Wolf-Heidegger

THE COLOR ATLAS
of HUMAN ANATOMY

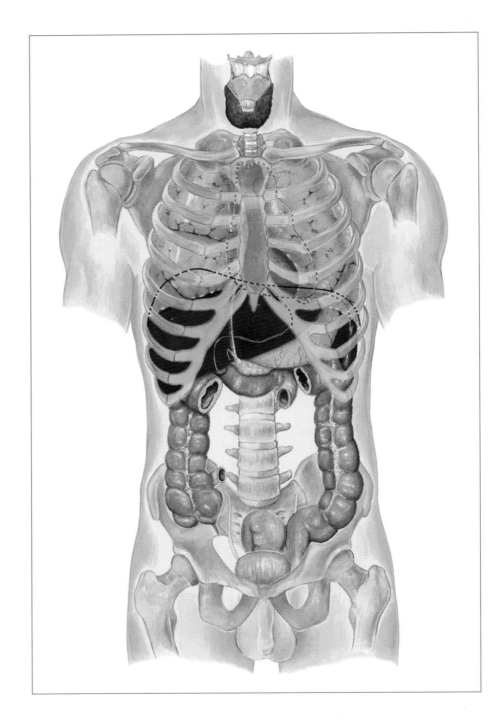

Wolf-Heidegger

THE COLOR ATLAS
of HUMAN ANATOMY

Dr. Petra Köpf-Maier, Editor

Dr. John C. Weber, Consulting Editor

Sterling Publishing Co., Inc.
New York

Library of Congress Cataloging-in-Publication Data Available

4 6 8 10 9 7 5

Published by Sterling Publishing Co., Inc.
387 Park Avenue South, New York, NY 10016

© 2004 by Karger AG, Basel
Paperback edition 2006 by Sterling Publishing Co., Inc.

Distributed in Canada by Sterling Publishing
c/o Canadian Manda Group, 165 Dufferin Street
Toronto, Ontario, Canada M6K 3H6

Printed in China
Sterling ISBN-13: 978-1-4027-4200-2
ISBN-10: 1-4027-4200-2

For information about custom editions, special sales, premium and
corporate purchases, please contact Sterling Special Sales
Department at 800-805-5489 or specialsales@sterlingpub.com.

To Those Who Bequeath Their Bodies to Science

Hic locus est ubi mors gaudet succurrere vitae

"This is the place where death delights in helping life"

—Inscription above the Anatomical Theatre of Bologna

This atlas of human anatomy begins by paying due homage and returning thanks to those who freely bequeath their bodies to anatomy. Such donations testify to an admirable, unselfish, and idealistic sense of sacrifice, and nothing can compensate for the invaluable service rendered to science and to society. Anatomy and medicine owe these individuals a tremendous debt of gratitude. By bequeathing their bodies, they enable students to learn through observation—even now, at the beginning of the twenty-first century, there is no substitute for this. Even beyond death, these altruistic people help the living—students, physicians, and patients alike. This is how the above inscription should be interpreted, and students should endeavor to be worthy of these voluntary and generous donations by respecting the dead as well as by working hard and learning eagerly.

Contents

Introduction

Dr. Gerhard Wolf-Heidegger first published his *Atlas of Human Anatomy* to international acclaim in 1953, aiming to give the medical student and the physician wishing to revise his or her anatomical knowledge as true a picture of the human body as possible. Since then, this venerable atlas has undergone several major revisions to keep pace with modern medical knowledge. The fifth edition published in the year 2000 has been completely revised and supplemented by Dr. Petra Köpf-Maier.

This classic medical reference is now available in a single concise volume designed to be accessible for general readers, students of anatomy, and allied health professionals such as nurses, physical therapists, sports trainers, hospital social workers, and medical secretaries and transcriptionists. In more than six hundred fifty superb anatomical drawings and paintings, *The Color Atlas of Human Anatomy* details the complex organ systems of the human body, offering clear and extensive labels and instructive color specifically designed to be easily understandable for beginners yet comprehensive enough for more advanced users.

As medical knowledge and technology advance, curricula in universities and medical schools have expanded to include the study of new diagnostic techniques, such as computed tomography, magnetic resonance imaging, and ultrasonography. In addition, the study of microscopic anatomy, physiological chemistry, and biochemistry have all taken on ever-greater importance over the past century, sometimes at the expense of traditional macroscopic anatomy. Yet it is only when students possess precise knowledge of human anatomical structures that they can properly interpret the information gathered by new imaging technologies. Students of anatomy at every level are thus challenged to build a solid understanding of human anatomy in order to maximize their skills within their chosen discipline.

To that end, *The Color Atlas of Human Anatomy* has been designed as a practical reference and as a supplement to anatomy texts. The volume is organized by anatomical region, echoing the method of study most common in biology and anatomy courses, and is complete with an exhaustive, twenty-page subject index. Nomenclature throughout the book conforms to the International Anatomical Terminology adopted in 1998 by the International Federation of Associations of Anatomists.

For those devoted to the study of human anatomy as well as those simply fascinated by the workings of the human body, this invaluable atlas maps the complexities of organ systems with clarity and accuracy.

Notes to the User

Anatomical Nomenclature

In this atlas, the designation of anatomical structures follows the most current international anatomical nomenclature, the *Terminologia Anatomica* (TA) adopted in 1998.

Abbreviations

Where required for space, the following abbreviations have been used:

Singular			**Plural**		
a.	=	artery	aa.	=	arteries
br.	=	branch	brr.	=	branches
cut.	=	cutaneous			
eth.	=	ethmoidal			
fem.	=	femoral			
inf.	=	inferior			
lig.	=	ligament	ligg.	=	ligaments
m.	=	muscle	mm.	=	muscles
n.	=	nerve	nn.	=	nerves
post.	=	posterior			
r.	=	ramus	rr.	=	rami
rad.	=	radiation			
rt.	=	right			
sup.	=	superior			
v.	=	vene	vv.	=	venes

Parentheses and Brackets

Parentheses ( ) are used to note terms also shown in parentheses in the *Terminologia Anatomica,* and for designating varieties, additional information, and explanations. In the legends, the relative size of images referred to the originals is given as a percentage in parentheses.

Terms commonly used but not present in the official *Terminologia Anatomica* are indicated in pointed parentheses ⟨ ⟩.

Numbers of vertebrae and cranial nerves appear in square brackets [ ], as in the *Terminologia Anatomica.*

Dashes

A dash following (in the left column) or preceding (in the right column) an entry indicates that one or several specific entries for the same body part will follow. The generic term is shown above it—usually without a pointer.

Pointers and Dots

If dots on a pointer identify two or more anatomical structures or if several dots appear on a pointer, the various designations are separated by a comma; their order follows that of the arrangement of the anatomical structures in the figure. In both columns, the labels are arranged according to the following principle: left first, then right; in the case of branched pointers, above first, then below.

Notation of Sizes

Unless otherwise indicated in the legends, the anatomical drawings in this atlas always represent the situation in adults; the percentages given in parentheses in the legends denote the relative size of the image referred to the original. With a view to the considerable biological variations in body size, the percentages have been rounded off and should only be considered as indicative.

Systemic Anatomy

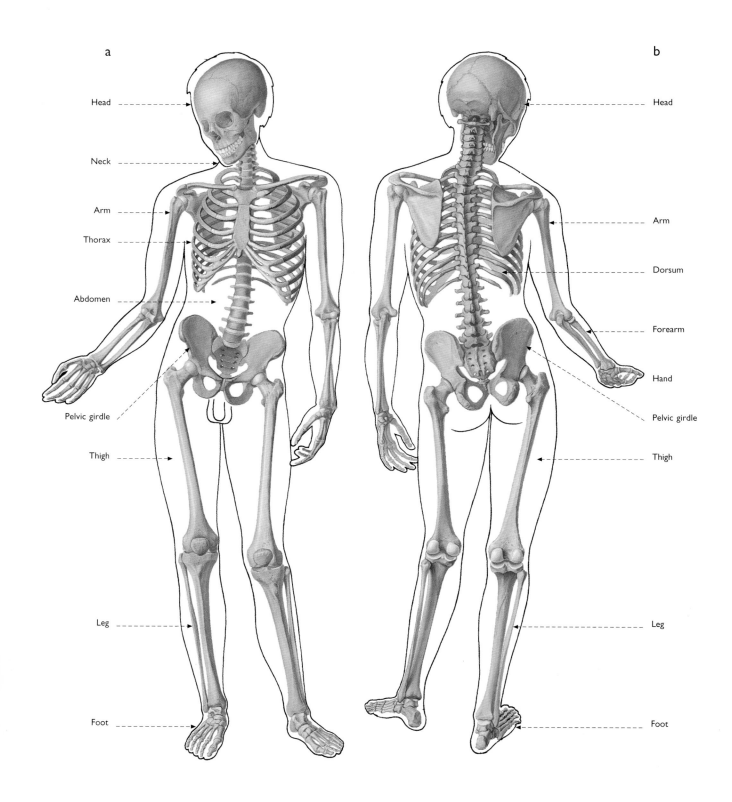

a

Head

Neck

Arm

Thorax

Abdomen

Pelvic girdle

Thigh

Leg

Foot

b

Head

Arm

Dorsum

Forearm

Hand

Pelvic girdle

Thigh

Leg

Foot

2 Skeleton and parts of the human body (10%)

Male skeleton

a Ventral aspect

b Dorsal aspect

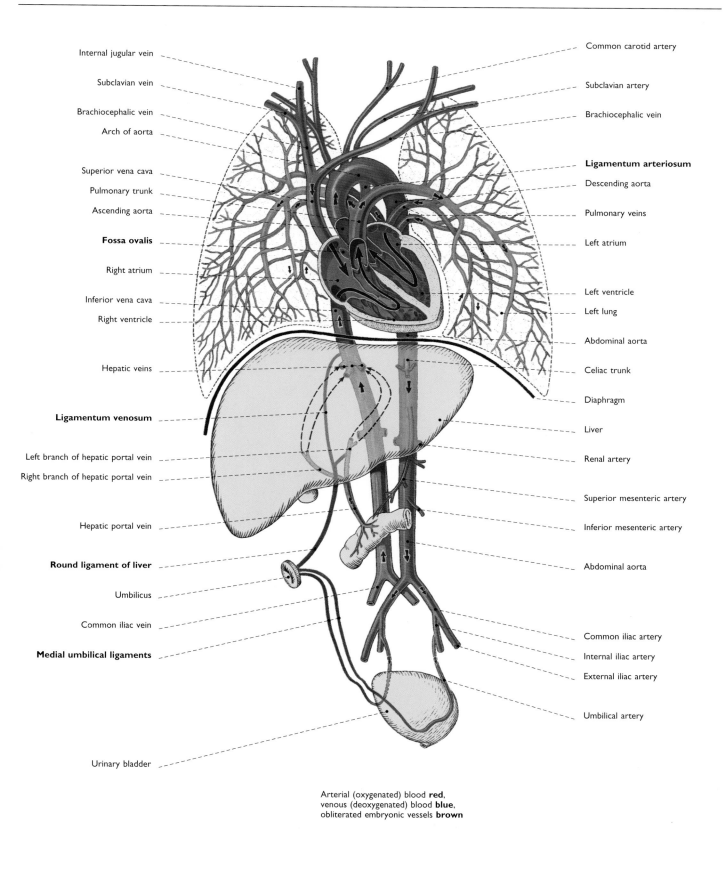

Internal jugular vein

Subclavian vein

Brachiocephalic vein

Arch of aorta

Superior vena cava

Pulmonary trunk

Ascending aorta

Fossa ovalis

Right atrium

Inferior vena cava

Right ventricle

Hepatic veins

Ligamentum venosum

Left branch of hepatic portal vein

Right branch of hepatic portal vein

Hepatic portal vein

Round ligament of liver

Umbilicus

Common iliac vein

Medial umbilical ligaments

Urinary bladder

Common carotid artery

Subclavian artery

Brachiocephalic vein

Ligamentum arteriosum

Descending aorta

Pulmonary veins

Left atrium

Left ventricle

Left lung

Abdominal aorta

Celiac trunk

Diaphragm

Liver

Renal artery

Superior mesenteric artery

Inferior mesenteric artery

Abdominal aorta

Common iliac artery

Internal iliac artery

External iliac artery

Umbilical artery

Arterial (oxygenated) blood **red**,
venous (deoxygenated) blood **blue**,
obliterated embryonic vessels **brown**

3　Adult cardiovascular system
Ventral aspect

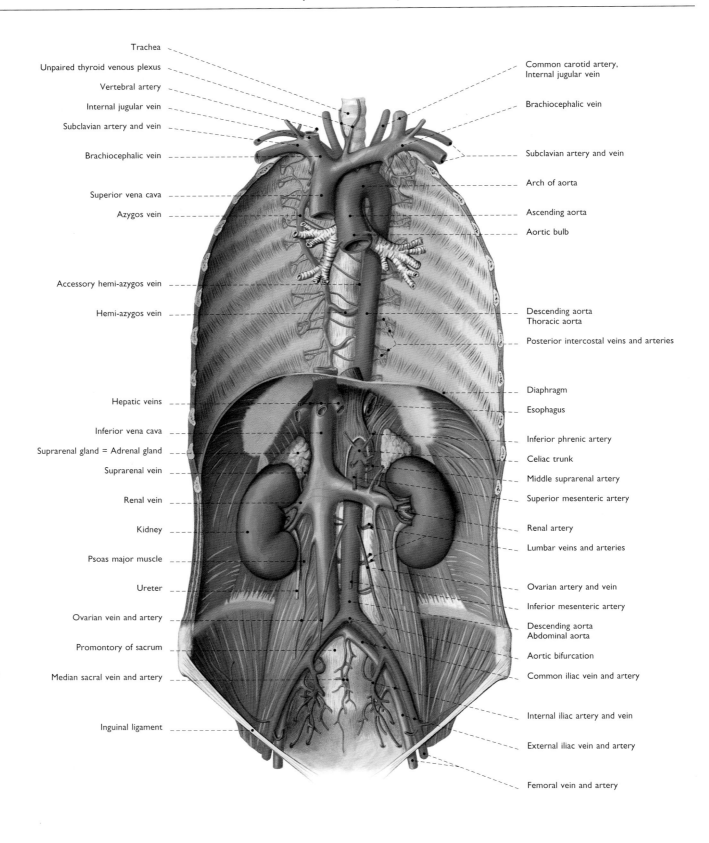

Trachea

Unpaired thyroid venous plexus

Vertebral artery

Internal jugular vein

Subclavian artery and vein

Brachiocephalic vein

Superior vena cava

Azygos vein

Accessory hemi-azygos vein

Hemi-azygos vein

Hepatic veins

Inferior vena cava

Suprarenal gland = Adrenal gland

Suprarenal vein

Renal vein

Kidney

Psoas major muscle

Ureter

Ovarian vein and artery

Promontory of sacrum

Median sacral vein and artery

Inguinal ligament

Common carotid artery,
Internal jugular vein

Brachiocephalic vein

Subclavian artery and vein

Arch of aorta

Ascending aorta

Aortic bulb

Descending aorta
Thoracic aorta

Posterior intercostal veins and arteries

Diaphragm

Esophagus

Inferior phrenic artery

Celiac trunk

Middle suprarenal artery

Superior mesenteric artery

Renal artery

Lumbar veins and arteries

Ovarian artery and vein

Inferior mesenteric artery

Descending aorta
Abdominal aorta

Aortic bifurcation

Common iliac vein and artery

Internal iliac artery and vein

External iliac vein and artery

Femoral vein and artery

4 Blood vessels of the trunk (30%)
Ventral aspect

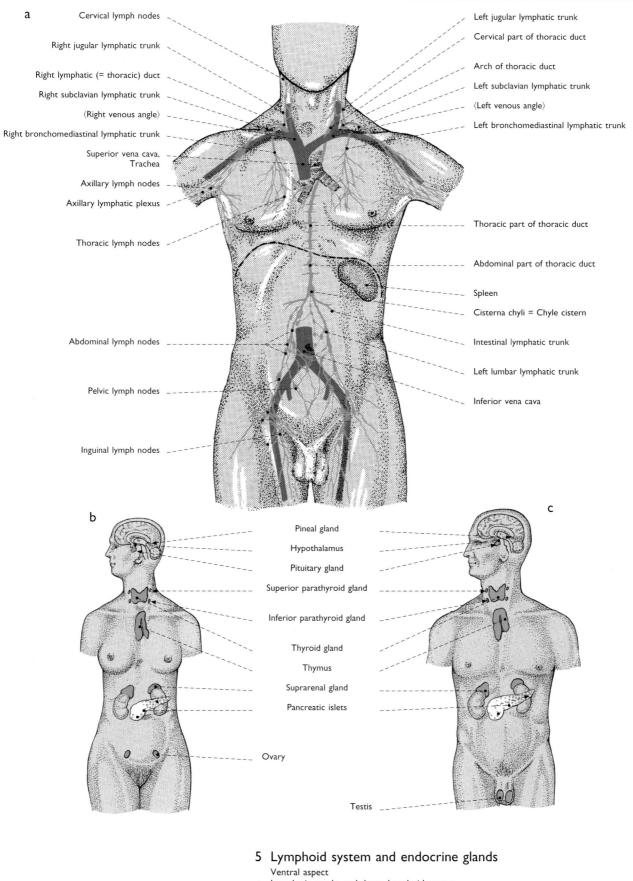

a

Cervical lymph nodes
Right jugular lymphatic trunk
Right lymphatic (= thoracic) duct
Right subclavian lymphatic trunk
⟨Right venous angle⟩
Right bronchomediastinal lymphatic trunk
Superior vena cava, Trachea
Axillary lymph nodes
Axillary lymphatic plexus
Thoracic lymph nodes
Abdominal lymph nodes
Pelvic lymph nodes
Inguinal lymph nodes

Left jugular lymphatic trunk
Cervical part of thoracic duct
Arch of thoracic duct
Left subclavian lymphatic trunk
⟨Left venous angle⟩
Left bronchomediastinal lymphatic trunk
Thoracic part of thoracic duct
Abdominal part of thoracic duct
Spleen
Cisterna chyli = Chyle cistern
Intestinal lymphatic trunk
Left lumbar lymphatic trunk
Inferior vena cava

b

c

Pineal gland
Hypothalamus
Pituitary gland
Superior parathyroid gland
Inferior parathyroid gland
Thyroid gland
Thymus
Suprarenal gland
Pancreatic islets
Ovary
Testis

5 Lymphoid system and endocrine glands
Ventral aspect
a Lymphatic trunks and ducts, lymphoid organs
b Female endocrine glands
c Male endocrine glands

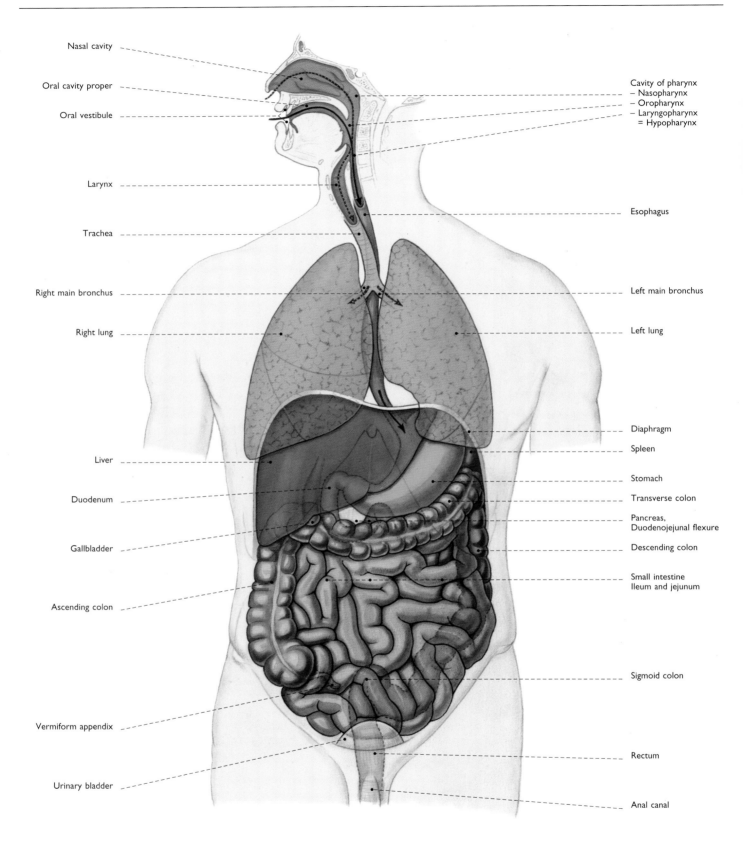

Nasal cavity

Oral cavity proper

Oral vestibule

Larynx

Trachea

Right main bronchus

Right lung

Liver

Duodenum

Gallbladder

Ascending colon

Vermiform appendix

Urinary bladder

Cavity of pharynx
– Nasopharynx
– Oropharynx
– Laryngopharynx
= Hypopharynx

Esophagus

Left main bronchus

Left lung

Diaphragm

Spleen

Stomach

Transverse colon

Pancreas,
Duodenojejunal flexure

Descending colon

Small intestine
Ileum and jejunum

Sigmoid colon

Rectum

Anal canal

6 Alimentary and respiratory systems (25%)
Ventral aspect

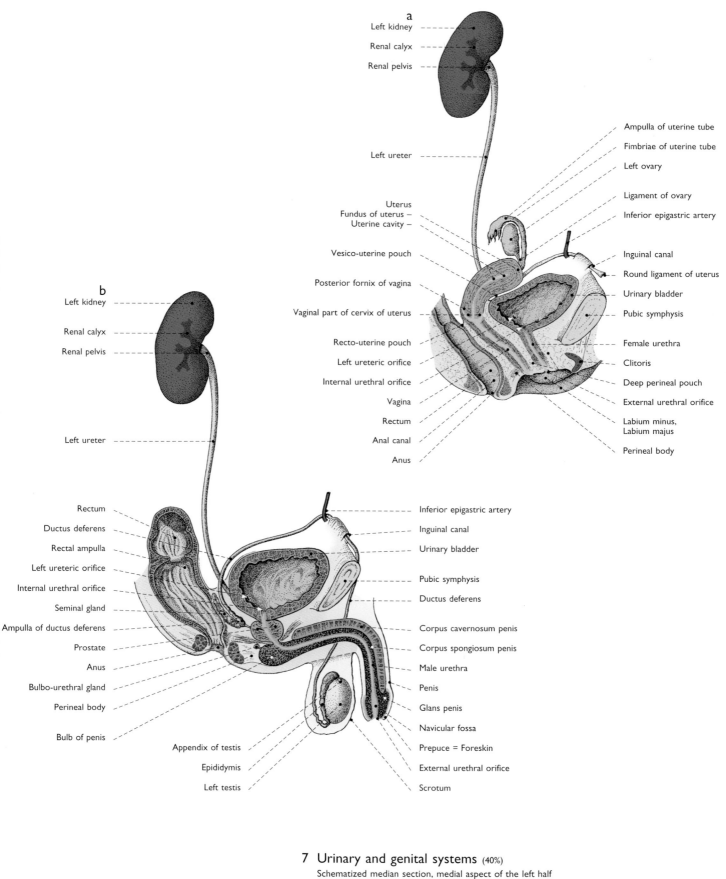

a
Left kidney
Renal calyx
Renal pelvis

Left ureter

Ampulla of uterine tube
Fimbriae of uterine tube
Left ovary

Ligament of ovary
Inferior epigastric artery

Uterus
Fundus of uterus —
Uterine cavity —

Vesico-uterine pouch

Posterior fornix of vagina

Vaginal part of cervix of uterus

Recto-uterine pouch
Left ureteric orifice
Internal urethral orifice
Vagina
Rectum
Anal canal
Anus

Inguinal canal
Round ligament of uterus
Urinary bladder
Pubic symphysis

Female urethra
Clitoris
Deep perineal pouch
External urethral orifice
Labium minus,
Labium majus
Perineal body

b
Left kidney
Renal calyx
Renal pelvis

Left ureter

Rectum
Ductus deferens
Rectal ampulla
Left ureteric orifice
Internal urethral orifice
Seminal gland
Ampulla of ductus deferens
Prostate
Anus
Bulbo-urethral gland
Perineal body
Bulb of penis

Appendix of testis
Epididymis
Left testis

Inferior epigastric artery
Inguinal canal
Urinary bladder

Pubic symphysis
Ductus deferens

Corpus cavernosum penis
Corpus spongiosum penis
Male urethra
Penis
Glans penis
Navicular fossa
Prepuce = Foreskin
External urethral orifice
Scrotum

7 Urinary and genital systems (40%)
   Schematized median section, medial aspect of the left half
   a Female
   b Male

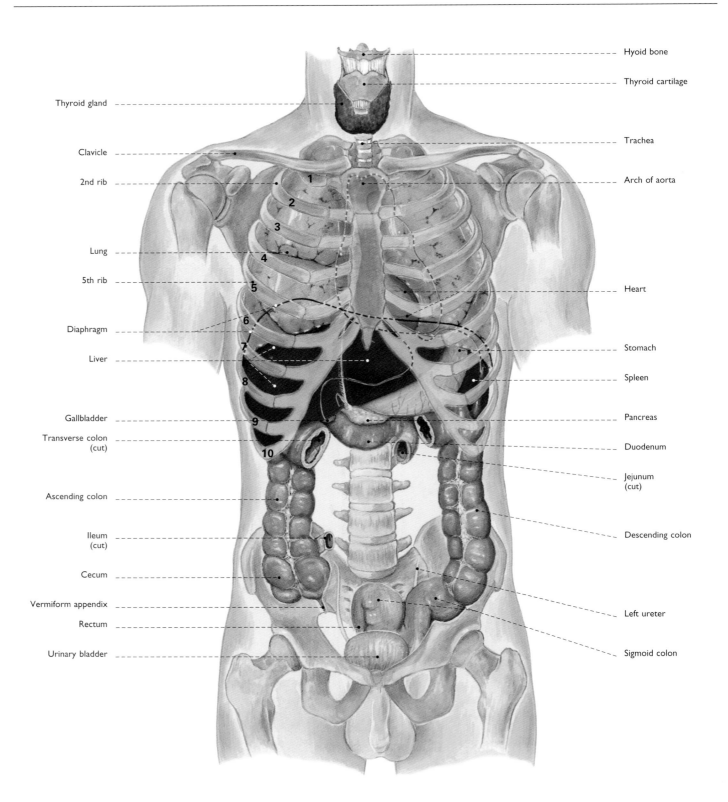

Hyoid bone

Thyroid cartilage

Thyroid gland

Trachea

Clavicle

2nd rib

Arch of aorta

Lung

5th rib

Heart

Diaphragm

Stomach

Liver

Spleen

Gallbladder

Pancreas

Transverse colon
(cut)

Duodenum

Jejunum
(cut)

Ascending colon

Descending colon

Ileum
(cut)

Cecum

Vermiform appendix

Left ureter

Rectum

Sigmoid colon

Urinary bladder

**8 Surface projections of
thoracic and abdominal viscera** (25%)
Jejunum, ileum, and transverse colon were removed. Ventral aspect

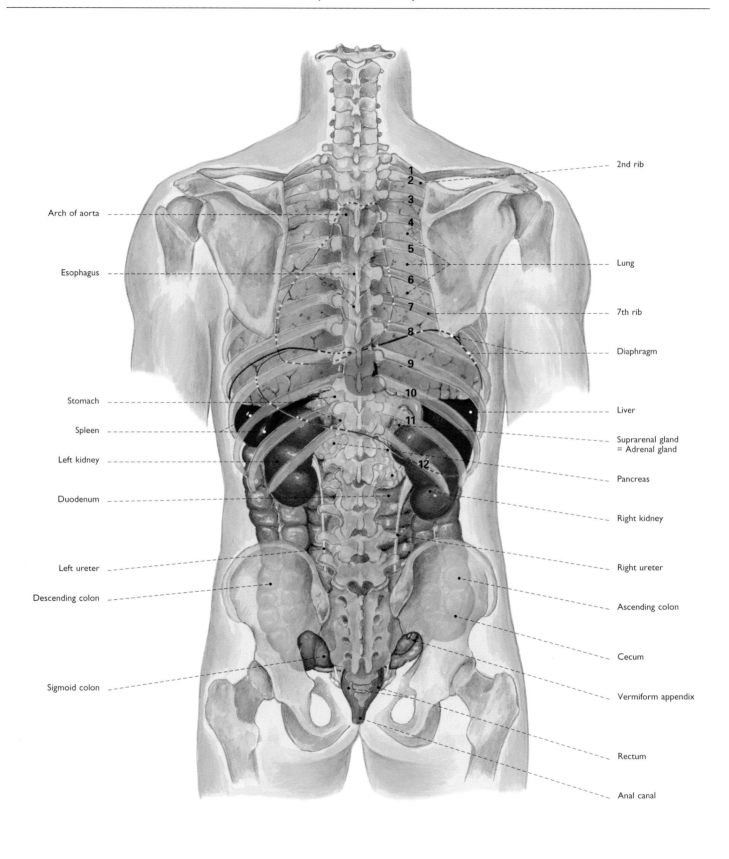

Arch of aorta

Esophagus

Stomach

Spleen

Left kidney

Duodenum

Left ureter

Descending colon

Sigmoid colon

2nd rib

Lung

7th rib

Diaphragm

Liver

Suprarenal gland = Adrenal gland

Pancreas

Right kidney

Right ureter

Ascending colon

Cecum

Vermiform appendix

Rectum

Anal canal

9 Surface projections of thoracic and abdominal viscera (25%)

Dorsal aspect

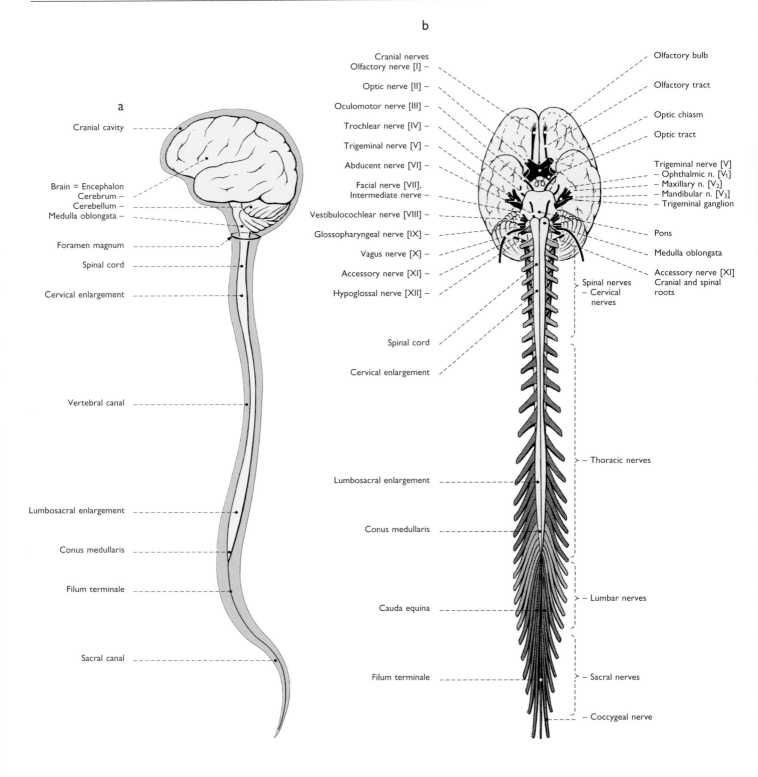

a

Cranial cavity

Brain = Encephalon
Cerebrum —
Cerebellum —
Medulla oblongata —

Foramen magnum

Spinal cord

Cervical enlargement

Vertebral canal

Lumbosacral enlargement

Conus medullaris

Filum terminale

Sacral canal

b

Cranial nerves
Olfactory nerve [I] –

Optic nerve [II] –

Oculomotor nerve [III] –

Trochlear nerve [IV] –

Trigeminal nerve [V] –

Abducent nerve [VI] –

Facial nerve [VII],
Intermediate nerve –

Vestibulocochlear nerve [VIII] –

Glossopharyngeal nerve [IX] –

Vagus nerve [X] –

Accessory nerve [XI] –

Hypoglossal nerve [XII] –

Spinal cord

Cervical enlargement

Lumbosacral enlargement

Conus medullaris

Cauda equina

Filum terminale

Olfactory bulb

Olfactory tract

Optic chiasm

Optic tract

Trigeminal nerve [V]
– Ophthalmic n. [V_1]
– Maxillary n. [V_2]
– Mandibular n. [V_3]
– Trigeminal ganglion

Pons

Medulla oblongata

Accessory nerve [XI]
Cranial and spinal
roots

Spinal nerves
– Cervical
nerves

> – Thoracic nerves

> – Lumbar nerves

> – Sacral nerves

– Coccygeal nerve

10 Central and peripheral nervous systems
a Central nervous system, left lateral aspect
b Cranial and spinal nerves, ventral aspect

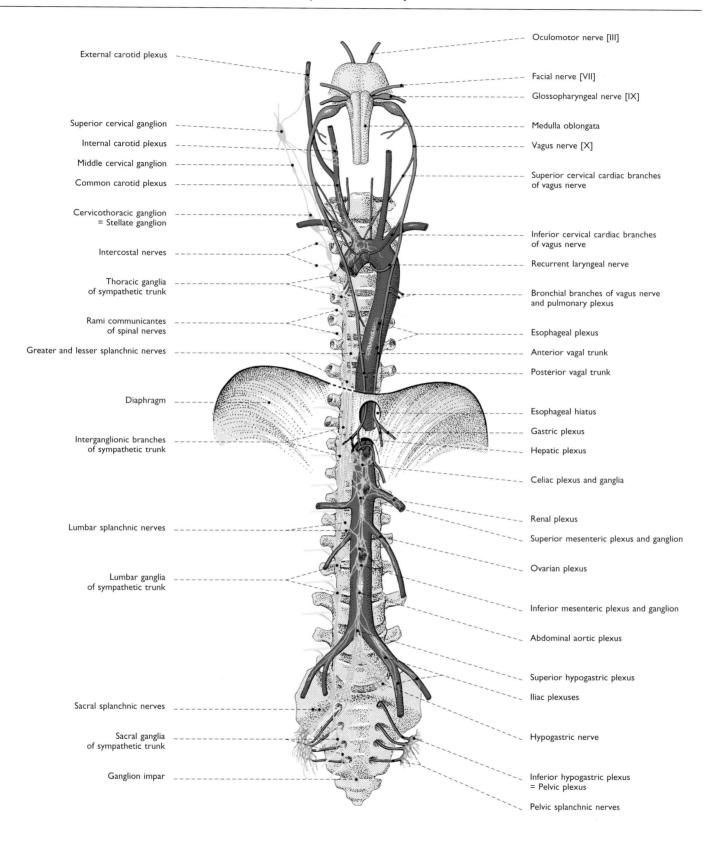

External carotid plexus

Superior cervical ganglion

Internal carotid plexus

Middle cervical ganglion

Common carotid plexus

Cervicothoracic ganglion
= Stellate ganglion

Intercostal nerves

Thoracic ganglia
of sympathetic trunk

Rami communicantes
of spinal nerves

Greater and lesser splanchnic nerves

Diaphragm

Interganglionic branches
of sympathetic trunk

Lumbar splanchnic nerves

Lumbar ganglia
of sympathetic trunk

Sacral splanchnic nerves

Sacral ganglia
of sympathetic trunk

Ganglion impar

Oculomotor nerve [III]

Facial nerve [VII]

Glossopharyngeal nerve [IX]

Medulla oblongata

Vagus nerve [X]

Superior cervical cardiac branches
of vagus nerve

Inferior cervical cardiac branches
of vagus nerve

Recurrent laryngeal nerve

Bronchial branches of vagus nerve
and pulmonary plexus

Esophageal plexus

Anterior vagal trunk

Posterior vagal trunk

Esophageal hiatus

Gastric plexus

Hepatic plexus

Celiac plexus and ganglia

Renal plexus

Superior mesenteric plexus and ganglion

Ovarian plexus

Inferior mesenteric plexus and ganglion

Abdominal aortic plexus

Superior hypogastric plexus

Iliac plexuses

Hypogastric nerve

Inferior hypogastric plexus
= Pelvic plexus

Pelvic splanchnic nerves

**11 Autonomic division
of the peripheral nervous system** (25%)
Peripheral sympathetic (**orange**) and parasympathetic (**brown**)
nerves and ganglia. The sympathetic components are shown
only on the left side of the picture. Ventral aspect

Body Wall

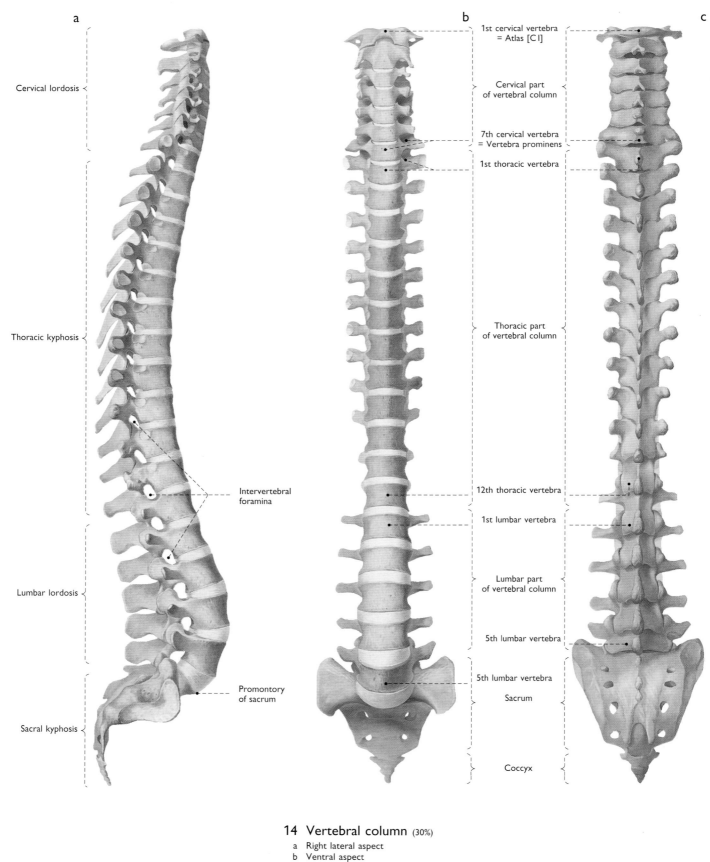

a

b

c

Cervical lordosis

Thoracic kyphosis

Lumbar lordosis

Sacral kyphosis

Intervertebral foramina

Promontory of sacrum

1st cervical vertebra = Atlas [C I]

Cervical part of vertebral column

7th cervical vertebra = Vertebra prominens

1st thoracic vertebra

Thoracic part of vertebral column

12th thoracic vertebra

1st lumbar vertebra

Lumbar part of vertebral column

5th lumbar vertebra

5th lumbar vertebra

Sacrum

Coccyx

14 Vertebral column (30%)

   a   Right lateral aspect
   b   Ventral aspect
   c   Dorsal aspect

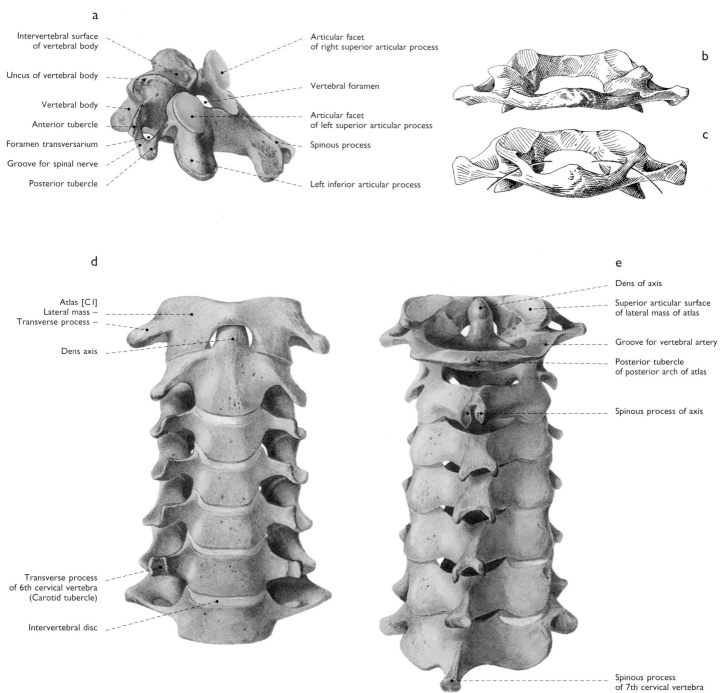

a

Intervertebral surface of vertebral body

Uncus of vertebral body

Vertebral body

Anterior tubercle

Foramen transversarium

Groove for spinal nerve

Posterior tubercle

Articular facet of right superior articular process

Vertebral foramen

Articular facet of left superior articular process

Spinous process

Left inferior articular process

b

c

d

Atlas [C I]
Lateral mass
Transverse process

Dens axis

Transverse process of 6th cervical vertebra (Carotid tubercle)

Intervertebral disc

e

Dens of axis

Superior articular surface of lateral mass of atlas

Groove for vertebral artery

Posterior tubercle of posterior arch of atlas

Spinous process of axis

Spinous process of 7th cervical vertebra = Spinous process of vertebra prominens

15   Cervical vertebrae and cervical spine

    a   Middle cervical vertebra (90%), left lateral aspect
b, c   First cervical vertebra = atlas [C I], dorsal aspect
    b   Deep groove for the vertebral artery on both sides
    c   Canal for the vertebral artery on both sides
d, e   Cervical spine with intervertebral discs (100%)
    d   Ventral aspect
    e   Dorsal aspect

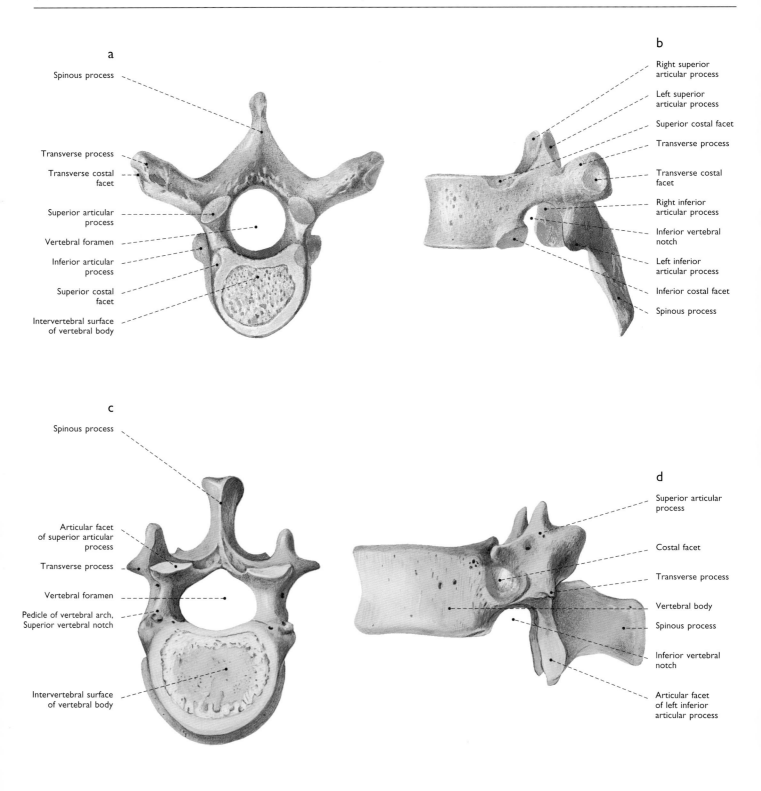

a

Spinous process

Transverse process

Transverse costal facet

Superior articular process

Vertebral foramen

Inferior articular process

Superior costal facet

Intervertebral surface of vertebral body

b

Right superior articular process

Left superior articular process

Superior costal facet

Transverse process

Transverse costal facet

Right inferior articular process

Inferior vertebral notch

Left inferior articular process

Inferior costal facet

Spinous process

c

Spinous process

Articular facet of superior articular process

Transverse process

Vertebral foramen

Pedicle of vertebral arch, Superior vertebral notch

Intervertebral surface of vertebral body

d

Superior articular process

Costal facet

Transverse process

Vertebral body

Spinous process

Inferior vertebral notch

Articular facet of left inferior articular process

16 Thoracic vertebrae (100%)

a, b   Sixth thoracic vertebra
   a   Cranial aspect
   b   Left lateral aspect
c, d   Twelfth thoracic vertebra
   c   Cranial aspect
   d   Left lateral aspect

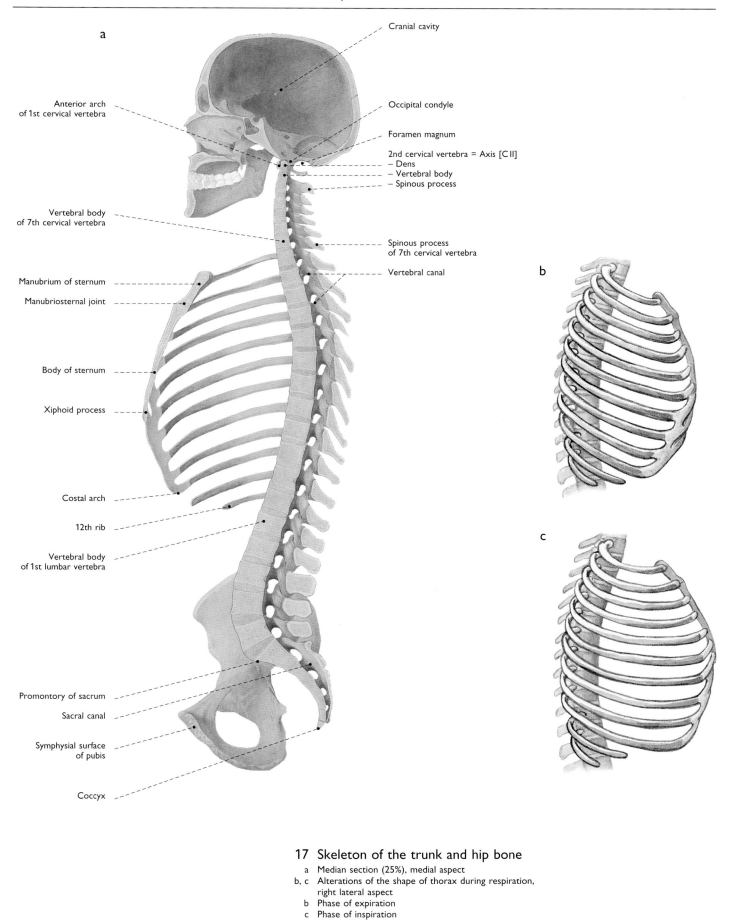

Cranial cavity

Occipital condyle

Anterior arch
of 1st cervical vertebra

Foramen magnum

2nd cervical vertebra = Axis [C II]
– Dens
– Vertebral body
– Spinous process

Vertebral body
of 7th cervical vertebra

Spinous process
of 7th cervical vertebra

Manubrium of sternum

Vertebral canal

Manubriosternal joint

Body of sternum

Xiphoid process

Costal arch

12th rib

Vertebral body
of 1st lumbar vertebra

Promontory of sacrum

Sacral canal

Symphysial surface
of pubis

Coccyx

b

c

17 Skeleton of the trunk and hip bone
a   Median section (25%), medial aspect
b, c   Alterations of the shape of thorax during respiration,
     right lateral aspect
b   Phase of expiration
c   Phase of inspiration

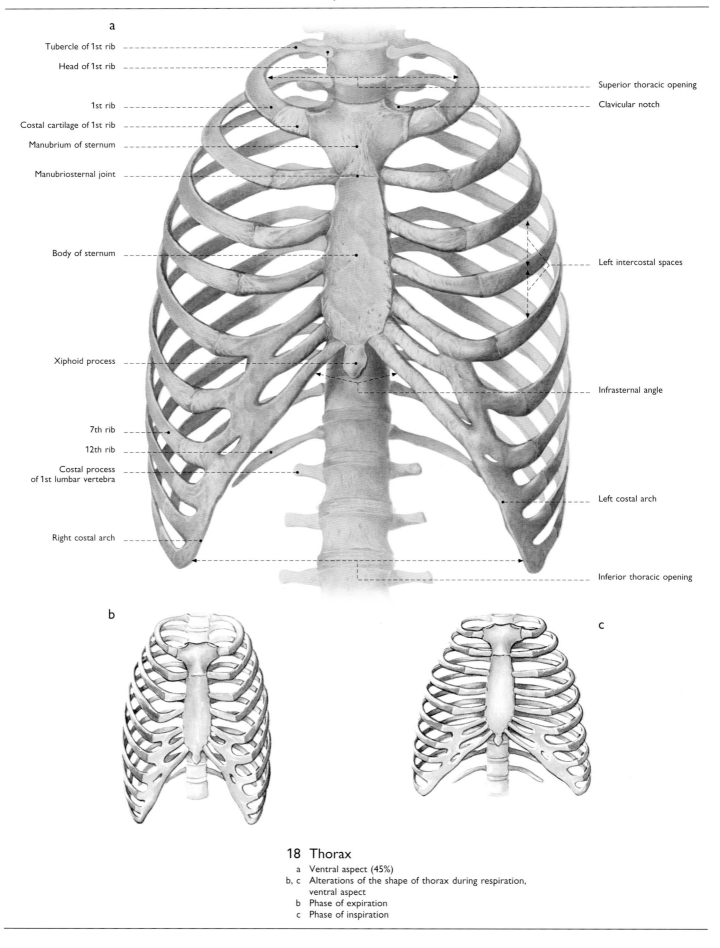

a

Tubercle of 1st rib

Head of 1st rib

Superior thoracic opening

Clavicular notch

1st rib

Costal cartilage of 1st rib

Manubrium of sternum

Manubriosternal joint

Body of sternum

Left intercostal spaces

Xiphoid process

Infrasternal angle

7th rib

12th rib

Costal process
of 1st lumbar vertebra

Left costal arch

Right costal arch

Inferior thoracic opening

b

c

18 Thorax

a Ventral aspect (45%)
b, c Alterations of the shape of thorax during respiration,
       ventral aspect
b Phase of expiration
c Phase of inspiration

1st thoracic vertebra
Articular facet of
superior articular process –
Transverse process –
Spinous process –

1st rib

Angle of 7th rib

12th rib

Spinous process
of 12th thoracic vertebra

Left costal arch

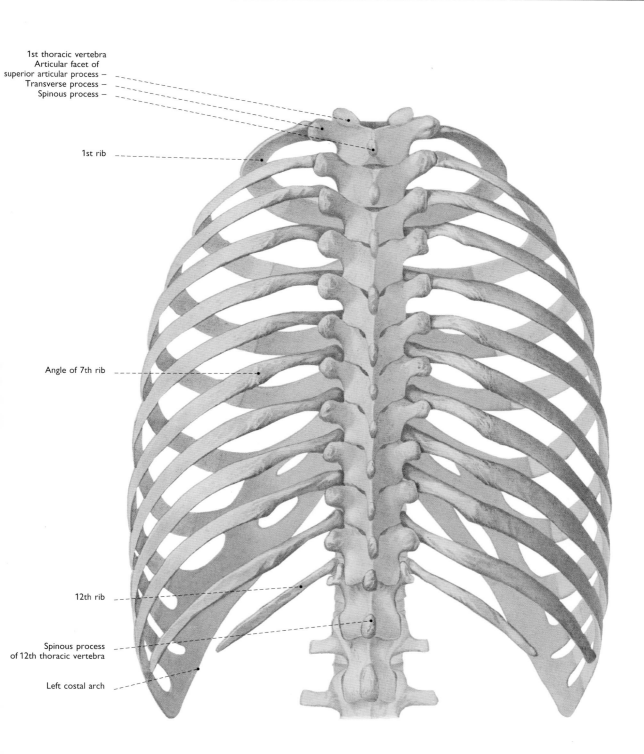

19 Thorax (50%)
Dorsal aspect

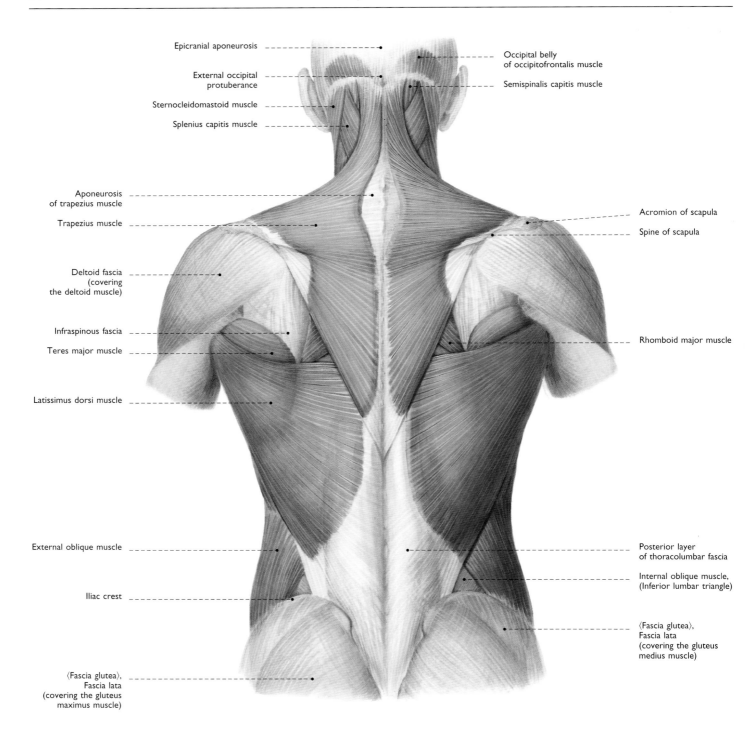

Epicranial aponeurosis

Occipital belly of occipitofrontalis muscle

External occipital protuberance

Semispinalis capitis muscle

Sternocleidomastoid muscle

Splenius capitis muscle

Aponeurosis of trapezius muscle

Acromion of scapula

Trapezius muscle

Spine of scapula

Deltoid fascia (covering the deltoid muscle)

Infraspinous fascia

Rhomboid major muscle

Teres major muscle

Latissimus dorsi muscle

External oblique muscle

Posterior layer of thoracolumbar fascia

Internal oblique muscle, (Inferior lumbar triangle)

Iliac crest

⟨Fascia glutea⟩, Fascia lata (covering the gluteus medius muscle)

⟨Fascia glutea⟩, Fascia lata (covering the gluteus maximus muscle)

20 Muscles of the back (25%)
Superficial layer

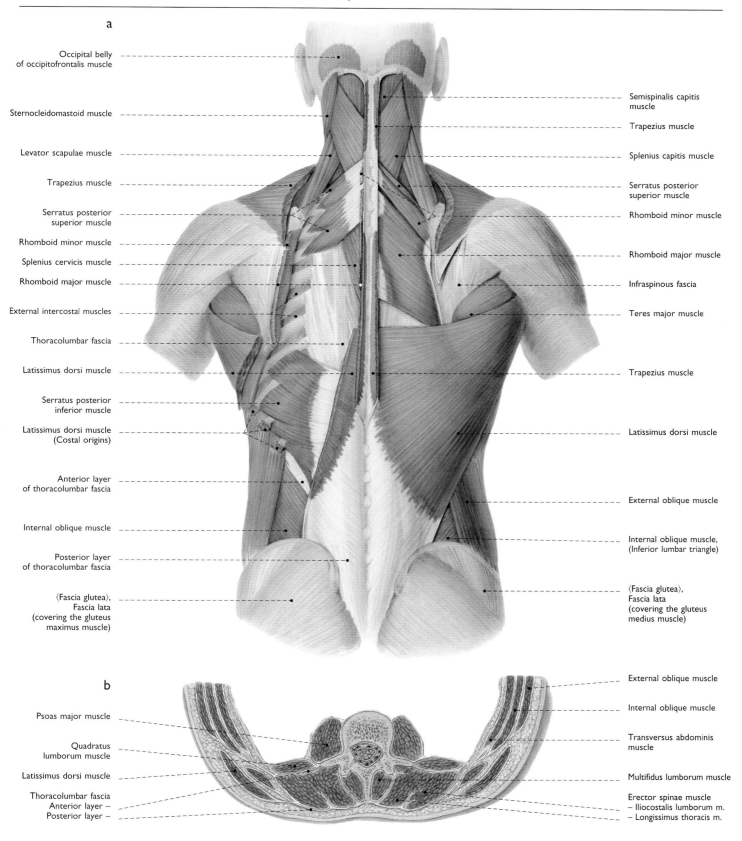

a

Occipital belly
of occipitofrontalis muscle

Sternocleidomastoid muscle

Levator scapulae muscle

Trapezius muscle

Serratus posterior
superior muscle

Rhomboid minor muscle

Splenius cervicis muscle

Rhomboid major muscle

External intercostal muscles

Thoracolumbar fascia

Latissimus dorsi muscle

Serratus posterior
inferior muscle

Latissimus dorsi muscle
(Costal origins)

Anterior layer
of thoracolumbar fascia

Internal oblique muscle

Posterior layer
of thoracolumbar fascia

⟨Fascia glutea⟩,
Fascia lata
(covering the gluteus
maximus muscle)

Semispinalis capitis
muscle

Trapezius muscle

Splenius capitis muscle

Serratus posterior
superior muscle

Rhomboid minor muscle

Rhomboid major muscle

Infraspinous fascia

Teres major muscle

Trapezius muscle

Latissimus dorsi muscle

External oblique muscle

Internal oblique muscle,
(Inferior lumbar triangle)

⟨Fascia glutea⟩,
Fascia lata
(covering the gluteus
medius muscle)

b

Psoas major muscle

Quadratus
lumborum muscle

Latissimus dorsi muscle

Thoracolumbar fascia
Anterior layer –
Posterior layer –

External oblique muscle

Internal oblique muscle

Transversus abdominis
muscle

Multifidus lumborum muscle

Erector spinae muscle
– Iliocostalis lumborum m.
– Longissimus thoracis m.

21   Muscles of the back

a   Deeper layer (25%)
b   Schematized transverse section through the posterior and
     lateral abdominal wall in the lumbar region (35%)

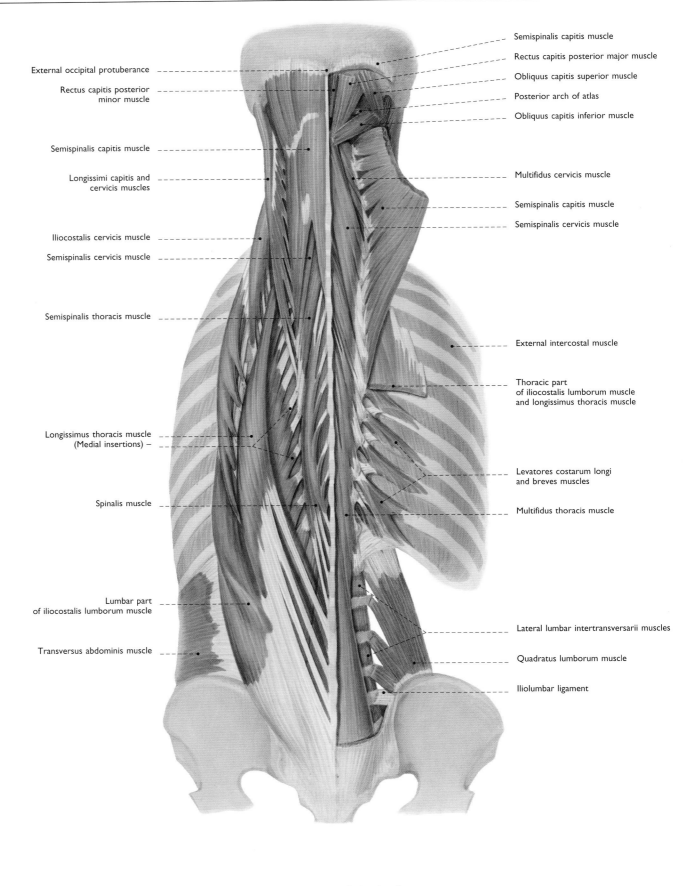

External occipital protuberance

Rectus capitis posterior minor muscle

Semispinalis capitis muscle

Longissimi capitis and cervicis muscles

Iliocostalis cervicis muscle

Semispinalis cervicis muscle

Semispinalis thoracis muscle

Longissimus thoracis muscle (Medial insertions)

Spinalis muscle

Lumbar part of iliocostalis lumborum muscle

Transversus abdominis muscle

Semispinalis capitis muscle

Rectus capitis posterior major muscle

Obliquus capitis superior muscle

Posterior arch of atlas

Obliquus capitis inferior muscle

Multifidus cervicis muscle

Semispinalis capitis muscle

Semispinalis cervicis muscle

External intercostal muscle

Thoracic part of iliocostalis lumborum muscle and longissimus thoracis muscle

Levatores costarum longi and breves muscles

Multifidus thoracis muscle

Lateral lumbar intertransversarii muscles

Quadratus lumborum muscle

Iliolumbar ligament

22 Muscles of the back proper (30%)
Deeper layer

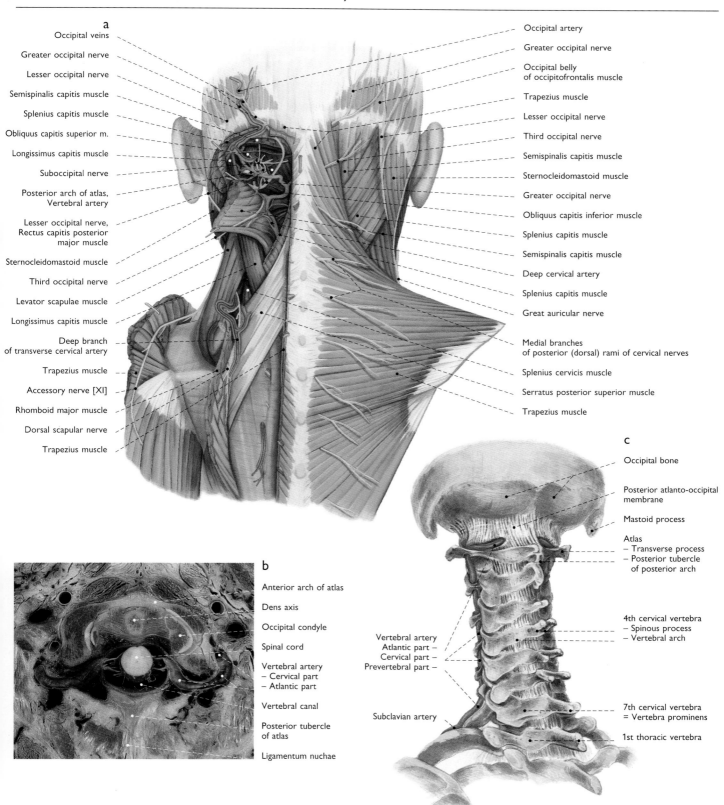

a

Occipital veins
Greater occipital nerve
Lesser occipital nerve
Semispinalis capitis muscle
Splenius capitis muscle
Obliquus capitis superior m.
Longissimus capitis muscle
Suboccipital nerve
Posterior arch of atlas, Vertebral artery
Lesser occipital nerve, Rectus capitis posterior major muscle
Sternocleidomastoid muscle
Third occipital nerve
Levator scapulae muscle
Longissimus capitis muscle
Deep branch of transverse cervical artery
Trapezius muscle
Accessory nerve [XI]
Rhomboid major muscle
Dorsal scapular nerve
Trapezius muscle

Occipital artery
Greater occipital nerve
Occipital belly of occipitofrontalis muscle
Trapezius muscle
Lesser occipital nerve
Third occipital nerve
Semispinalis capitis muscle
Sternocleidomastoid muscle
Greater occipital nerve
Obliquus capitis inferior muscle
Splenius capitis muscle
Semispinalis capitis muscle
Deep cervical artery
Splenius capitis muscle
Great auricular nerve
Medial branches of posterior (dorsal) rami of cervical nerves
Splenius cervicis muscle
Serratus posterior superior muscle
Trapezius muscle

b

Anterior arch of atlas
Dens axis
Occipital condyle
Spinal cord
Vertebral artery – Cervical part – Atlantic part
Vertebral canal
Posterior tubercle of atlas
Ligamentum nuchae

c

Occipital bone
Posterior atlanto-occipital membrane
Mastoid process
Atlas – Transverse process – Posterior tubercle of posterior arch
Vertebral artery Atlantic part – Cervical part – Prevertebral part –
4th cervical vertebra – Spinous process – Vertebral arch
Subclavian artery
7th cervical vertebra = Vertebra prominens
1st thoracic vertebra

23 Neck and shoulder regions

a Right, superficial layer; left, deeper layer (40%). Dorsal aspect
b Horizontal section at the level of the first cervical vertebra (= atlas) (60%). Cranial aspect
c Course of the vertebral artery (60%). Left dorsolateral aspect

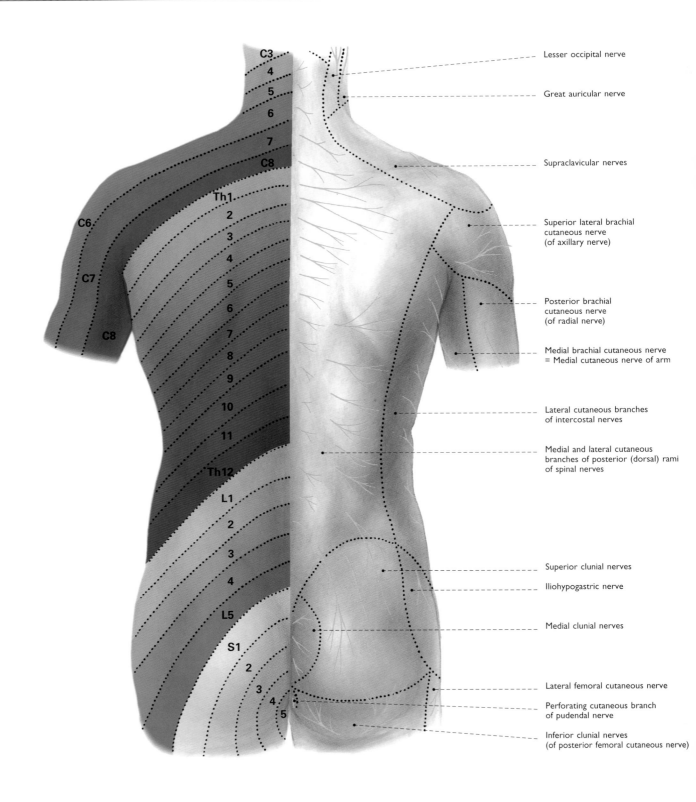

C3
4
5
6
7
C8
Th1
C6
2
3
C7
4
5
6
C8
7
8
9
10
11
Th12
L1
2
3
4
L5
S1
2
3
4
5

Lesser occipital nerve

Great auricular nerve

Supraclavicular nerves

Superior lateral brachial
cutaneous nerve
(of axillary nerve)

Posterior brachial
cutaneous nerve
(of radial nerve)

Medial brachial cutaneous nerve
= Medial cutaneous nerve of arm

Lateral cutaneous branches
of intercostal nerves

Medial and lateral cutaneous
branches of posterior (dorsal) rami
of spinal nerves

Superior clunial nerves

Iliohypogastric nerve

Medial clunial nerves

Lateral femoral cutaneous nerve

Perforating cutaneous branch
of pudendal nerve

Inferior clunial nerves
(of posterior femoral cutaneous nerve)

**24 Cutaneous and segmental innervation
of the dorsal body wall (25%)**
Schematic representation

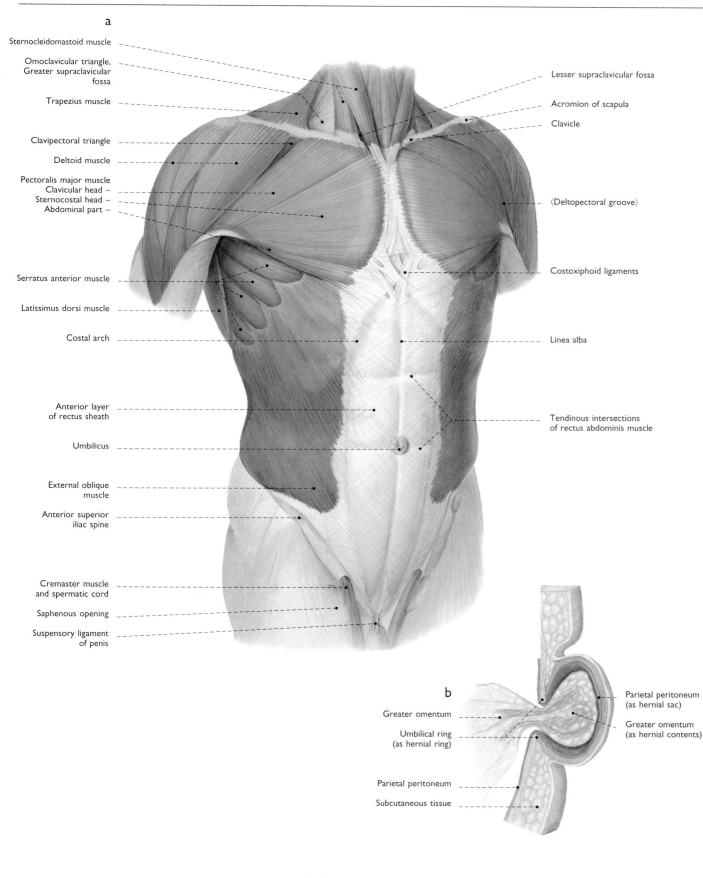

a

Sternocleidomastoid muscle

Omoclavicular triangle, Greater supraclavicular fossa

Trapezius muscle

Clavipectoral triangle

Deltoid muscle

Pectoralis major muscle
Clavicular head –
Sternocostal head –
Abdominal part –

Serratus anterior muscle

Latissimus dorsi muscle

Costal arch

Anterior layer of rectus sheath

Umbilicus

External oblique muscle

Anterior superior iliac spine

Cremaster muscle and spermatic cord

Saphenous opening

Suspensory ligament of penis

Lesser supraclavicular fossa

Acromion of scapula

Clavicle

⟨Deltopectoral groove⟩

Costoxiphoid ligaments

Linea alba

Tendinous intersections of rectus abdominis muscle

b

Greater omentum

Umbilical ring (as hernial ring)

Parietal peritoneum

Subcutaneous tissue

Parietal peritoneum (as hernial sac)

Greater omentum (as hernial contents)

25   Ventral muscles of the trunk
a   Superficial layer (25%)
b   Schematic representation of an umbilical hernia

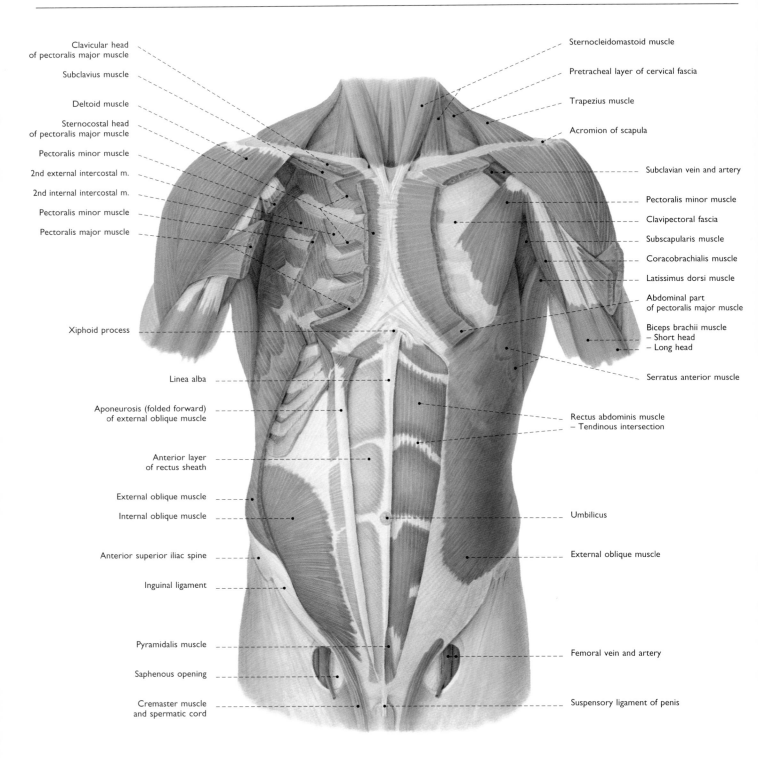

Clavicular head
of pectoralis major muscle

Subclavius muscle

Deltoid muscle

Sternocostal head
of pectoralis major muscle

Pectoralis minor muscle

2nd external intercostal m.

2nd internal intercostal m.

Pectoralis minor muscle

Pectoralis major muscle

Xiphoid process

Linea alba

Aponeurosis (folded forward)
of external oblique muscle

Anterior layer
of rectus sheath

External oblique muscle

Internal oblique muscle

Anterior superior iliac spine

Inguinal ligament

Pyramidalis muscle

Saphenous opening

Cremaster muscle
and spermatic cord

Sternocleidomastoid muscle

Pretracheal layer of cervical fascia

Trapezius muscle

Acromion of scapula

Subclavian vein and artery

Pectoralis minor muscle

Clavipectoral fascia

Subscapularis muscle

Coracobrachialis muscle

Latissimus dorsi muscle

Abdominal part
of pectoralis major muscle

Biceps brachii muscle
– Short head
– Long head

Serratus anterior muscle

Rectus abdominis muscle
– Tendinous intersection

Umbilicus

External oblique muscle

Femoral vein and artery

Suspensory ligament of penis

26 Ventral muscles of the trunk (25%)
Deeper layer

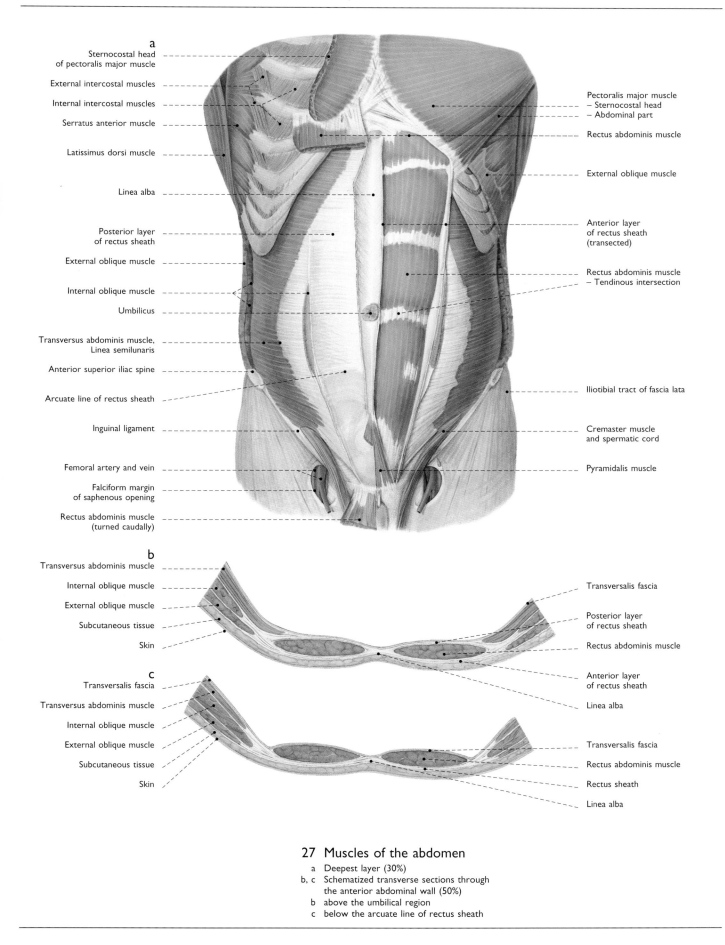

a

Sternocostal head of pectoralis major muscle

External intercostal muscles

Internal intercostal muscles

Serratus anterior muscle

Latissimus dorsi muscle

Linea alba

Posterior layer of rectus sheath

External oblique muscle

Internal oblique muscle

Umbilicus

Transversus abdominis muscle, Linea semilunaris

Anterior superior iliac spine

Arcuate line of rectus sheath

Inguinal ligament

Femoral artery and vein

Falciform margin of saphenous opening

Rectus abdominis muscle (turned caudally)

Pectoralis major muscle
– Sternocostal head
– Abdominal part

Rectus abdominis muscle

External oblique muscle

Anterior layer of rectus sheath (transected)

Rectus abdominis muscle – Tendinous intersection

Iliotibial tract of fascia lata

Cremaster muscle and spermatic cord

Pyramidalis muscle

b

Transversus abdominis muscle

Internal oblique muscle

External oblique muscle

Subcutaneous tissue

Skin

Transversalis fascia

Posterior layer of rectus sheath

Rectus abdominis muscle

Anterior layer of rectus sheath

Linea alba

c

Transversalis fascia

Transversus abdominis muscle

Internal oblique muscle

External oblique muscle

Subcutaneous tissue

Skin

Transversalis fascia

Rectus abdominis muscle

Rectus sheath

Linea alba

27 Muscles of the abdomen

a   Deepest layer (30%)

b, c   Schematized transverse sections through the anterior abdominal wall (50%)

b   above the umbilical region

c   below the arcuate line of rectus sheath

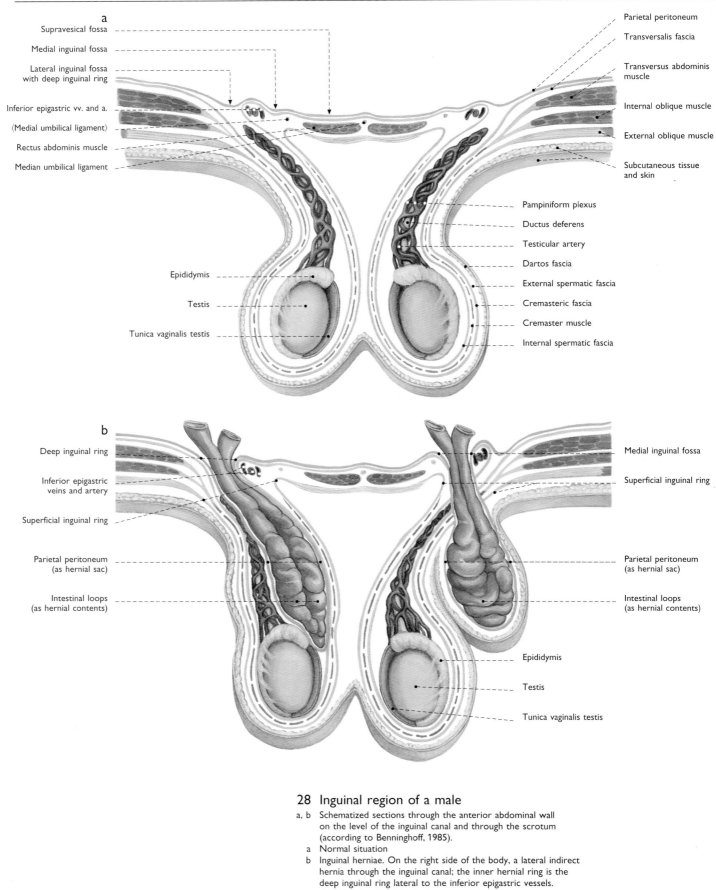

Supravesical fossa

Medial inguinal fossa

Lateral inguinal fossa
with deep inguinal ring

Inferior epigastric vv. and a.
⟨Medial umbilical ligament⟩

Rectus abdominis muscle

Median umbilical ligament

Parietal peritoneum

Transversalis fascia

Transversus abdominis
muscle

Internal oblique muscle

External oblique muscle

Subcutaneous tissue
and skin

Pampiniform plexus

Ductus deferens

Testicular artery

Dartos fascia

External spermatic fascia

Cremasteric fascia

Cremaster muscle

Internal spermatic fascia

Epididymis

Testis

Tunica vaginalis testis

Deep inguinal ring

Inferior epigastric
veins and artery

Superficial inguinal ring

Parietal peritoneum
(as hernial sac)

Intestinal loops
(as hernial contents)

Medial inguinal fossa

Superficial inguinal ring

Parietal peritoneum
(as hernial sac)

Intestinal loops
(as hernial contents)

Epididymis

Testis

Tunica vaginalis testis

28 Inguinal region of a male

a, b   Schematized sections through the anterior abdominal wall
         on the level of the inguinal canal and through the scrotum
         (according to Benninghoff, 1985).
   a   Normal situation
   b   Inguinal herniae. On the right side of the body, a lateral indirect
         hernia through the inguinal canal; the inner hernial ring is the
         deep inguinal ring lateral to the inferior epigastric vessels.
         On the left side of the body, a medial direct hernia; the inner hernial ring
         is the medial inguinal fossa medial to the inferior epigastric vessels.

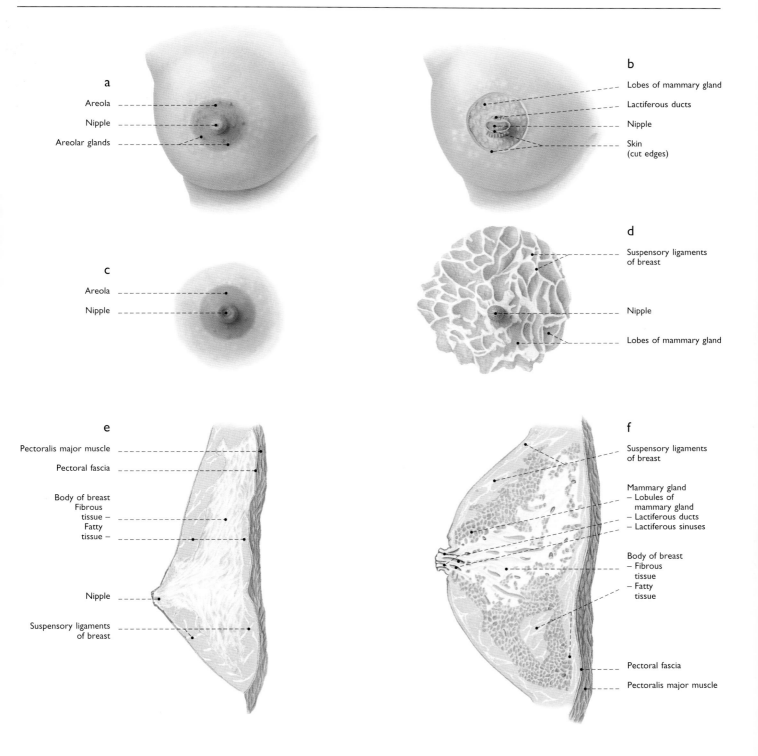

a

Areola

Nipple

Areolar glands

b

Lobes of mammary gland

Lactiferous ducts

Nipple

Skin
(cut edges)

c

Areola

Nipple

d

Suspensory ligaments
of breast

Nipple

Lobes of mammary gland

e

Pectoralis major muscle

Pectoral fascia

Body of breast
Fibrous
tissue –
Fatty
tissue –

Nipple

Suspensory ligaments
of breast

f

Suspensory ligaments
of breast

Mammary gland
– Lobules of
mammary gland
– Lactiferous ducts
– Lactiferous sinuses

Body of breast
– Fibrous
tissue
– Fatty
tissue

Pectoral fascia

Pectoralis major muscle

29 Breast

a   Ventral aspect (40%)
b   Ventral aspect (40%). The skin around the nipple was removed.
c   Depressed nipple (70%)
d   Parenchyma of the mammary gland after removal of
    the skin and the subcutaneous tissue (40%)
e   Sagittal section through the breast of a 16-year-old
    non-pregnant nullipara (60%)
f   Sagittal section through the breast of a 28-year-old
    woman immediately before lactation (60%)

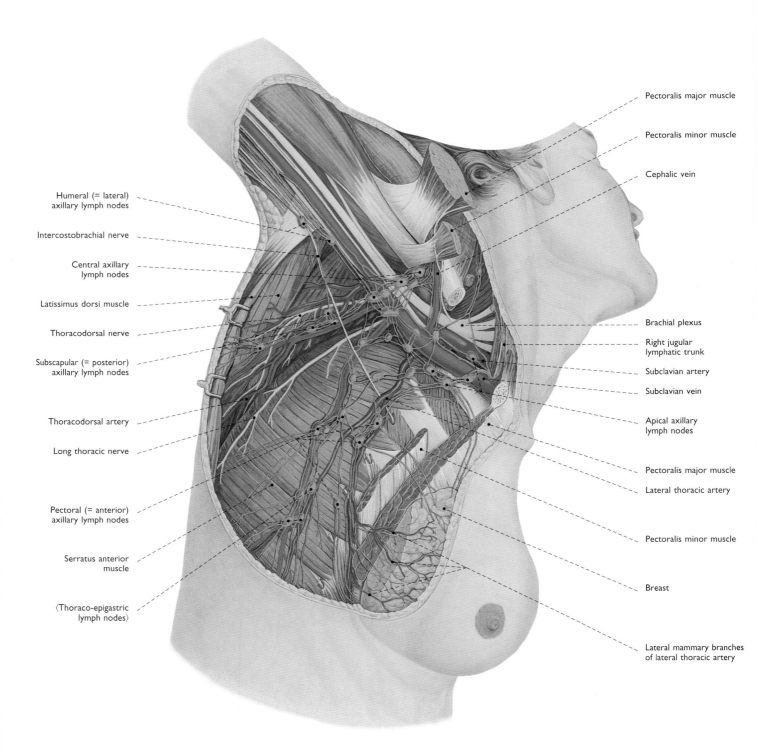

Pectoralis major muscle

Pectoralis minor muscle

Cephalic vein

Humeral (= lateral) axillary lymph nodes

Intercostobrachial nerve

Central axillary lymph nodes

Latissimus dorsi muscle

Thoracodorsal nerve

Subscapular (= posterior) axillary lymph nodes

Thoracodorsal artery

Long thoracic nerve

Pectoral (= anterior) axillary lymph nodes

Serratus anterior muscle

⟨Thoraco-epigastric lymph nodes⟩

Brachial plexus

Right jugular lymphatic trunk

Subclavian artery

Subclavian vein

Apical axillary lymph nodes

Pectoralis major muscle

Lateral thoracic artery

Pectoralis minor muscle

Breast

Lateral mammary branches of lateral thoracic artery

30 Lymphatic vessels and lymph nodes of the axilla and the anterior chest wall (50%)
Lateral aspect

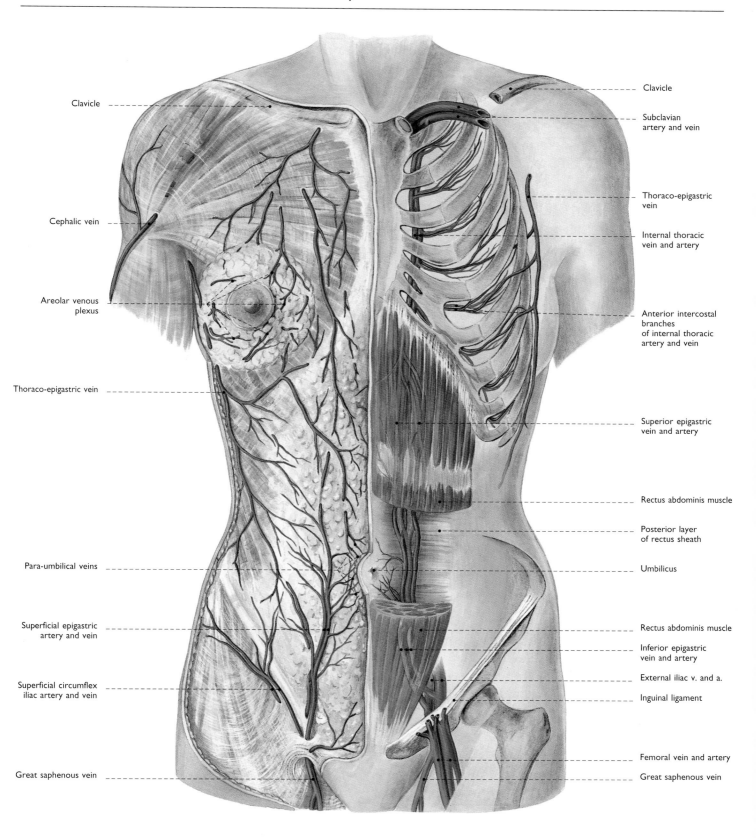

Clavicle

Cephalic vein

Areolar venous
plexus

Thoraco-epigastric vein

Para-umbilical veins

Superficial epigastric
artery and vein

Superficial circumflex
iliac artery and vein

Great saphenous vein

Clavicle

Subclavian
artery and vein

Thoraco-epigastric
vein

Internal thoracic
vein and artery

Anterior intercostal
branches
of internal thoracic
artery and vein

Superior epigastric
vein and artery

Rectus abdominis muscle

Posterior layer
of rectus sheath

Umbilicus

Rectus abdominis muscle

Inferior epigastric
vein and artery

External iliac v. and a.

Inguinal ligament

Femoral vein and artery

Great saphenous vein

31 Blood vessels of the ventral body wall (35%)
On the right side of the body, superficial vessels in the subcutaneous fatty tissue;
on the left side of the body, deep vessels shining through the covering layers
(the rectus abdominis muscle is cut above and below the umbilical region)

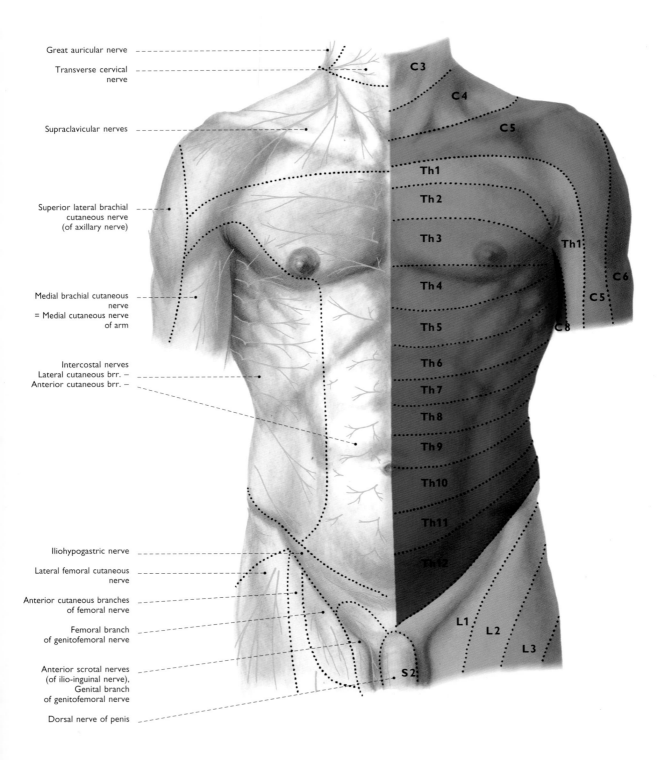

Great auricular nerve

Transverse cervical nerve

Supraclavicular nerves

Superior lateral brachial cutaneous nerve (of axillary nerve)

Medial brachial cutaneous nerve = Medial cutaneous nerve of arm

Intercostal nerves
Lateral cutaneous brr. —
Anterior cutaneous brr. —

Iliohypogastric nerve

Lateral femoral cutaneous nerve

Anterior cutaneous branches of femoral nerve

Femoral branch of genitofemoral nerve

Anterior scrotal nerves (of ilio-inguinal nerve), Genital branch of genitofemoral nerve

Dorsal nerve of penis

C 3
C 4
C 5
Th 1
Th 2
Th 3
Th 1
C 6
C 5
Th 4
C 8
Th 5
Th 6
Th 7
Th 8
Th 9
Th 10
Th 11
Th 12
L 1
L 2
L 3
S 2

32 Cutaneous and segmental innervation of the ventral body wall (25%)
Schematic representation

Upper Limb

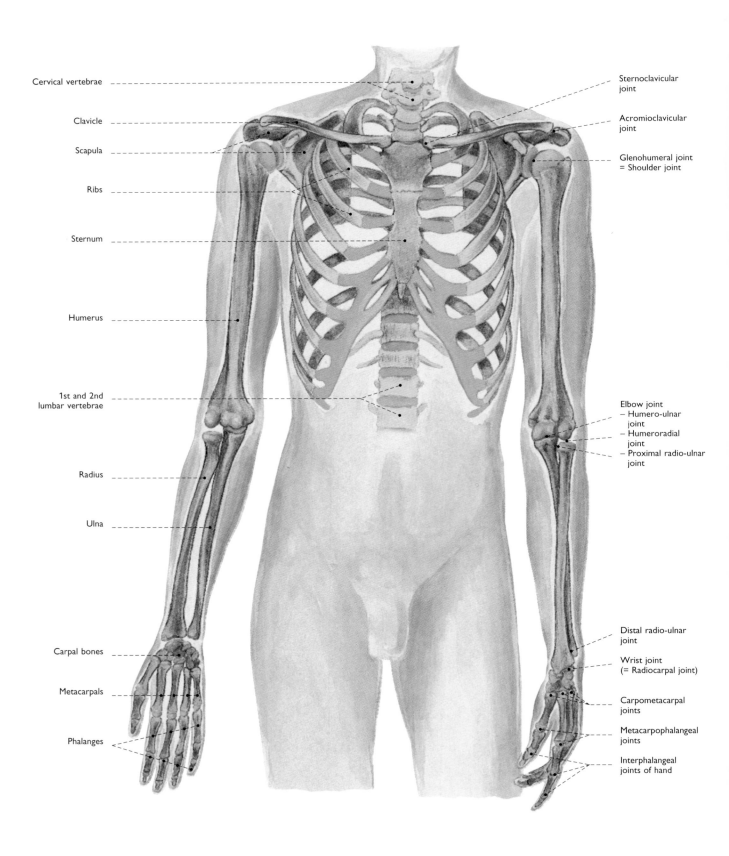

Cervical vertebrae

Clavicle

Scapula

Ribs

Sternum

Humerus

1st and 2nd
lumbar vertebrae

Radius

Ulna

Carpal bones

Metacarpals

Phalanges

Sternoclavicular
joint

Acromioclavicular
joint

Glenohumeral joint
= Shoulder joint

Elbow joint
– Humero-ulnar
  joint
– Humeroradial
  joint
– Proximal radio-ulnar
  joint

Distal radio-ulnar
joint

Wrist joint
(= Radiocarpal joint)

Carpometacarpal
joints

Metacarpophalangeal
joints

Interphalangeal
joints of hand

34 Upper limbs and thorax (25%)
Ventral aspect

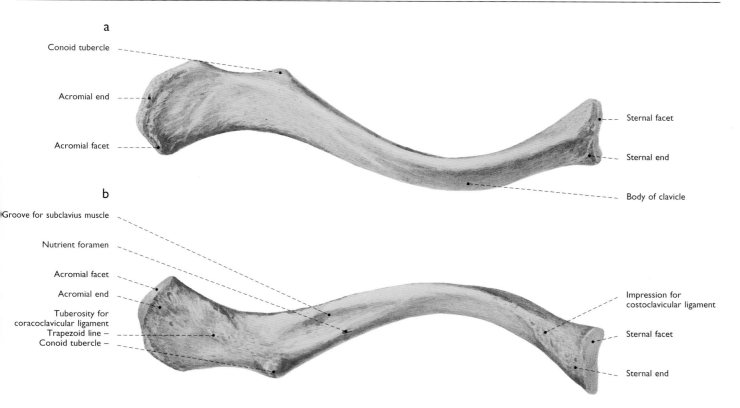

a

Conoid tubercle

Acromial end

Acromial facet

Sternal facet

Sternal end

Body of clavicle

b

Groove for subclavius muscle

Nutrient foramen

Acromial facet

Acromial end

Tuberosity for coracoclavicular ligament

Trapezoid line

Conoid tubercle

Impression for costoclavicular ligament

Sternal facet

Sternal end

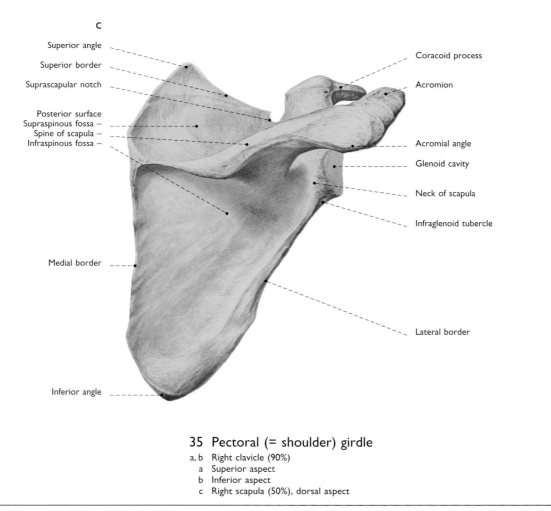

c

Superior angle

Superior border

Suprascapular notch

Posterior surface

Supraspinous fossa

Spine of scapula

Infraspinous fossa

Medial border

Inferior angle

Coracoid process

Acromion

Acromial angle

Glenoid cavity

Neck of scapula

Infraglenoid tubercle

Lateral border

35 Pectoral (= shoulder) girdle

a, b Right clavicle (90%)
  a Superior aspect
  b Inferior aspect
  c Right scapula (50%), dorsal aspect

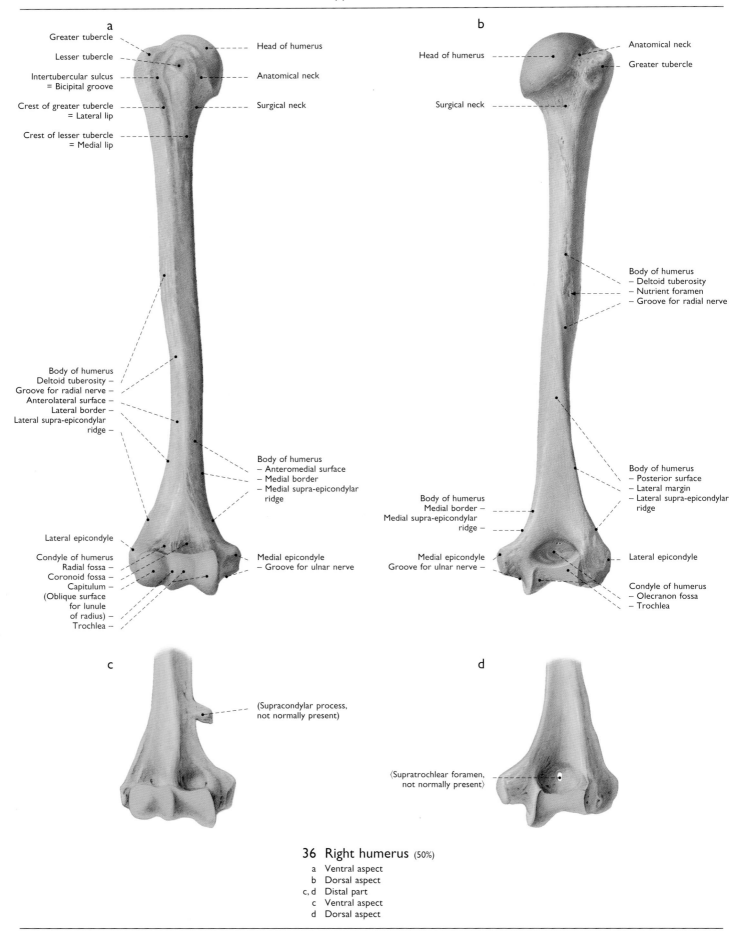

a

Greater tubercle
Lesser tubercle
Intertubercular sulcus = Bicipital groove
Crest of greater tubercle = Lateral lip
Crest of lesser tubercle = Medial lip

Head of humerus
Anatomical neck
Surgical neck

Body of humerus
Deltoid tuberosity –
Groove for radial nerve –
Anterolateral surface –
Lateral border –
Lateral supra-epicondylar ridge –

Body of humerus
– Anteromedial surface
– Medial border
– Medial supra-epicondylar ridge

Lateral epicondyle
Condyle of humerus
Radial fossa –
Coronoid fossa –
Capitulum –
(Oblique surface for lunule of radius) –
Trochlea –

Medial epicondyle
– Groove for ulnar nerve

b

Head of humerus
Surgical neck

Anatomical neck
Greater tubercle

Body of humerus
– Deltoid tuberosity
– Nutrient foramen
– Groove for radial nerve

Body of humerus
Medial border –
Medial supra-epicondylar ridge –

Body of humerus
– Posterior surface
– Lateral margin
– Lateral supra-epicondylar ridge

Medial epicondyle
Groove for ulnar nerve –

Lateral epicondyle

Condyle of humerus
– Olecranon fossa
– Trochlea

c

(Supracondylar process, not normally present)

d

(Supratrochlear foramen, not normally present)

36 Right humerus (50%)
a   Ventral aspect
b   Dorsal aspect
c, d Distal part
c   Ventral aspect
d   Dorsal aspect

a   b   c

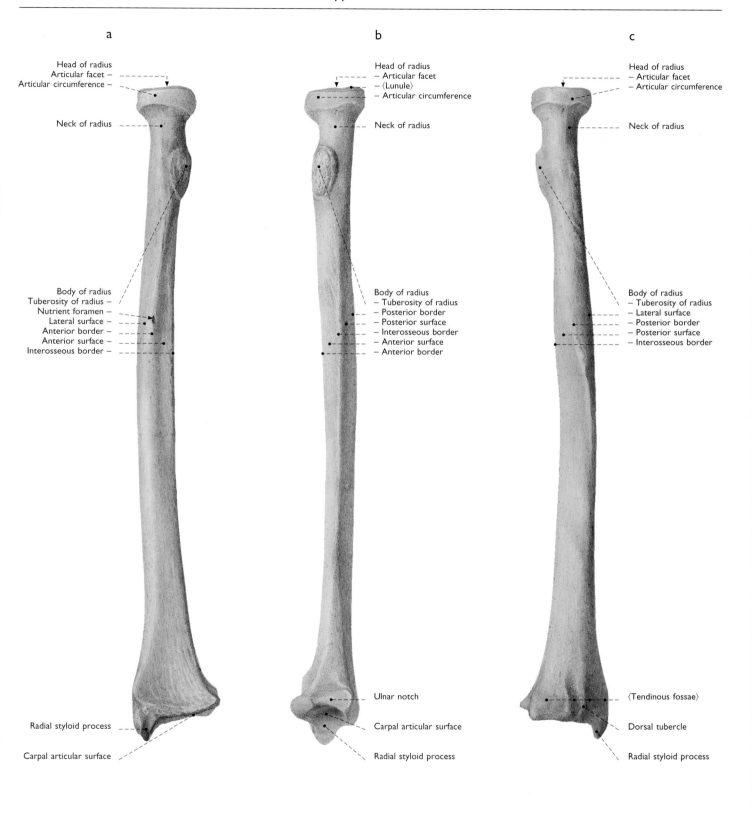

Head of radius
Articular facet —
Articular circumference —

Neck of radius —

Body of radius
Tuberosity of radius —
Nutrient foramen —
Lateral surface —
Anterior border —
Anterior surface —
Interosseous border —

Radial styloid process

Carpal articular surface

Head of radius
— Articular facet
— ⟨Lunule⟩
— Articular circumference

Neck of radius

Body of radius
— Tuberosity of radius
— Posterior border
— Posterior surface
— Interosseous border
— Anterior surface
— Anterior border

Ulnar notch

Carpal articular surface

Radial styloid process

Head of radius
— Articular facet
— Articular circumference

Neck of radius

Body of radius
— Tuberosity of radius
— Lateral surface
— Posterior border
— Posterior surface
— Interosseous border

⟨Tendinous fossae⟩

Dorsal tubercle

Radial styloid process

37 Right radius (70%)

a Ventral aspect
b Medial aspect
c Dorsal aspect

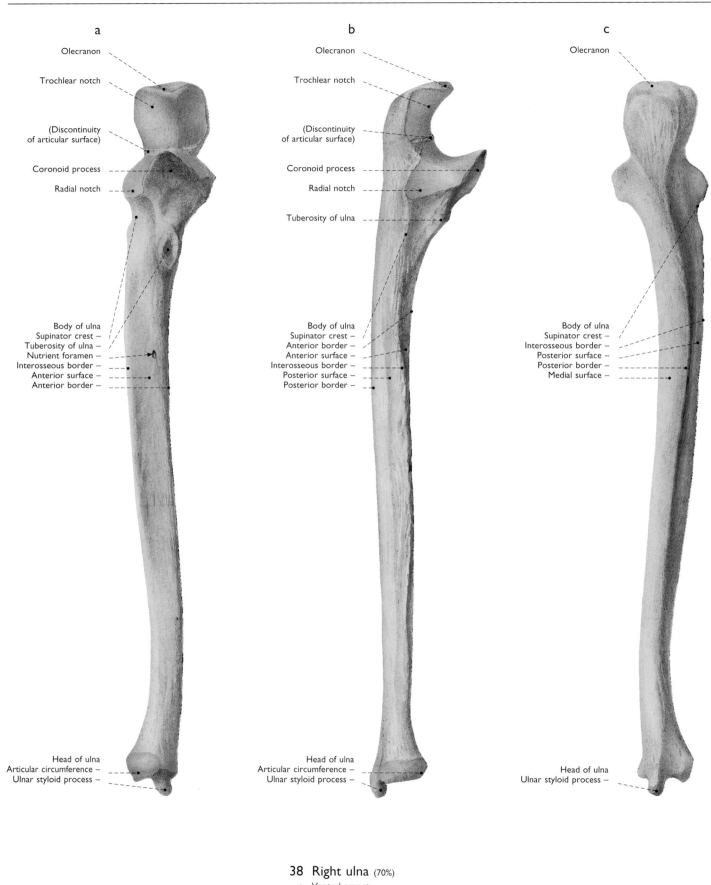

a

Olecranon
Trochlear notch
(Discontinuity of articular surface)
Coronoid process
Radial notch

Body of ulna
Supinator crest –
Tuberosity of ulna –
Nutrient foramen –
Interosseous border –
Anterior surface –
Anterior border –

Head of ulna
Articular circumference –
Ulnar styloid process –

b

Olecranon
Trochlear notch
(Discontinuity of articular surface)
Coronoid process
Radial notch
Tuberosity of ulna

Body of ulna
Supinator crest –
Anterior border –
Anterior surface –
Interosseous border –
Posterior surface –
Posterior border –

Head of ulna
Articular circumference –
Ulnar styloid process –

c

Olecranon

Body of ulna
Supinator crest –
Interosseous border –
Posterior surface –
Posterior border –
Medial surface –

Head of ulna
Ulnar styloid process –

38 Right ulna (70%)

a Ventral aspect
b Lateral aspect
c Dorsal aspect

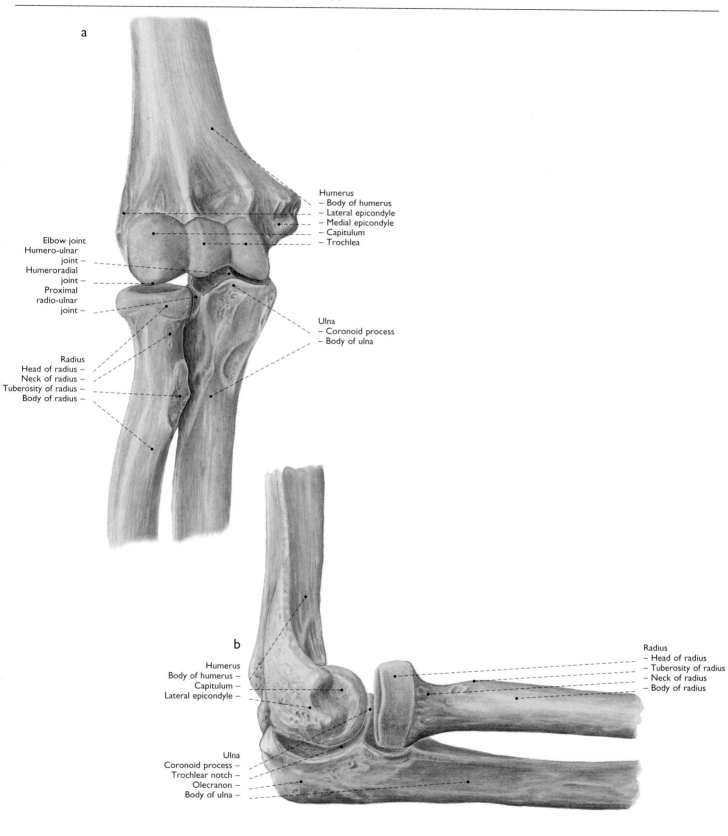

a

Humerus
– Body of humerus
– Lateral epicondyle
– Medial epicondyle
– Capitulum
– Trochlea

Elbow joint
Humero-ulnar
joint –
Humeroradial
joint –
Proximal
radio-ulnar
joint –

Ulna
– Coronoid process
– Body of ulna

Radius
Head of radius –
Neck of radius –
Tuberosity of radius –
Body of radius –

b

Humerus
Body of humerus –
Capitulum –
Lateral epicondyle –

Ulna
Coronoid process –
Trochlear notch –
Olecranon –
Body of ulna –

Radius
– Head of radius
– Tuberosity of radius
– Neck of radius
– Body of radius

39 Bones of the right elbow joint (90%)

a   Ventral aspect
b   Lateral (radial) aspect

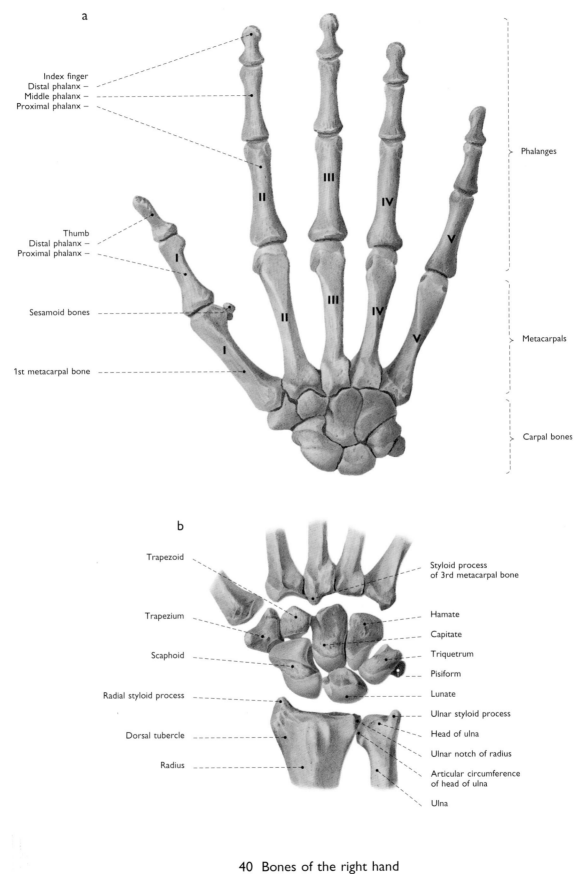

a

Index finger
Distal phalanx –
Middle phalanx –
Proximal phalanx –

Thumb
Distal phalanx –
Proximal phalanx –

Sesamoid bones

1st metacarpal bone

Phalanges

Metacarpals

Carpal bones

b

Trapezoid

Trapezium

Scaphoid

Radial styloid process

Dorsal tubercle

Radius

Styloid process
of 3rd metacarpal bone

Hamate

Capitate

Triquetrum

Pisiform

Lunate

Ulnar styloid process

Head of ulna

Ulnar notch of radius

Articular circumference
of head of ulna

Ulna

40   Bones of the right hand
    a   Dorsal aspect (60%)
    b   Carpal bones (70%), dorsal aspect

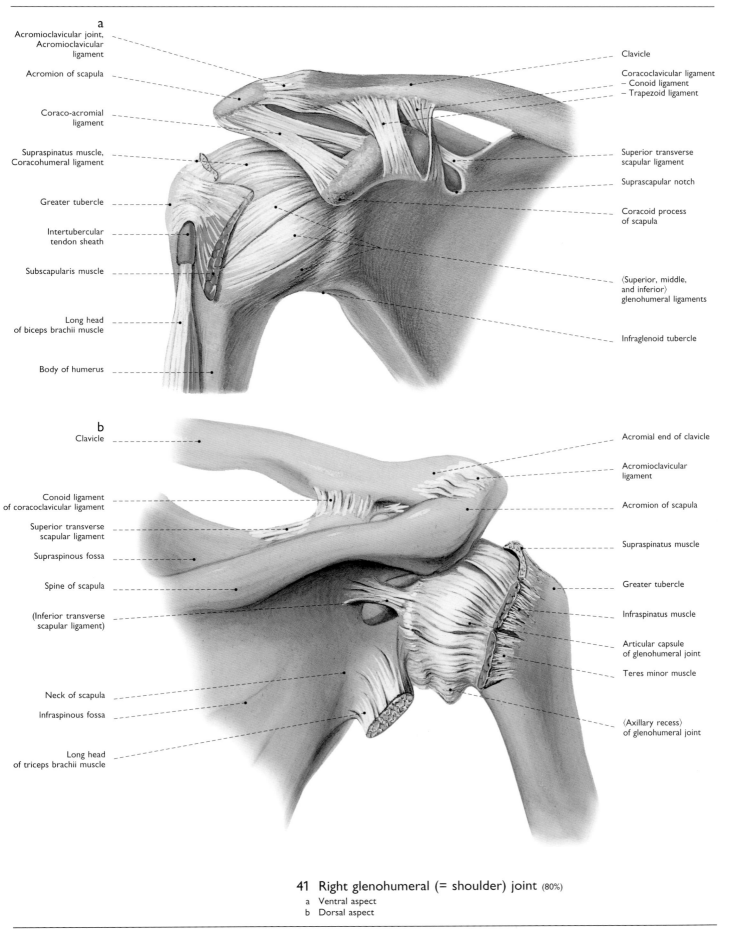

a

Acromioclavicular joint, Acromioclavicular ligament

Acromion of scapula

Coraco-acromial ligament

Supraspinatus muscle, Coracohumeral ligament

Greater tubercle

Intertubercular tendon sheath

Subscapularis muscle

Long head of biceps brachii muscle

Body of humerus

Clavicle

Coracoclavicular ligament
– Conoid ligament
– Trapezoid ligament

Superior transverse scapular ligament

Suprascapular notch

Coracoid process of scapula

⟨Superior, middle, and inferior⟩ glenohumeral ligaments

Infraglenoid tubercle

b

Clavicle

Conoid ligament of coracoclavicular ligament

Superior transverse scapular ligament

Supraspinous fossa

Spine of scapula

(Inferior transverse scapular ligament)

Neck of scapula

Infraspinous fossa

Long head of triceps brachii muscle

Acromial end of clavicle

Acromioclavicular ligament

Acromion of scapula

Supraspinatus muscle

Greater tubercle

Infraspinatus muscle

Articular capsule of glenohumeral joint

Teres minor muscle

⟨Axillary recess⟩ of glenohumeral joint

41 Right glenohumeral (= shoulder) joint (80%)
   a Ventral aspect
   b Dorsal aspect

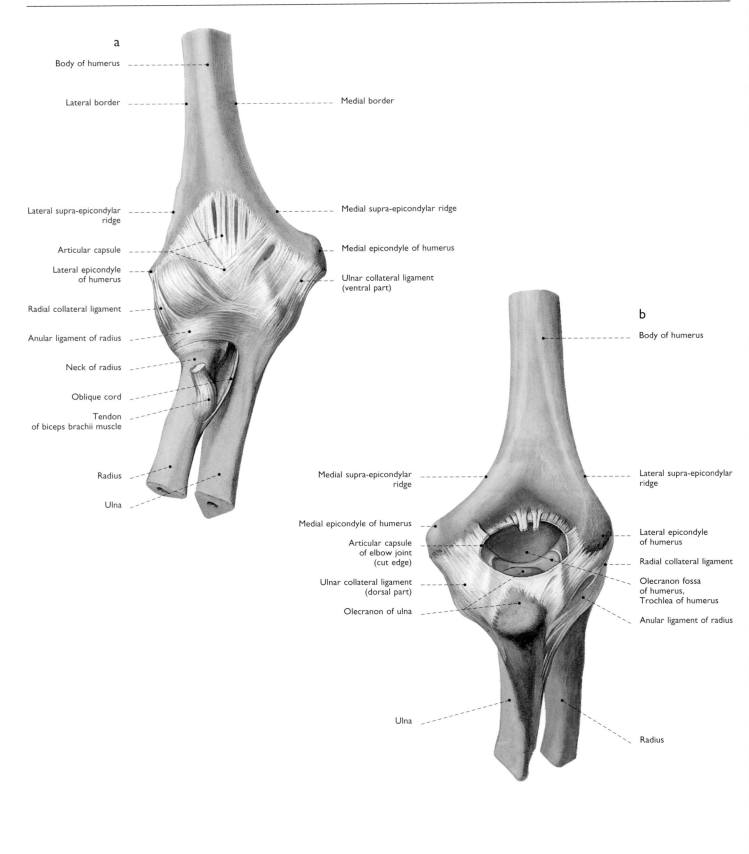

a

Body of humerus

Lateral border

Medial border

Lateral supra-epicondylar
ridge

Medial supra-epicondylar ridge

Articular capsule

Medial epicondyle of humerus

Lateral epicondyle
of humerus

Radial collateral ligament

Ulnar collateral ligament
(ventral part)

Anular ligament of radius

Neck of radius

Oblique cord

Tendon
of biceps brachii muscle

Radius

Ulna

b

Body of humerus

Medial supra-epicondylar
ridge

Lateral supra-epicondylar
ridge

Medial epicondyle of humerus

Articular capsule
of elbow joint
(cut edge)

Lateral epicondyle
of humerus

Radial collateral ligament

Ulnar collateral ligament
(dorsal part)

Olecranon fossa
of humerus,
Trochlea of humerus

Olecranon of ulna

Anular ligament of radius

Ulna

Radius

42 Right elbow joint (85%)
a Ventral aspect
b Dorsal aspect

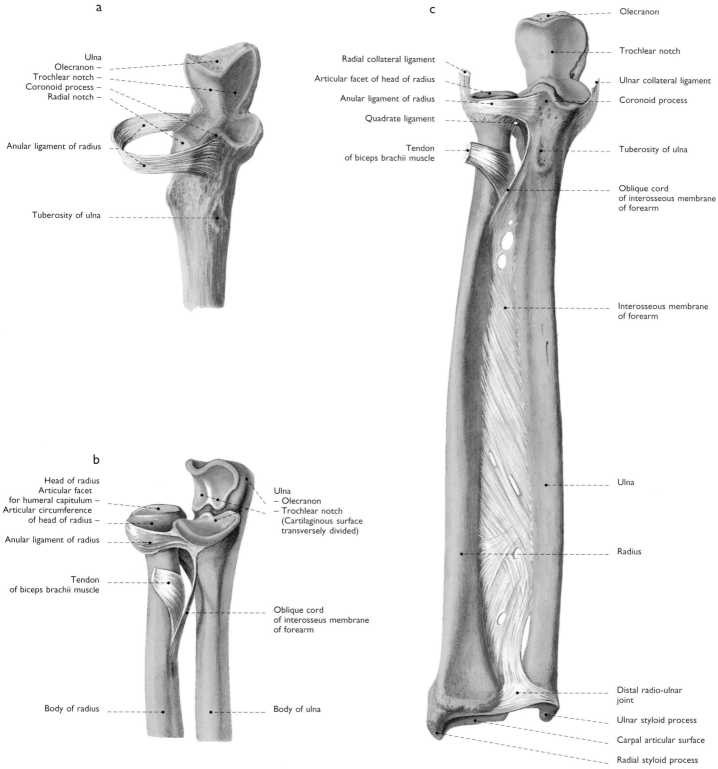

a

Ulna
Olecranon –
Trochlear notch –
Coronoid process –
Radial notch –

Anular ligament of radius

Tuberosity of ulna

b

Head of radius
Articular facet
for humeral capitulum –
Articular circumference
of head of radius –

Anular ligament of radius

Tendon
of biceps brachii muscle

Body of radius

Ulna
– Olecranon
– Trochlear notch
(Cartilaginous surface
transversely divided)

Oblique cord
of interosseus membrane
of forearm

Body of ulna

c

Radial collateral ligament

Articular facet of head of radius

Anular ligament of radius

Quadrate ligament

Tendon
of biceps brachii muscle

Olecranon

Trochlear notch

Ulnar collateral ligament

Coronoid process

Tuberosity of ulna

Oblique cord
of interosseous membrane
of forearm

Interosseous membrane
of forearm

Ulna

Radius

Distal radio-ulnar
joint

Ulnar styloid process

Carpal articular surface

Radial styloid process

43 Radio-ulnar joints of the right forearm
a Proximal end of ulna and anular ligament of radius (80%),
   ventral aspect
b Proximal radio-ulnar joint (70%), ventral aspect
c Forearm bones in supinated position (70%), ventral aspect

a

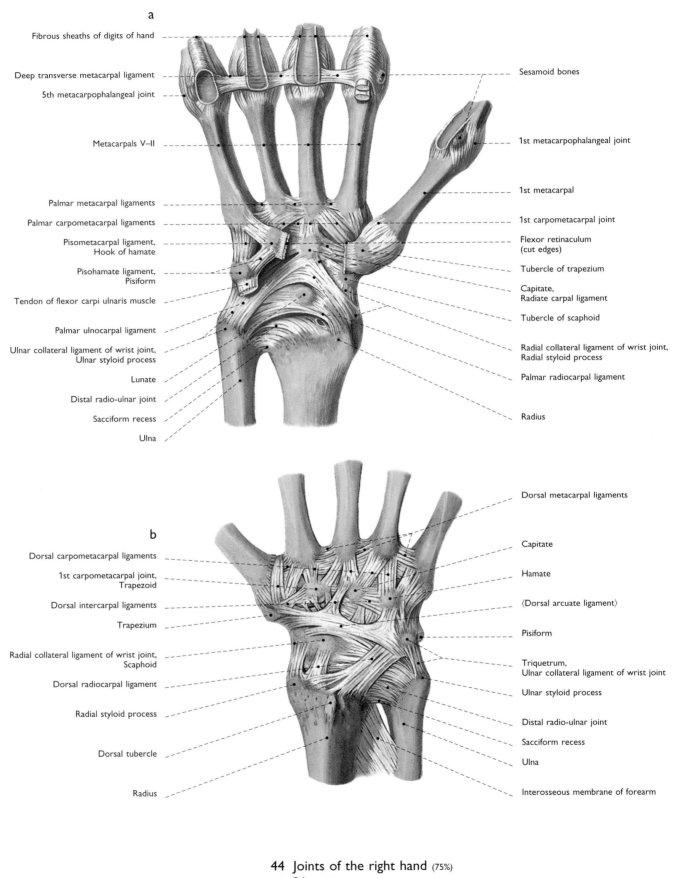

Fibrous sheaths of digits of hand

Deep transverse metacarpal ligament

5th metacarpophalangeal joint

Metacarpals V–II

Palmar metacarpal ligaments

Palmar carpometacarpal ligaments

Pisometacarpal ligament,
Hook of hamate

Pisohamate ligament,
Pisiform

Tendon of flexor carpi ulnaris muscle

Palmar ulnocarpal ligament

Ulnar collateral ligament of wrist joint,
Ulnar styloid process

Lunate

Distal radio-ulnar joint

Sacciform recess

Ulna

Sesamoid bones

1st metacarpophalangeal joint

1st metacarpal

1st carpometacarpal joint

Flexor retinaculum
(cut edges)

Tubercle of trapezium

Capitate,
Radiate carpal ligament

Tubercle of scaphoid

Radial collateral ligament of wrist joint,
Radial styloid process

Palmar radiocarpal ligament

Radius

b

Dorsal carpometacarpal ligaments

1st carpometacarpal joint,
Trapezoid

Dorsal intercarpal ligaments

Trapezium

Radial collateral ligament of wrist joint,
Scaphoid

Dorsal radiocarpal ligament

Radial styloid process

Dorsal tubercle

Radius

Dorsal metacarpal ligaments

Capitate

Hamate

⟨Dorsal arcuate ligament⟩

Pisiform

Triquetrum,
Ulnar collateral ligament of wrist joint

Ulnar styloid process

Distal radio-ulnar joint

Sacciform recess

Ulna

Interosseous membrane of forearm

44 Joints of the right hand (75%)
 a   Palmar aspect
 b   Dorsal aspect

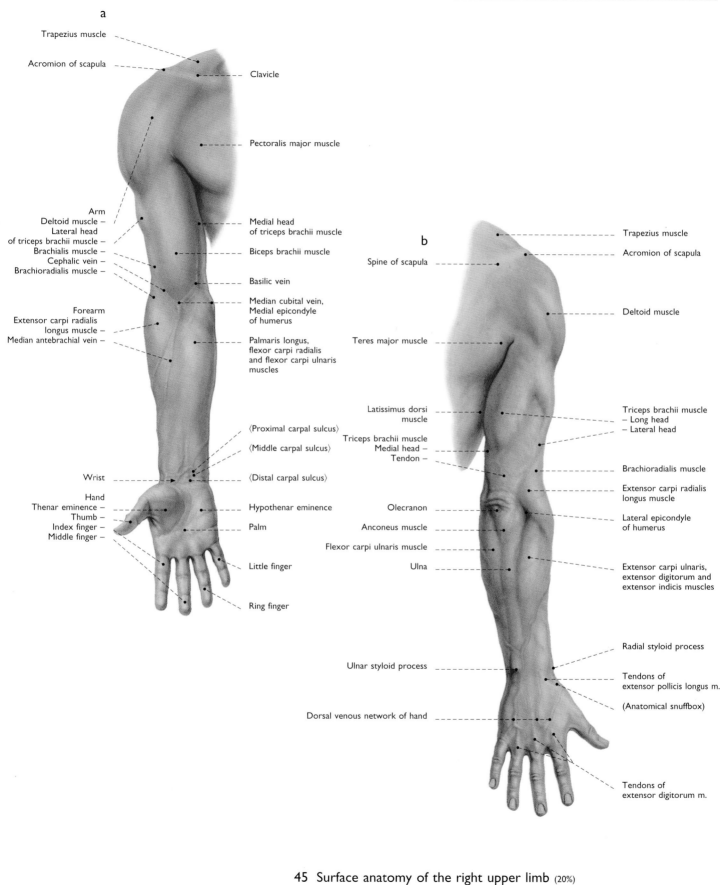

a

Trapezius muscle

Acromion of scapula

Clavicle

Pectoralis major muscle

Arm
Deltoid muscle –
Lateral head
of triceps brachii muscle –
Brachialis muscle –
Cephalic vein –
Brachioradialis muscle –

Medial head
of triceps brachii muscle

Biceps brachii muscle

Basilic vein

Median cubital vein,
Medial epicondyle
of humerus

Forearm
Extensor carpi radialis
longus muscle –
Median antebrachial vein –

Palmaris longus,
flexor carpi radialis
and flexor carpi ulnaris
muscles

⟨Proximal carpal sulcus⟩

⟨Middle carpal sulcus⟩

Wrist

⟨Distal carpal sulcus⟩

Hand
Thenar eminence –
Thumb –
Index finger –
Middle finger –

Hypothenar eminence

Palm

Little finger

Ring finger

b

Spine of scapula

Trapezius muscle

Acromion of scapula

Deltoid muscle

Teres major muscle

Latissimus dorsi
muscle

Triceps brachii muscle
– Long head
– Lateral head

Triceps brachii muscle
Medial head –
Tendon –

Brachioradialis muscle

Extensor carpi radialis
longus muscle

Olecranon

Lateral epicondyle
of humerus

Anconeus muscle

Flexor carpi ulnaris muscle

Ulna

Extensor carpi ulnaris,
extensor digitorum and
extensor indicis muscles

Radial styloid process

Ulnar styloid process

Tendons of
extensor pollicis longus m.

(Anatomical snuffbox)

Dorsal venous network of hand

Tendons of
extensor digitorum m.

45 Surface anatomy of the right upper limb (20%)
a Ventral aspect
b Dorsal aspect

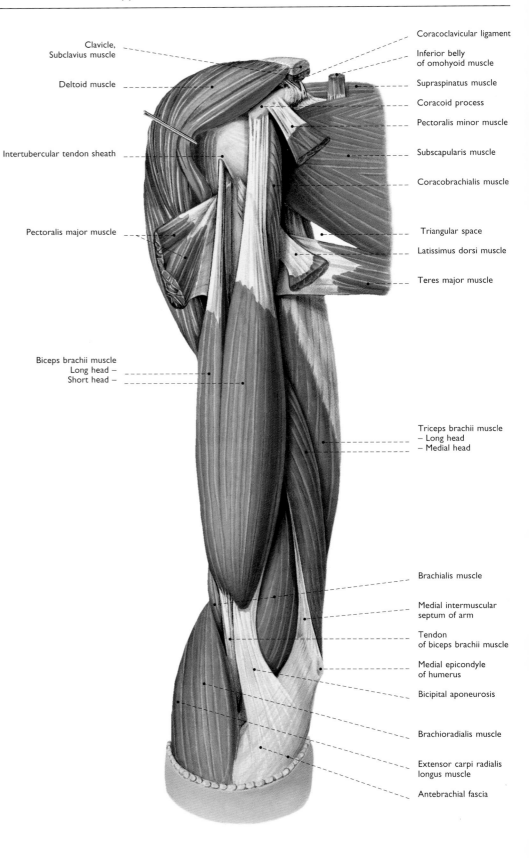

Clavicle,
Subclavius muscle

Deltoid muscle

Intertubercular tendon sheath

Pectoralis major muscle

Biceps brachii muscle
Long head –
Short head –

Coracoclavicular ligament

Inferior belly
of omohyoid muscle

Supraspinatus muscle

Coracoid process

Pectoralis minor muscle

Subscapularis muscle

Coracobrachialis muscle

Triangular space

Latissimus dorsi muscle

Teres major muscle

Triceps brachii muscle
– Long head
– Medial head

Brachialis muscle

Medial intermuscular
septum of arm

Tendon
of biceps brachii muscle

Medial epicondyle
of humerus

Bicipital aponeurosis

Brachioradialis muscle

Extensor carpi radialis
longus muscle

Antebrachial fascia

**46 Muscles of the right shoulder
and the right arm** (50%)
Ventral aspect

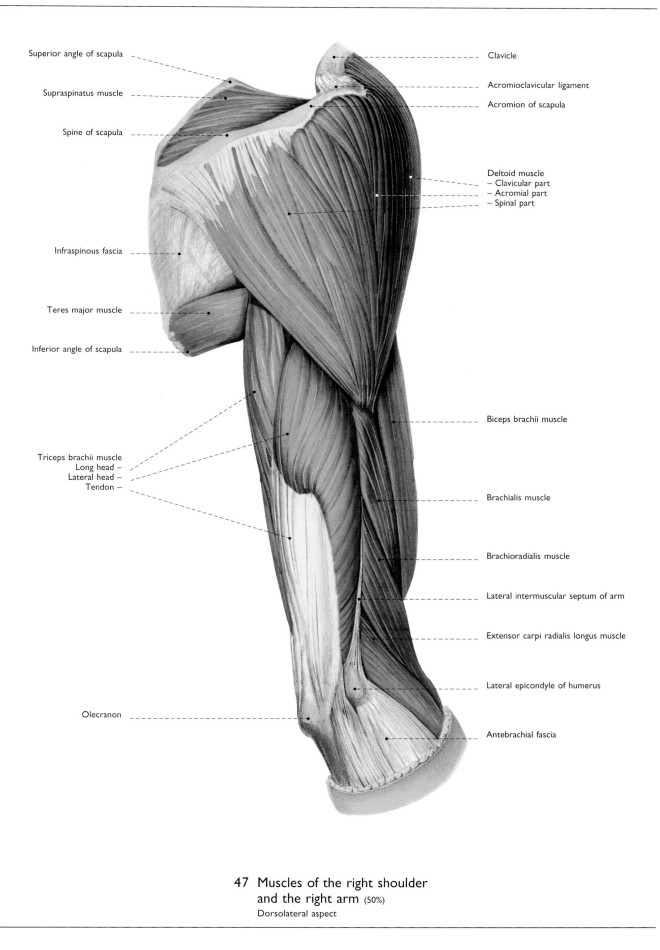

Superior angle of scapula

Supraspinatus muscle

Spine of scapula

Infraspinous fascia

Teres major muscle

Inferior angle of scapula

Triceps brachii muscle
Long head –
Lateral head –
Tendon –

Olecranon

Clavicle

Acromioclavicular ligament

Acromion of scapula

Deltoid muscle
– Clavicular part
– Acromial part
– Spinal part

Biceps brachii muscle

Brachialis muscle

Brachioradialis muscle

Lateral intermuscular septum of arm

Extensor carpi radialis longus muscle

Lateral epicondyle of humerus

Antebrachial fascia

**47 Muscles of the right shoulder
and the right arm** (50%)
Dorsolateral aspect

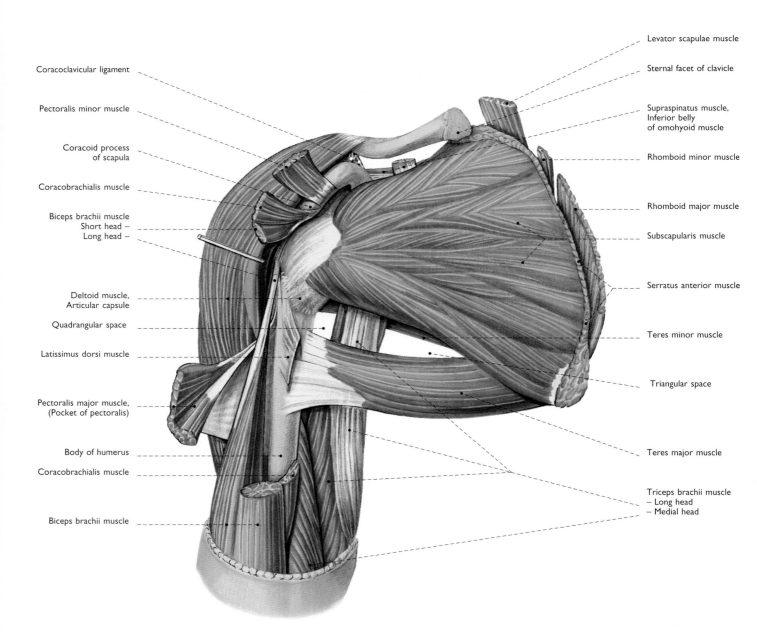

Coracoclavicular ligament

Pectoralis minor muscle

Coracoid process
of scapula

Coracobrachialis muscle

Biceps brachii muscle
Short head –
Long head –

Deltoid muscle,
Articular capsule

Quadrangular space

Latissimus dorsi muscle

Pectoralis major muscle,
(Pocket of pectoralis)

Body of humerus

Coracobrachialis muscle

Biceps brachii muscle

Levator scapulae muscle

Sternal facet of clavicle

Supraspinatus muscle,
Inferior belly
of omohyoid muscle

Rhomboid minor muscle

Rhomboid major muscle

Subscapularis muscle

Serratus anterior muscle

Teres minor muscle

Triangular space

Teres major muscle

Triceps brachii muscle
– Long head
– Medial head

**48 Muscles of the right shoulder
and the right arm** (60%)
Ventromedial aspect

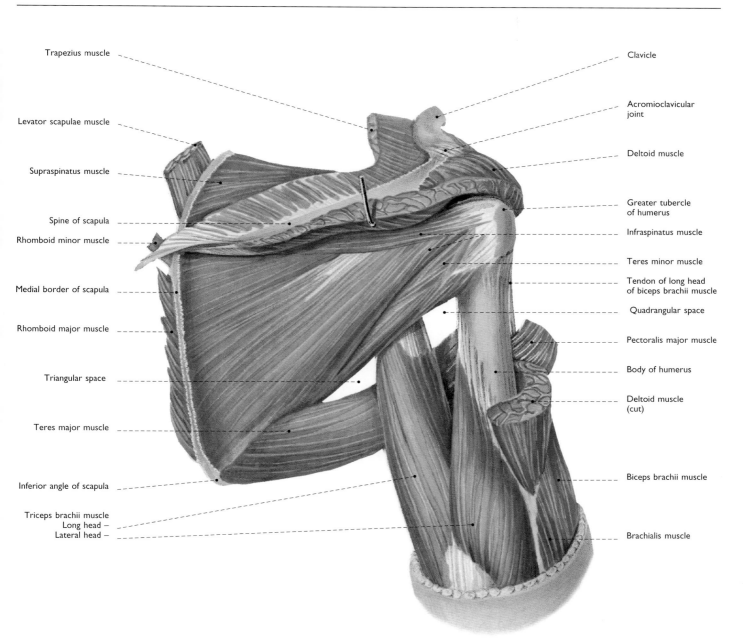

Trapezius muscle

Levator scapulae muscle

Supraspinatus muscle

Spine of scapula

Rhomboid minor muscle

Medial border of scapula

Rhomboid major muscle

Triangular space

Teres major muscle

Inferior angle of scapula

Triceps brachii muscle
Long head –
Lateral head –

Clavicle

Acromioclavicular
joint

Deltoid muscle

Greater tubercle
of humerus

Infraspinatus muscle

Teres minor muscle

Tendon of long head
of biceps brachii muscle

Quadrangular space

Pectoralis major muscle

Body of humerus

Deltoid muscle
(cut)

Biceps brachii muscle

Brachialis muscle

**49 Muscles of the right shoulder
and the right arm** (60%)
The deltoid muscle was partially removed.
Dorsal aspect

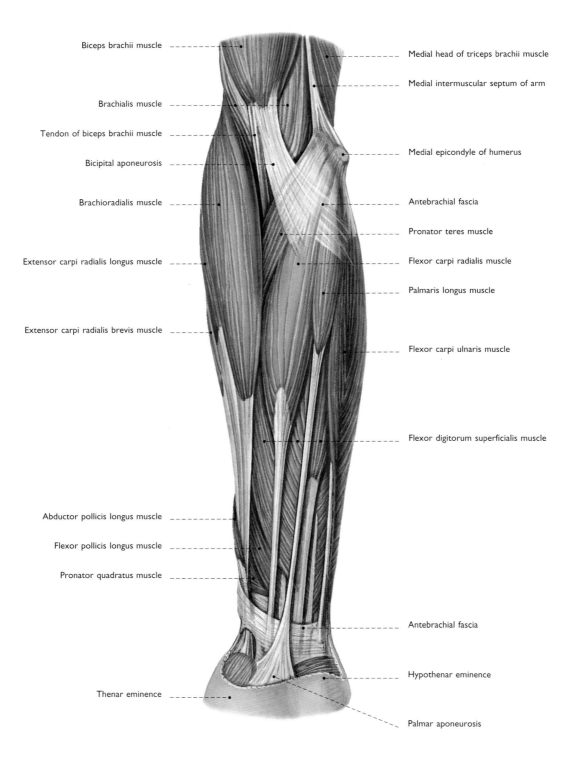

Biceps brachii muscle

Brachialis muscle

Tendon of biceps brachii muscle

Bicipital aponeurosis

Brachioradialis muscle

Extensor carpi radialis longus muscle

Extensor carpi radialis brevis muscle

Abductor pollicis longus muscle

Flexor pollicis longus muscle

Pronator quadratus muscle

Thenar eminence

Medial head of triceps brachii muscle

Medial intermuscular septum of arm

Medial epicondyle of humerus

Antebrachial fascia

Pronator teres muscle

Flexor carpi radialis muscle

Palmaris longus muscle

Flexor carpi ulnaris muscle

Flexor digitorum superficialis muscle

Antebrachial fascia

Hypothenar eminence

Palmar aponeurosis

50 Muscles of the right forearm (50%)
Superficial layer, ventral aspect

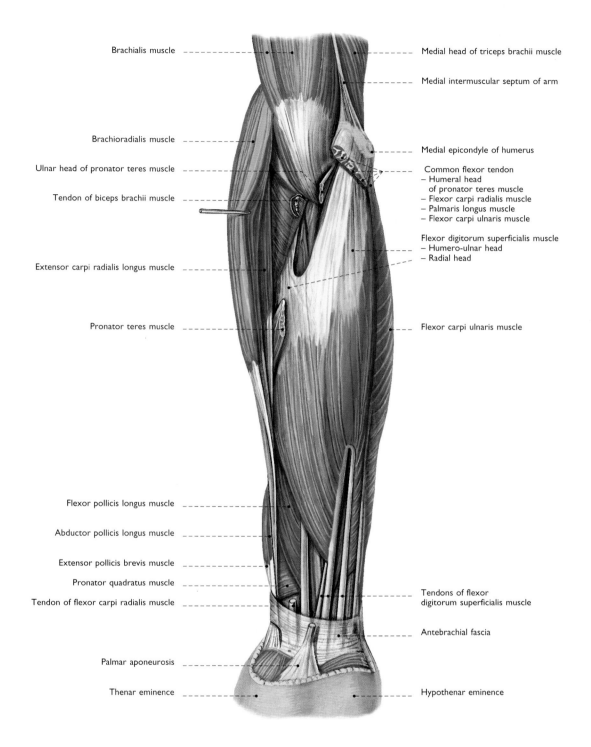

Brachialis muscle

Brachioradialis muscle

Ulnar head of pronator teres muscle

Tendon of biceps brachii muscle

Extensor carpi radialis longus muscle

Pronator teres muscle

Flexor pollicis longus muscle

Abductor pollicis longus muscle

Extensor pollicis brevis muscle

Pronator quadratus muscle

Tendon of flexor carpi radialis muscle

Palmar aponeurosis

Thenar eminence

Medial head of triceps brachii muscle

Medial intermuscular septum of arm

Medial epicondyle of humerus

Common flexor tendon
– Humeral head
  of pronator teres muscle
– Flexor carpi radialis muscle
– Palmaris longus muscle
– Flexor carpi ulnaris muscle

Flexor digitorum superficialis muscle
– Humero-ulnar head
– Radial head

Flexor carpi ulnaris muscle

Tendons of flexor
digitorum superficialis muscle

Antebrachial fascia

Hypothenar eminence

51 Muscles of the right forearm (50%)
Superficial layer. Some superficial forearm flexors were removed.
Ventral aspect

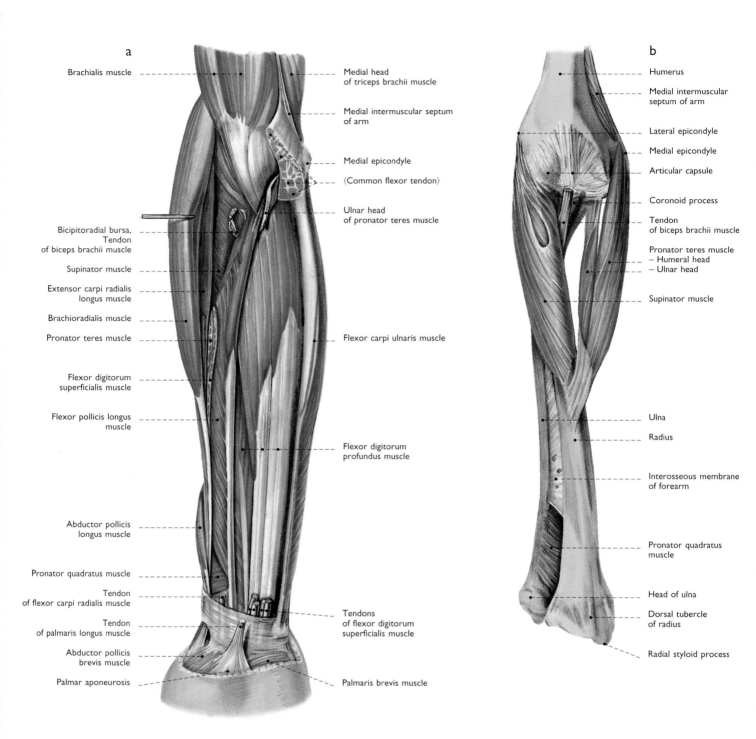

a

Brachialis muscle

Medial head
of triceps brachii muscle

Medial intermuscular septum
of arm

Medial epicondyle

⟨Common flexor tendon⟩

Ulnar head
of pronator teres muscle

Bicipitoradial bursa,
Tendon
of biceps brachii muscle

Supinator muscle

Extensor carpi radialis
longus muscle

Brachioradialis muscle

Pronator teres muscle

Flexor carpi ulnaris muscle

Flexor digitorum
superficialis muscle

Flexor pollicis longus
muscle

Flexor digitorum
profundus muscle

Abductor pollicis
longus muscle

Pronator quadratus muscle

Tendon
of flexor carpi radialis muscle

Tendon
of palmaris longus muscle

Abductor pollicis
brevis muscle

Tendons
of flexor digitorum
superficialis muscle

Palmar aponeurosis

Palmaris brevis muscle

b

Humerus

Medial intermuscular
septum of arm

Lateral epicondyle

Medial epicondyle

Articular capsule

Coronoid process

Tendon
of biceps brachii muscle

Pronator teres muscle
– Humeral head
– Ulnar head

Supinator muscle

Ulna

Radius

Interosseous membrane
of forearm

Pronator quadratus
muscle

Head of ulna

Dorsal tubercle
of radius

Radial styloid process

52 Muscles of the right forearm (50%)
Ventral aspect
a Deep layer
b Supinator and pronator muscles in pronated position
of the forearm

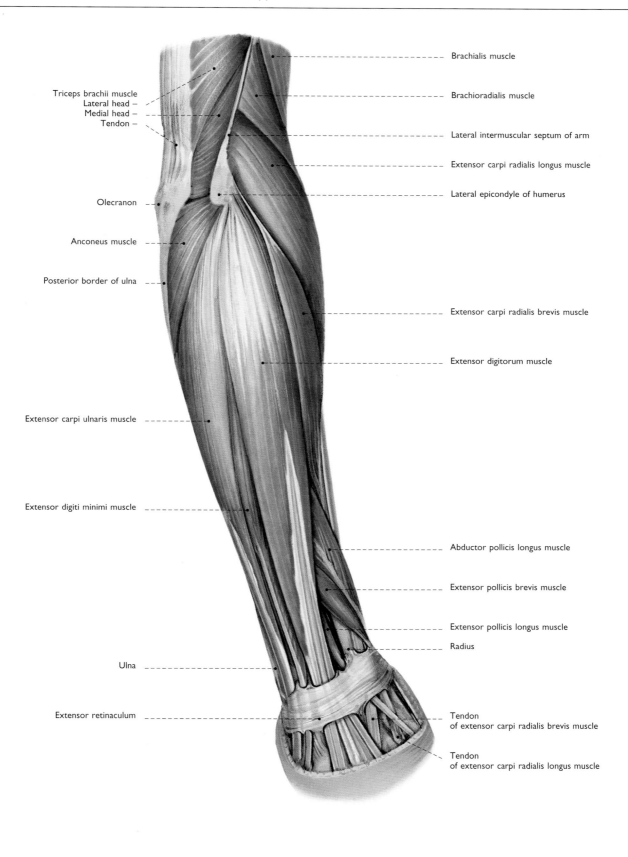

Triceps brachii muscle
Lateral head –
Medial head –
Tendon –

Olecranon –

Anconeus muscle –

Posterior border of ulna –

Extensor carpi ulnaris muscle –

Extensor digiti minimi muscle –

Ulna –

Extensor retinaculum –

Brachialis muscle

Brachioradialis muscle

Lateral intermuscular septum of arm

Extensor carpi radialis longus muscle

Lateral epicondyle of humerus

Extensor carpi radialis brevis muscle

Extensor digitorum muscle

Abductor pollicis longus muscle

Extensor pollicis brevis muscle

Extensor pollicis longus muscle

Radius

Tendon
of extensor carpi radialis brevis muscle

Tendon
of extensor carpi radialis longus muscle

53 Muscles of the right forearm (50%)
Superficial layer. The forearm is slightly pronated.
Dorsolateral aspect

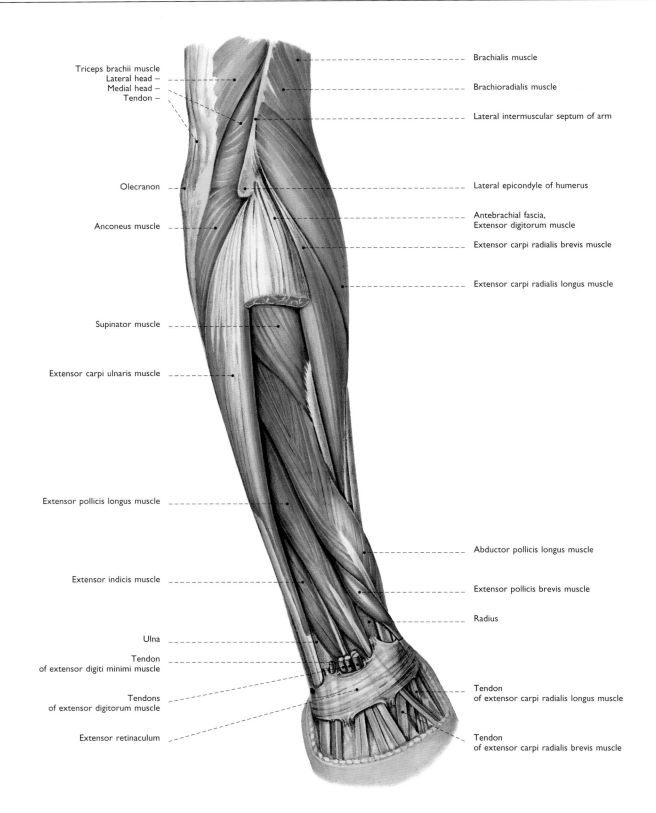

Triceps brachii muscle
Lateral head –
Medial head –
Tendon –

Olecranon

Anconeus muscle

Supinator muscle

Extensor carpi ulnaris muscle

Extensor pollicis longus muscle

Extensor indicis muscle

Ulna

Tendon
of extensor digiti minimi muscle

Tendons
of extensor digitorum muscle

Extensor retinaculum

Brachialis muscle

Brachioradialis muscle

Lateral intermuscular septum of arm

Lateral epicondyle of humerus

Antebrachial fascia,
Extensor digitorum muscle

Extensor carpi radialis brevis muscle

Extensor carpi radialis longus muscle

Abductor pollicis longus muscle

Extensor pollicis brevis muscle

Radius

Tendon
of extensor carpi radialis longus muscle

Tendon
of extensor carpi radialis brevis muscle

54 Muscles of the right forearm (50%)
Deep layer. The forearm is slightly pronated.
Dorsolateral aspect

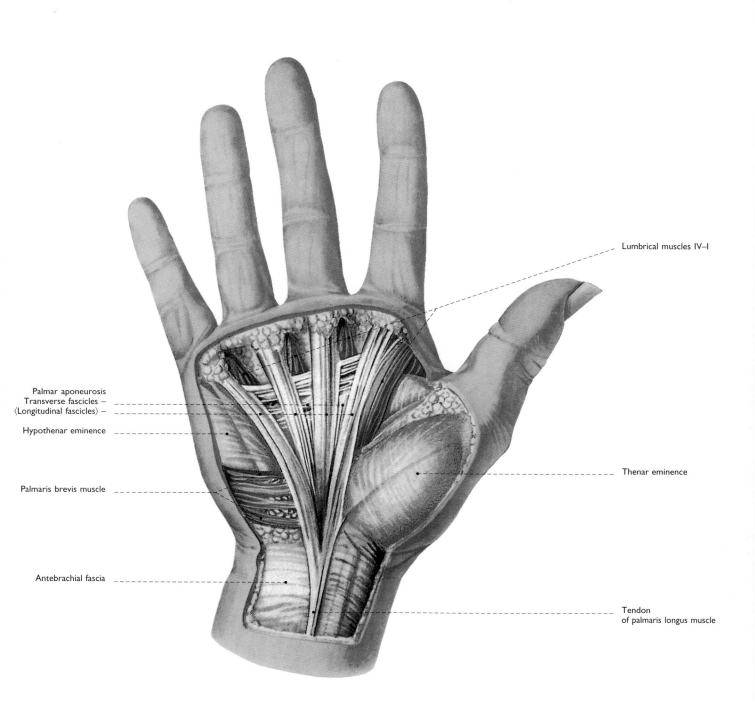

Lumbrical muscles IV–I

Palmar aponeurosis
Transverse fascicles –
(Longitudinal fascicles) –

Hypothenar eminence

Palmaris brevis muscle

Antebrachial fascia

Thenar eminence

Tendon
of palmaris longus muscle

55 Palmar aponeurosis of the right hand (75%)
Palmar aspect

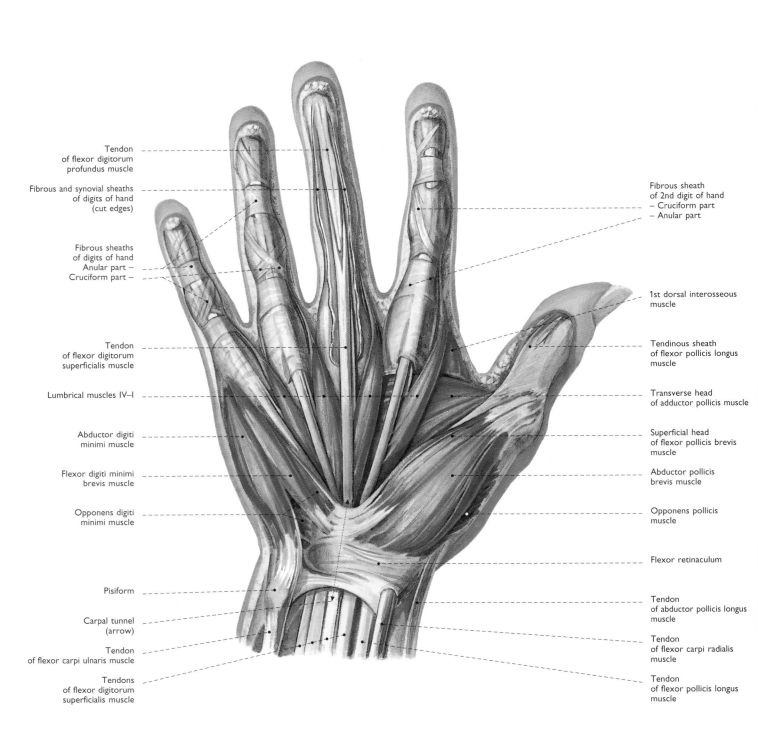

Tendon
of flexor digitorum
profundus muscle

Fibrous and synovial sheaths
of digits of hand
(cut edges)

Fibrous sheaths
of digits of hand
Anular part –
Cruciform part –

Tendon
of flexor digitorum
superficialis muscle

Lumbrical muscles IV–I

Abductor digiti
minimi muscle

Flexor digiti minimi
brevis muscle

Opponens digiti
minimi muscle

Pisiform

Carpal tunnel
(arrow)

Tendon
of flexor carpi ulnaris muscle

Tendons
of flexor digitorum
superficialis muscle

Fibrous sheath
of 2nd digit of hand
– Cruciform part
– Anular part

1st dorsal interosseous
muscle

Tendinous sheath
of flexor pollicis longus
muscle

Transverse head
of adductor pollicis muscle

Superficial head
of flexor pollicis brevis
muscle

Abductor pollicis
brevis muscle

Opponens pollicis
muscle

Flexor retinaculum

Tendon
of abductor pollicis longus
muscle

Tendon
of flexor carpi radialis
muscle

Tendon
of flexor pollicis longus
muscle

56 Muscles of the right hand (75%)
Superficial layer. The palmar aponeurosis was removed.
Palmar aspect

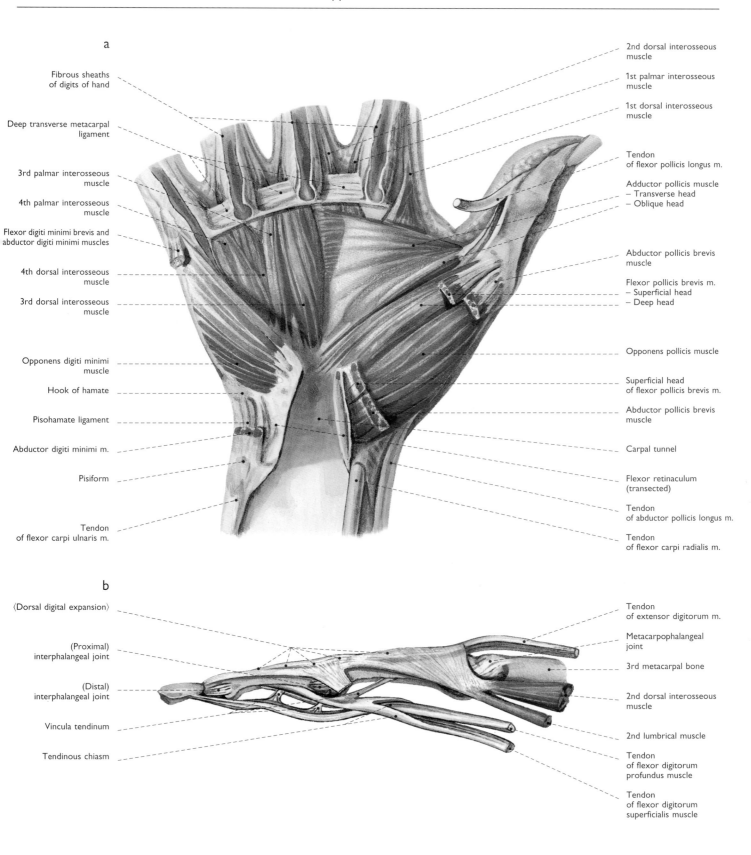

a

Fibrous sheaths
of digits of hand

Deep transverse metacarpal
ligament

3rd palmar interosseous
muscle

4th palmar interosseous
muscle

Flexor digiti minimi brevis and
abductor digiti minimi muscles

4th dorsal interosseous
muscle

3rd dorsal interosseous
muscle

Opponens digiti minimi
muscle

Hook of hamate

Pisohamate ligament

Abductor digiti minimi m.

Pisiform

Tendon
of flexor carpi ulnaris m.

2nd dorsal interosseous
muscle

1st palmar interosseous
muscle

1st dorsal interosseous
muscle

Tendon
of flexor pollicis longus m.

Adductor pollicis muscle
– Transverse head
– Oblique head

Abductor pollicis brevis
muscle

Flexor pollicis brevis m.
– Superficial head
– Deep head

Opponens pollicis muscle

Superficial head
of flexor pollicis brevis m.

Abductor pollicis brevis
muscle

Carpal tunnel

Flexor retinaculum
(transected)

Tendon
of abductor pollicis longus m.

Tendon
of flexor carpi radialis m.

b

⟨Dorsal digital expansion⟩

(Proximal)
interphalangeal joint

(Distal)
interphalangeal joint

Vincula tendinum

Tendinous chiasm

Tendon
of extensor digitorum m.

Metacarpophalangeal
joint

3rd metacarpal bone

2nd dorsal interosseous
muscle

2nd lumbrical muscle

Tendon
of flexor digitorum
profundus muscle

Tendon
of flexor digitorum
superficialis muscle

57 Muscles of the right hand (75%)

a Deep layer, palmar aspect
b Middle finger with the dorsal digital expansion,
radial aspect

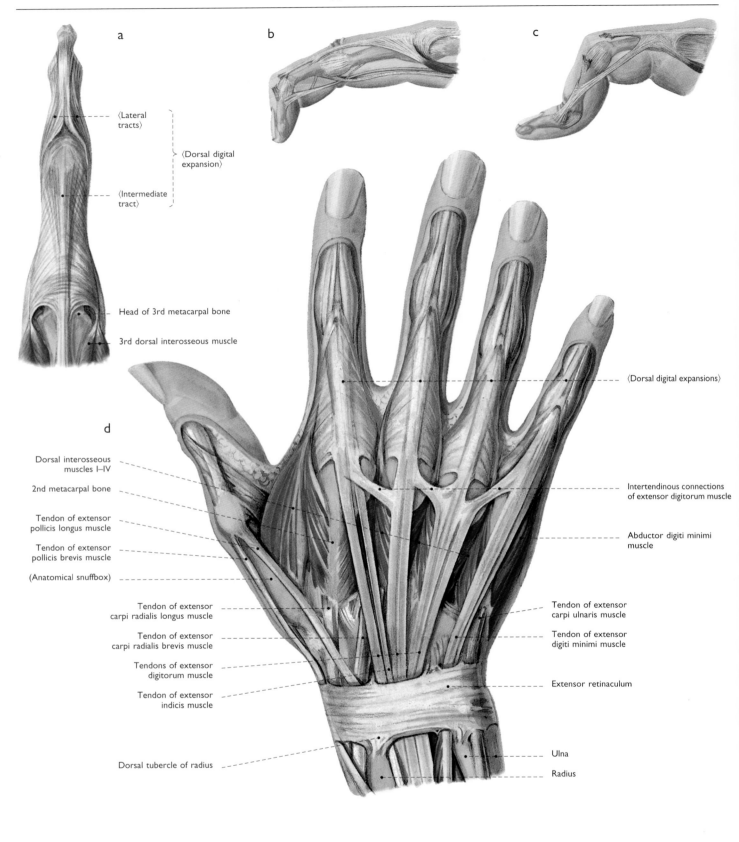

a

⟨Lateral tracts⟩

⟨Dorsal digital expansion⟩

⟨Intermediate tract⟩

Head of 3rd metacarpal bone

3rd dorsal interosseous muscle

b

c

⟨Dorsal digital expansions⟩

d

Dorsal interosseous muscles I–IV

2nd metacarpal bone

Tendon of extensor pollicis longus muscle

Tendon of extensor pollicis brevis muscle

(Anatomical snuffbox)

Tendon of extensor carpi radialis longus muscle

Tendon of extensor carpi radialis brevis muscle

Tendons of extensor digitorum muscle

Tendon of extensor indicis muscle

Dorsal tubercle of radius

Intertendinous connections of extensor digitorum muscle

Abductor digiti minimi muscle

Tendon of extensor carpi ulnaris muscle

Tendon of extensor digiti minimi muscle

Extensor retinaculum

Ulna

Radius

58 Muscles of the right hand

a Muscles of the dorsum and dorsal digital expansion of the middle finger (75%), dorsal aspect

b, c Ruptures of the dorsal digital expansion above the distal (b) and proximal (c) interphalangeal joints (50%)

d Muscles of the dorsum of hand (75%), dorsal aspect

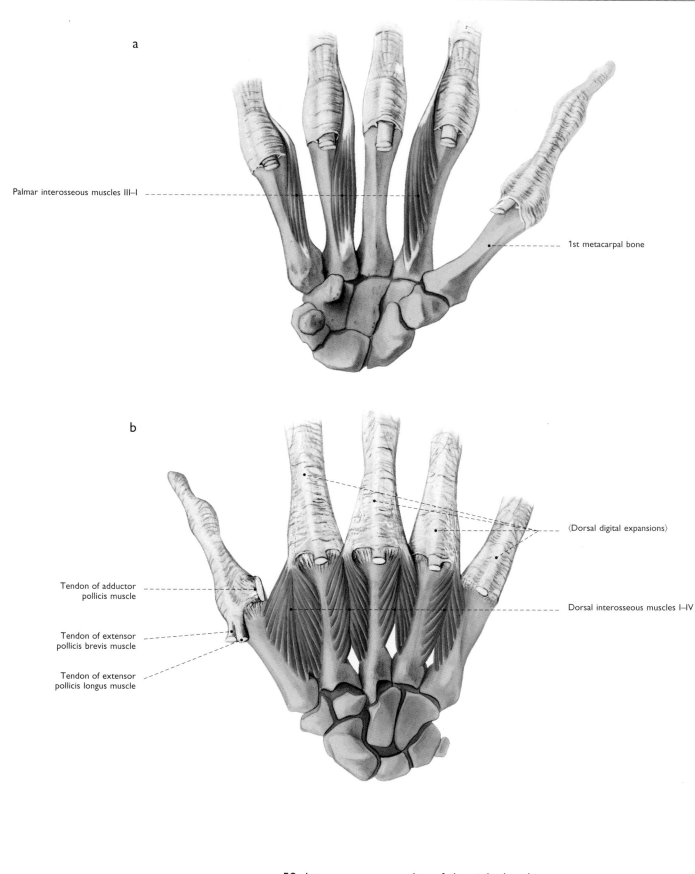

a

Palmar interosseous muscles III–I

1st metacarpal bone

b

Tendon of adductor
pollicis muscle

Tendon of extensor
pollicis brevis muscle

Tendon of extensor
pollicis longus muscle

⟨Dorsal digital expansions⟩

Dorsal interosseous muscles I–IV

59 Interosseous muscles of the right hand (75%)
a Palmar interosseous muscles, palmar aspect
b Dorsal interosseous muscles, dorsal aspect

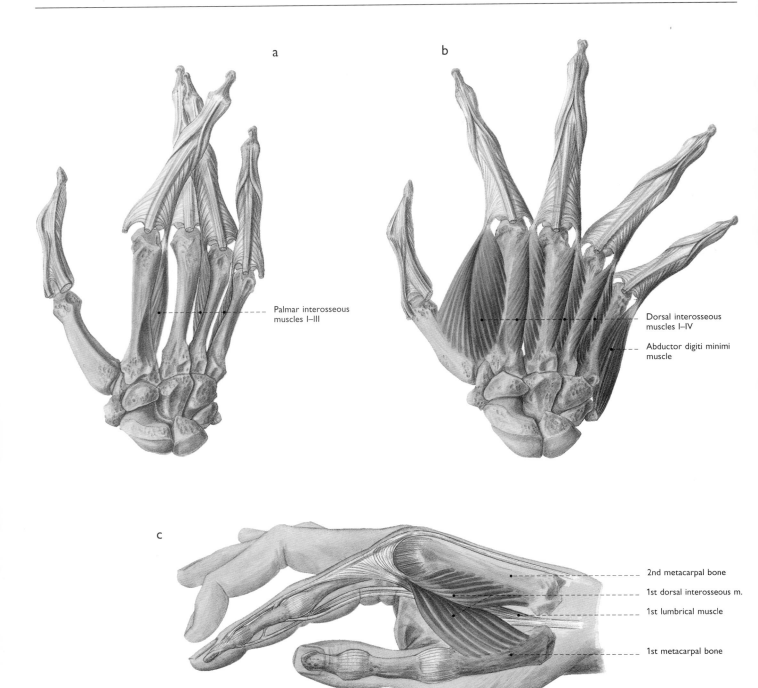

a

b

Palmar interosseous
muscles I–III

Dorsal interosseous
muscles I–IV

Abductor digiti minimi
muscle

c

2nd metacarpal bone

1st dorsal interosseous m.

1st lumbrical muscle

1st metacarpal bone

**60 Interosseous and lumbrical muscles
of the right hand** (60%)

a Function of the palmar interosseous muscles, dorsal aspect
b Function of the dorsal interosseous muscles, dorsal aspect
c Function of the first lumbrical and first dorsal interosseous muscles,
radial aspect

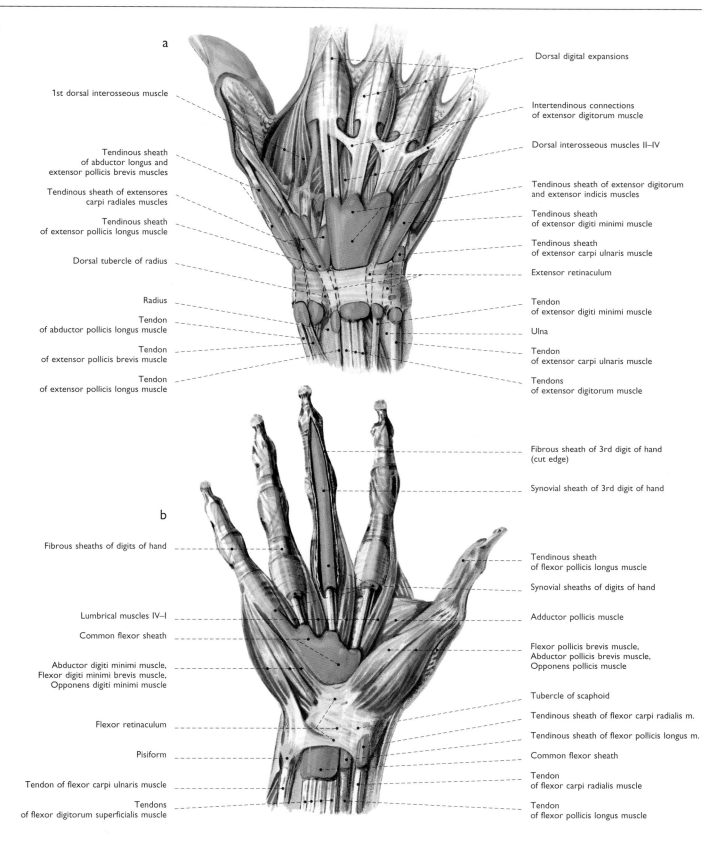

a

1st dorsal interosseous muscle

Tendinous sheath
of abductor longus and
extensor pollicis brevis muscles

Tendinous sheath of extensores
carpi radiales muscles

Tendinous sheath
of extensor pollicis longus muscle

Dorsal tubercle of radius

Radius

Tendon
of abductor pollicis longus muscle

Tendon
of extensor pollicis brevis muscle

Tendon
of extensor pollicis longus muscle

Dorsal digital expansions

Intertendinous connections
of extensor digitorum muscle

Dorsal interosseous muscles II–IV

Tendinous sheath of extensor digitorum
and extensor indicis muscles

Tendinous sheath
of extensor digiti minimi muscle

Tendinous sheath
of extensor carpi ulnaris muscle

Extensor retinaculum

Tendon
of extensor digiti minimi muscle

Ulna

Tendon
of extensor carpi ulnaris muscle

Tendons
of extensor digitorum muscle

b

Fibrous sheaths of digits of hand

Lumbrical muscles IV–I

Common flexor sheath

Abductor digiti minimi muscle,
Flexor digiti minimi brevis muscle,
Opponens digiti minimi muscle

Flexor retinaculum

Pisiform

Tendon of flexor carpi ulnaris muscle

Tendons
of flexor digitorum superficialis muscle

Fibrous sheath of 3rd digit of hand
(cut edge)

Synovial sheath of 3rd digit of hand

Tendinous sheath
of flexor pollicis longus muscle

Synovial sheaths of digits of hand

Adductor pollicis muscle

Flexor pollicis brevis muscle,
Abductor pollicis brevis muscle,
Opponens pollicis muscle

Tubercle of scaphoid

Tendinous sheath of flexor carpi radialis m.

Tendinous sheath of flexor pollicis longus m.

Common flexor sheath

Tendon
of flexor carpi radialis muscle

Tendon
of flexor pollicis longus muscle

**61 Tendinous sheaths on the wrist
and the fingers of the right hand** (50%)
  a Dorsal aspect
  b Palmar aspect

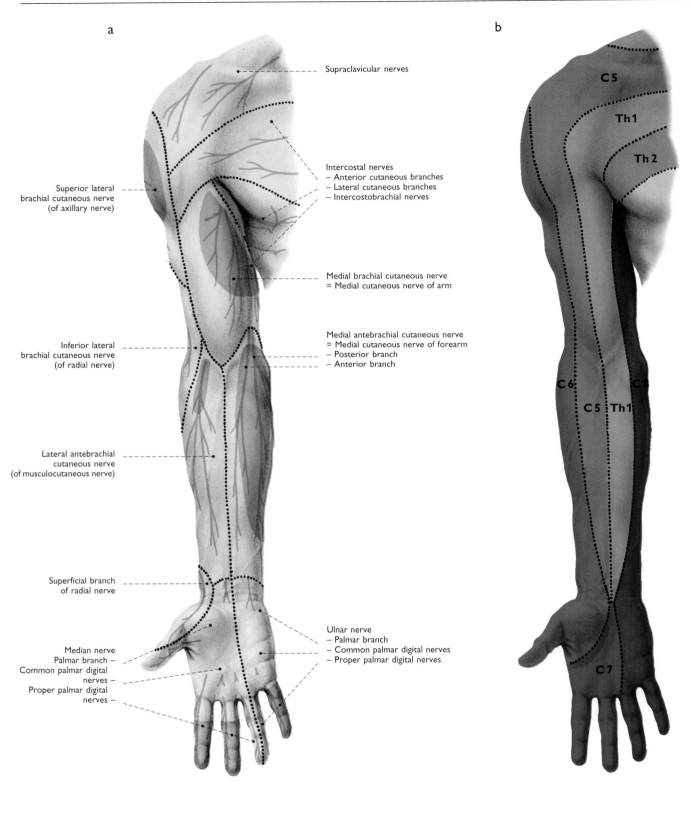

a

Supraclavicular nerves

Intercostal nerves
– Anterior cutaneous branches
– Lateral cutaneous branches
– Intercostobrachial nerves

Superior lateral
brachial cutaneous nerve
(of axillary nerve)

Medial brachial cutaneous nerve
= Medial cutaneous nerve of arm

Inferior lateral
brachial cutaneous nerve
(of radial nerve)

Medial antebrachial cutaneous nerve
= Medial cutaneous nerve of forearm
– Posterior branch
– Anterior branch

Lateral antebrachial
cutaneous nerve
(of musculocutaneous nerve)

Superficial branch
of radial nerve

Median nerve
Palmar branch –
Common palmar digital
nerves –
Proper palmar digital
nerves –

Ulnar nerve
– Palmar branch
– Common palmar digital nerves
– Proper palmar digital nerves

b

C 5

Th 1

Th 2

C 6   C 8

C 5   Th 1

C 7

**62 Cutaneous and segmental innervation
of the right upper limb** (25%)
Schematic representations, ventral aspect
a Cutaneous nerves and areas of distribution,
the autonomic areas of the different nerves
are given in a darker gray.
b Segmental innervation (dermatomes)

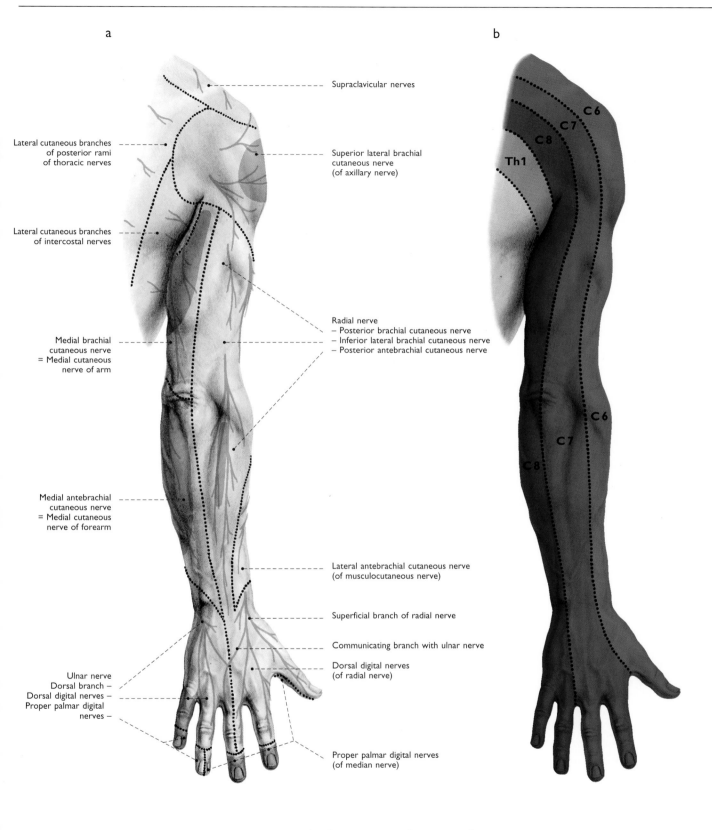

Supraclavicular nerves

Superior lateral brachial cutaneous nerve (of axillary nerve)

Lateral cutaneous branches of posterior rami of thoracic nerves

Lateral cutaneous branches of intercostal nerves

Radial nerve
– Posterior brachial cutaneous nerve
– Inferior lateral brachial cutaneous nerve
– Posterior antebrachial cutaneous nerve

Medial brachial cutaneous nerve = Medial cutaneous nerve of arm

Medial antebrachial cutaneous nerve = Medial cutaneous nerve of forearm

Lateral antebrachial cutaneous nerve (of musculocutaneous nerve)

Superficial branch of radial nerve

Communicating branch with ulnar nerve

Dorsal digital nerves (of radial nerve)

Ulnar nerve
Dorsal branch –
Dorsal digital nerves –
Proper palmar digital nerves –

Proper palmar digital nerves (of median nerve)

C 6
C 7
C 8
Th1

C 6
C 7
C 8

a

b

63 Cutaneous and segmental innervation of the right upper limb (25%)

Schematic representations, dorsal aspect
a Cutaneous nerves and areas of distribution, the autonomic areas of the different nerves are given in a darker gray.
b Segmental innervation (dermatomes)

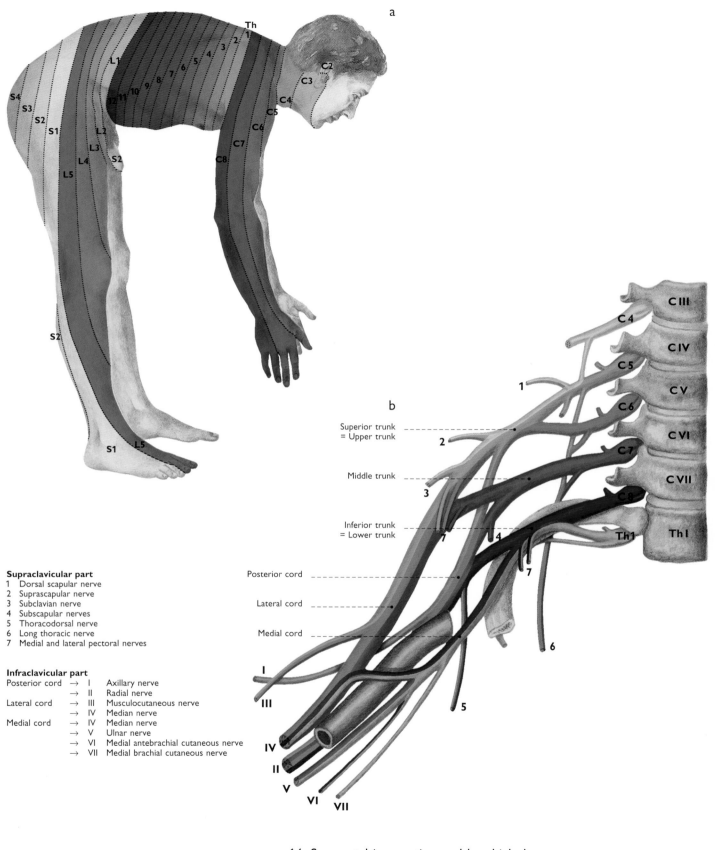

Supraclavicular part
1 Dorsal scapular nerve
2 Suprascapular nerve
3 Subclavian nerve
4 Subscapular nerves
5 Thoracodorsal nerve
6 Long thoracic nerve
7 Medial and lateral pectoral nerves

Infraclavicular part
Posterior cord → I Axillary nerve
→ II Radial nerve
Lateral cord → III Musculocutaneous nerve
→ IV Median nerve
Medial cord → IV Median nerve
→ V Ulnar nerve
→ VI Medial antebrachial cutaneous nerve
→ VII Medial brachial cutaneous nerve

Superior trunk = Upper trunk
Middle trunk
Inferior trunk = Lower trunk
Posterior cord
Lateral cord
Medial cord

64 Segmental innervation and brachial plexus
a Segmental innervation (dermatomes) of the upper limb, trunk, and lower limb (according to von Lanz and Wachsmuth, 1959)
b Plan of the brachial plexus

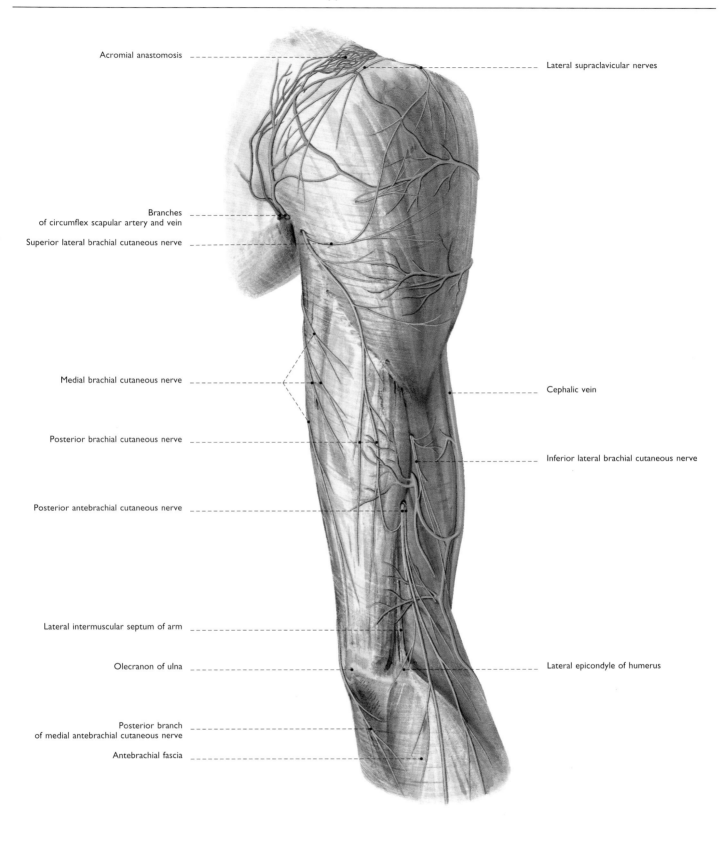

Acromial anastomosis

Lateral supraclavicular nerves

Branches
of circumflex scapular artery and vein

Superior lateral brachial cutaneous nerve

Medial brachial cutaneous nerve

Cephalic vein

Posterior brachial cutaneous nerve

Inferior lateral brachial cutaneous nerve

Posterior antebrachial cutaneous nerve

Lateral intermuscular septum of arm

Olecranon of ulna

Lateral epicondyle of humerus

Posterior branch
of medial antebrachial cutaneous nerve

Antebrachial fascia

**65 Subcutaneous veins and nerves of the
right shoulder and the right arm** (50%)
In this case the inferior lateral brachial cutaneous nerve
originates from the axillary nerve. Dorsolateral aspect

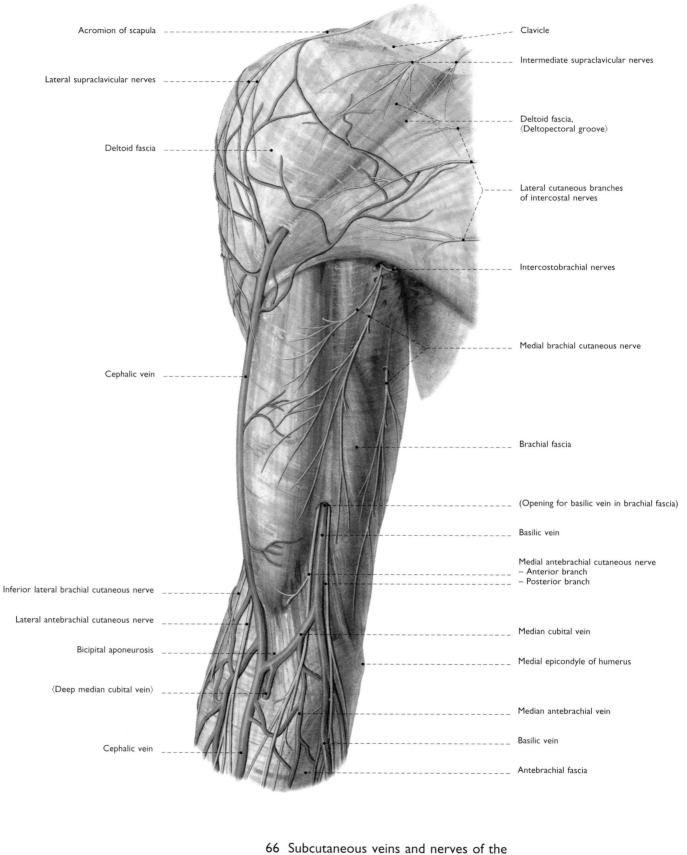

Acromion of scapula
Lateral supraclavicular nerves
Deltoid fascia
Cephalic vein
Inferior lateral brachial cutaneous nerve
Lateral antebrachial cutaneous nerve
Bicipital aponeurosis
⟨Deep median cubital vein⟩
Cephalic vein

Clavicle
Intermediate supraclavicular nerves
Deltoid fascia, ⟨Deltopectoral groove⟩
Lateral cutaneous branches of intercostal nerves
Intercostobrachial nerves
Medial brachial cutaneous nerve
Brachial fascia
(Opening for basilic vein in brachial fascia)
Basilic vein
Medial antebrachial cutaneous nerve – Anterior branch – Posterior branch
Median cubital vein
Medial epicondyle of humerus
Median antebrachial vein
Basilic vein
Antebrachial fascia

66 Subcutaneous veins and nerves of the right shoulder and the right arm (50%)
Ventral aspect

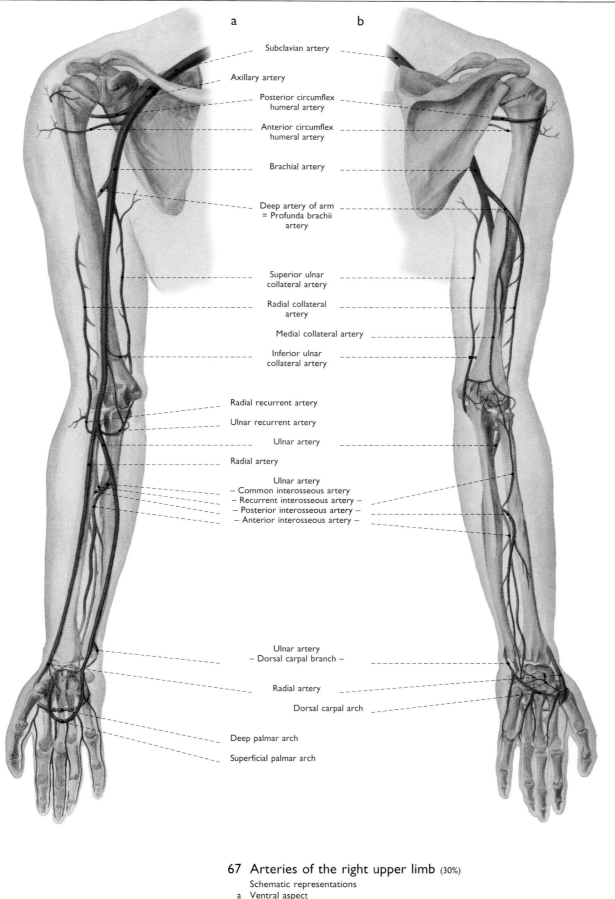

a   b

Subclavian artery

Axillary artery

Posterior circumflex
humeral artery

Anterior circumflex
humeral artery

Brachial artery

Deep artery of arm
= Profunda brachii
artery

Superior ulnar
collateral artery

Radial collateral
artery

Medial collateral artery

Inferior ulnar
collateral artery

Radial recurrent artery

Ulnar recurrent artery

Ulnar artery

Radial artery

Ulnar artery
– Common interosseous artery
– Recurrent interosseous artery –
– Posterior interosseous artery –
– Anterior interosseous artery –

Ulnar artery
– Dorsal carpal branch –

Radial artery

Dorsal carpal arch

Deep palmar arch

Superficial palmar arch

67 Arteries of the right upper limb (30%)
Schematic representations
a Ventral aspect
b Dorsal aspect

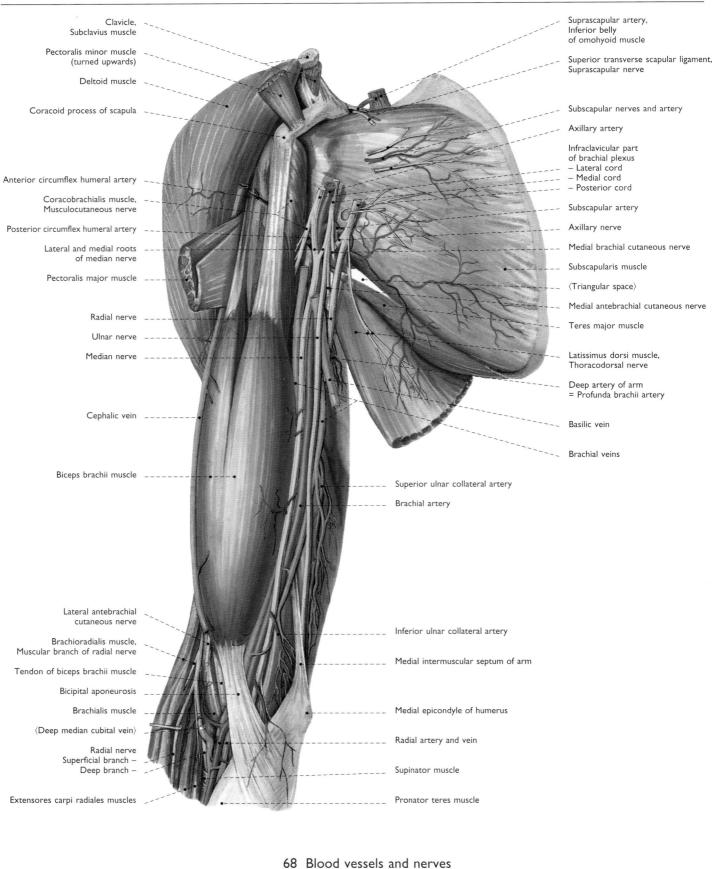

Clavicle,
Subclavius muscle

Pectoralis minor muscle
(turned upwards)

Deltoid muscle

Coracoid process of scapula

Anterior circumflex humeral artery

Coracobrachialis muscle,
Musculocutaneous nerve

Posterior circumflex humeral artery

Lateral and medial roots
of median nerve

Pectoralis major muscle

Radial nerve

Ulnar nerve

Median nerve

Cephalic vein

Biceps brachii muscle

Lateral antebrachial
cutaneous nerve

Brachioradialis muscle,
Muscular branch of radial nerve

Tendon of biceps brachii muscle

Bicipital aponeurosis

Brachialis muscle

⟨Deep median cubital vein⟩

Radial nerve
Superficial branch –
Deep branch –

Extensores carpi radiales muscles

Suprascapular artery,
Inferior belly
of omohyoid muscle

Superior transverse scapular ligament,
Suprascapular nerve

Subscapular nerves and artery

Axillary artery

Infraclavicular part
of brachial plexus
– Lateral cord
– Medial cord
– Posterior cord

Subscapular artery

Axillary nerve

Medial brachial cutaneous nerve

Subscapularis muscle

⟨Triangular space⟩

Medial antebrachial cutaneous nerve

Teres major muscle

Latissimus dorsi muscle,
Thoracodorsal nerve

Deep artery of arm
= Profunda brachii artery

Basilic vein

Brachial veins

Superior ulnar collateral artery

Brachial artery

Inferior ulnar collateral artery

Medial intermuscular septum of arm

Medial epicondyle of humerus

Radial artery and vein

Supinator muscle

Pronator teres muscle

**68 Blood vessels and nerves
of the right shoulder, the arm,
and the anterior region of elbow** (50%)
Ventral aspect

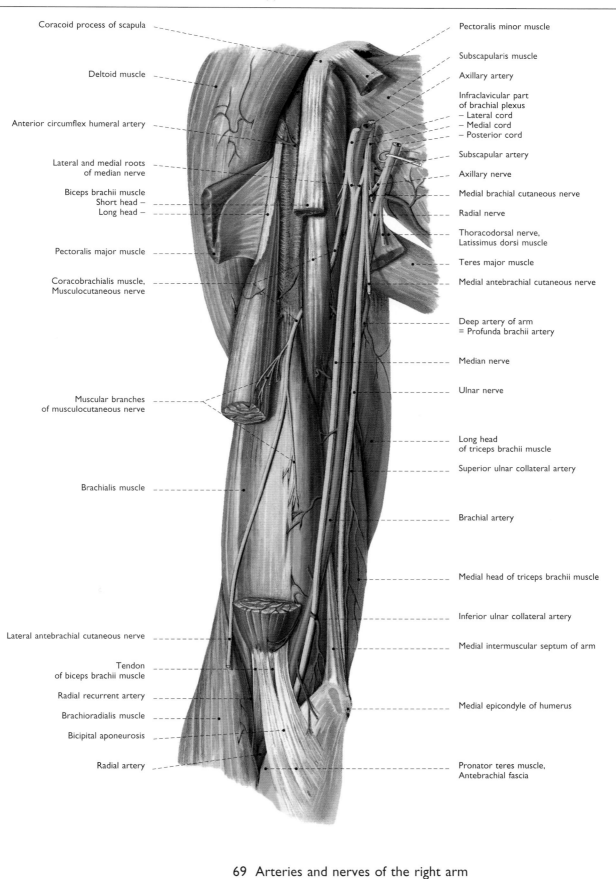

Coracoid process of scapula

Deltoid muscle

Anterior circumflex humeral artery

Lateral and medial roots
of median nerve

Biceps brachii muscle
Short head –
Long head –

Pectoralis major muscle

Coracobrachialis muscle,
Musculocutaneous nerve

Muscular branches
of musculocutaneous nerve

Brachialis muscle

Lateral antebrachial cutaneous nerve

Tendon
of biceps brachii muscle

Radial recurrent artery

Brachioradialis muscle

Bicipital aponeurosis

Radial artery

Pectoralis minor muscle

Subscapularis muscle

Axillary artery

Infraclavicular part
of brachial plexus
– Lateral cord
– Medial cord
– Posterior cord

Subscapular artery

Axillary nerve

Medial brachial cutaneous nerve

Radial nerve

Thoracodorsal nerve,
Latissimus dorsi muscle

Teres major muscle

Medial antebrachial cutaneous nerve

Deep artery of arm
= Profunda brachii artery

Median nerve

Ulnar nerve

Long head
of triceps brachii muscle

Superior ulnar collateral artery

Brachial artery

Medial head of triceps brachii muscle

Inferior ulnar collateral artery

Medial intermuscular septum of arm

Medial epicondyle of humerus

Pronator teres muscle,
Antebrachial fascia

**69 Arteries and nerves of the right arm
and the anterior region of elbow** (50%)
The biceps brachii muscle was partially removed.
Ventral aspect

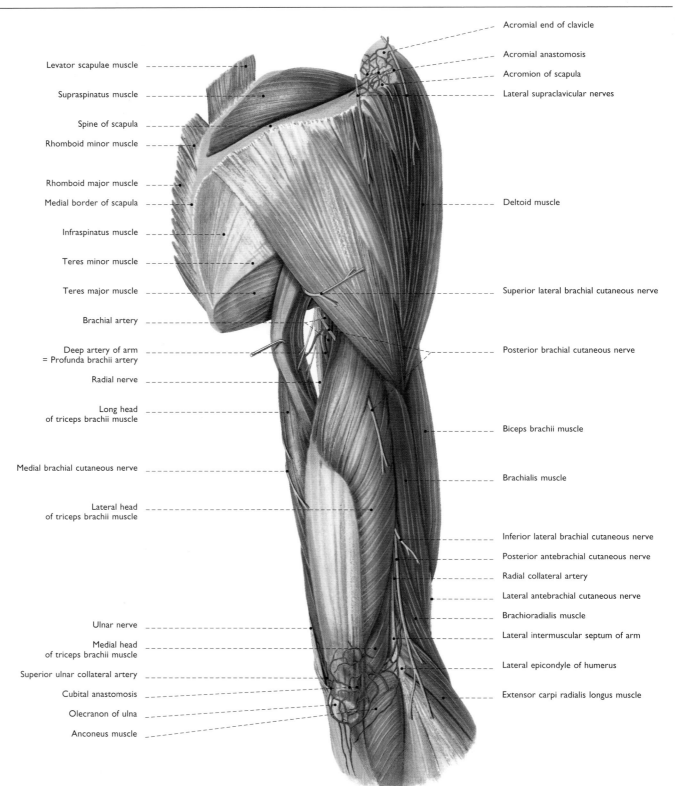

Levator scapulae muscle

Supraspinatus muscle

Spine of scapula

Rhomboid minor muscle

Rhomboid major muscle

Medial border of scapula

Infraspinatus muscle

Teres minor muscle

Teres major muscle

Brachial artery

Deep artery of arm
= Profunda brachii artery

Radial nerve

Long head
of triceps brachii muscle

Medial brachial cutaneous nerve

Lateral head
of triceps brachii muscle

Ulnar nerve

Medial head
of triceps brachii muscle

Superior ulnar collateral artery

Cubital anastomosis

Olecranon of ulna

Anconeus muscle

Acromial end of clavicle

Acromial anastomosis

Acromion of scapula

Lateral supraclavicular nerves

Deltoid muscle

Superior lateral brachial cutaneous nerve

Posterior brachial cutaneous nerve

Biceps brachii muscle

Brachialis muscle

Inferior lateral brachial cutaneous nerve

Posterior antebrachial cutaneous nerve

Radial collateral artery

Lateral antebrachial cutaneous nerve

Brachioradialis muscle

Lateral intermuscular septum of arm

Lateral epicondyle of humerus

Extensor carpi radialis longus muscle

70 Arteries and nerves of the right shoulder,
the right arm, and the posterior region of elbow (50%)
Dorsolateral aspect

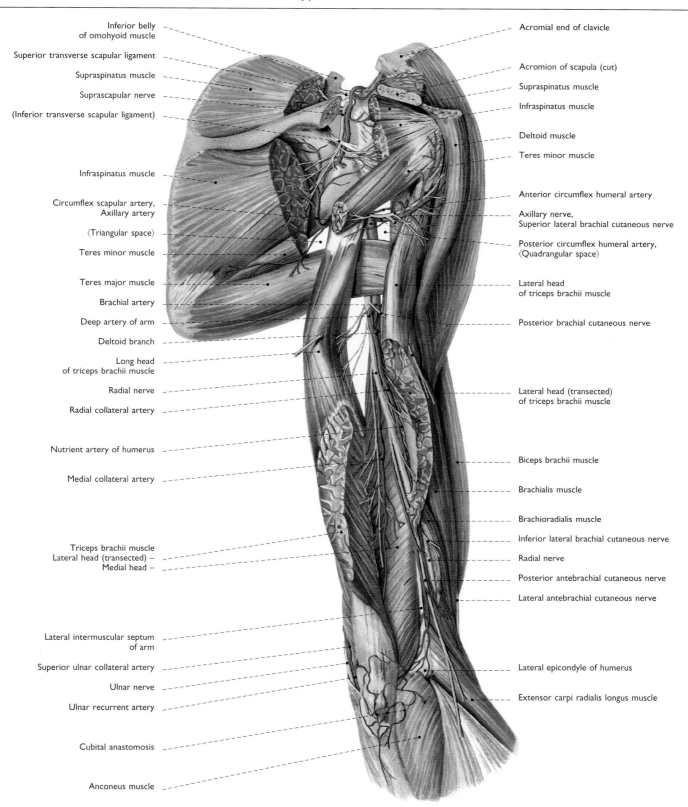

Inferior belly
of omohyoid muscle

Superior transverse scapular ligament

Supraspinatus muscle

Suprascapular nerve

(Inferior transverse scapular ligament)

Infraspinatus muscle

Circumflex scapular artery,
Axillary artery

⟨Triangular space⟩

Teres minor muscle

Teres major muscle

Brachial artery

Deep artery of arm

Deltoid branch

Long head
of triceps brachii muscle

Radial nerve

Radial collateral artery

Nutrient artery of humerus

Medial collateral artery

Triceps brachii muscle
Lateral head (transected) –
Medial head –

Lateral intermuscular septum
of arm

Superior ulnar collateral artery

Ulnar nerve

Ulnar recurrent artery

Cubital anastomosis

Anconeus muscle

Acromial end of clavicle

Acromion of scapula (cut)

Supraspinatus muscle

Infraspinatus muscle

Deltoid muscle

Teres minor muscle

Anterior circumflex humeral artery

Axillary nerve,
Superior lateral brachial cutaneous nerve

Posterior circumflex humeral artery,
⟨Quadrangular space⟩

Lateral head
of triceps brachii muscle

Posterior brachial cutaneous nerve

Lateral head (transected)
of triceps brachii muscle

Biceps brachii muscle

Brachialis muscle

Brachioradialis muscle

Inferior lateral brachial cutaneous nerve

Radial nerve

Posterior antebrachial cutaneous nerve

Lateral antebrachial cutaneous nerve

Lateral epicondyle of humerus

Extensor carpi radialis longus muscle

**71 Arteries and nerves of the right shoulder,
      the right arm, and the posterior region of elbow** (50%)
The lateral head of the triceps brachii muscle was divided,
the radial nerve channel opened. Dorsolateral aspect

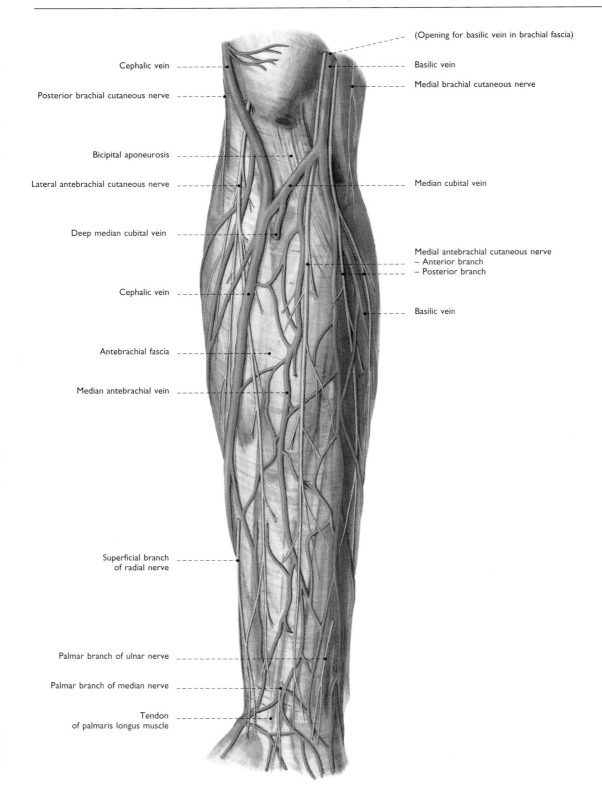

Cephalic vein

Posterior brachial cutaneous nerve

Bicipital aponeurosis

Lateral antebrachial cutaneous nerve

Deep median cubital vein

Cephalic vein

Antebrachial fascia

Median antebrachial vein

Superficial branch of radial nerve

Palmar branch of ulnar nerve

Palmar branch of median nerve

Tendon of palmaris longus muscle

(Opening for basilic vein in brachial fascia)

Basilic vein

Medial brachial cutaneous nerve

Median cubital vein

Medial antebrachial cutaneous nerve
– Anterior branch
– Posterior branch

Basilic vein

72 Subcutaneous veins and nerves of the anterior region of elbow and the anterior (flexor) region of the right forearm (50%)
Ventral aspect

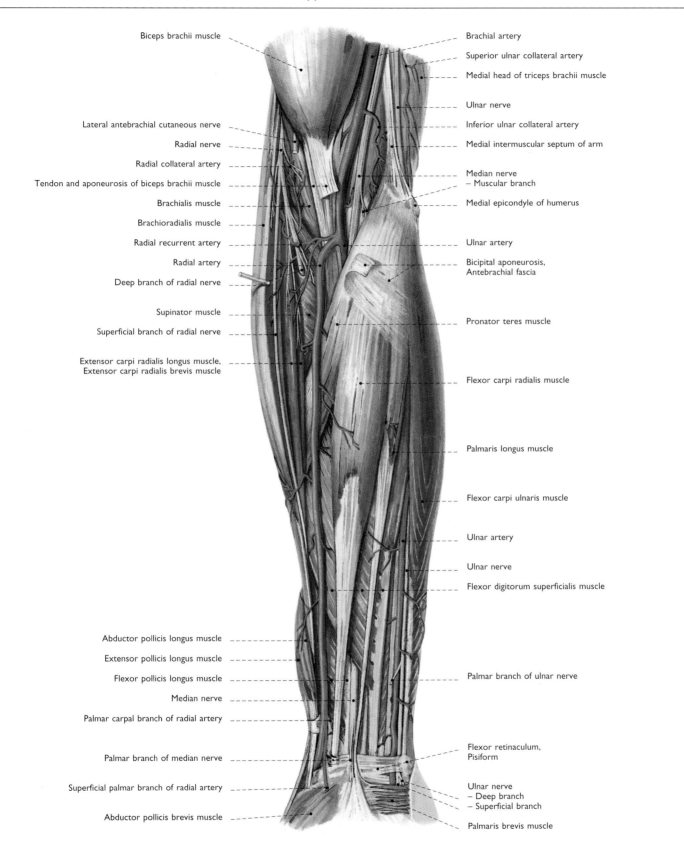

Biceps brachii muscle

Lateral antebrachial cutaneous nerve

Radial nerve

Radial collateral artery

Tendon and aponeurosis of biceps brachii muscle

Brachialis muscle

Brachioradialis muscle

Radial recurrent artery

Radial artery

Deep branch of radial nerve

Supinator muscle

Superficial branch of radial nerve

Extensor carpi radialis longus muscle,
Extensor carpi radialis brevis muscle

Abductor pollicis longus muscle

Extensor pollicis longus muscle

Flexor pollicis longus muscle

Median nerve

Palmar carpal branch of radial artery

Palmar branch of median nerve

Superficial palmar branch of radial artery

Abductor pollicis brevis muscle

Brachial artery

Superior ulnar collateral artery

Medial head of triceps brachii muscle

Ulnar nerve

Inferior ulnar collateral artery

Medial intermuscular septum of arm

Median nerve
– Muscular branch

Medial epicondyle of humerus

Ulnar artery

Bicipital aponeurosis,
Antebrachial fascia

Pronator teres muscle

Flexor carpi radialis muscle

Palmaris longus muscle

Flexor carpi ulnaris muscle

Ulnar artery

Ulnar nerve

Flexor digitorum superficialis muscle

Palmar branch of ulnar nerve

Flexor retinaculum,
Pisiform

Ulnar nerve
– Deep branch
– Superficial branch

Palmaris brevis muscle

73 Arteries and nerves of the right forearm (50%)
Ventral aspect

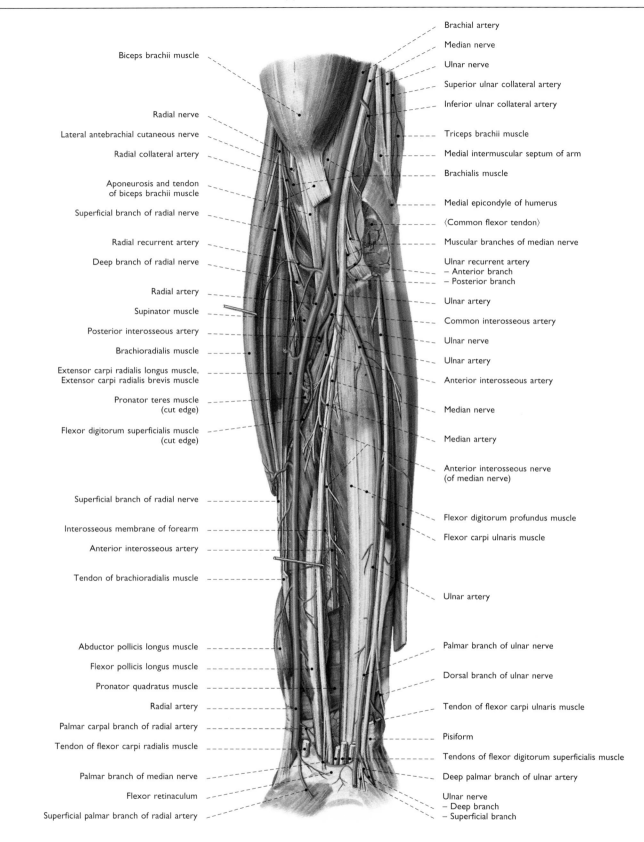

Biceps brachii muscle

Radial nerve
Lateral antebrachial cutaneous nerve
Radial collateral artery

Aponeurosis and tendon
of biceps brachii muscle

Superficial branch of radial nerve

Radial recurrent artery
Deep branch of radial nerve

Radial artery
Supinator muscle
Posterior interosseous artery
Brachioradialis muscle
Extensor carpi radialis longus muscle,
Extensor carpi radialis brevis muscle
Pronator teres muscle
(cut edge)
Flexor digitorum superficialis muscle
(cut edge)

Superficial branch of radial nerve

Interosseous membrane of forearm

Anterior interosseous artery

Tendon of brachioradialis muscle

Abductor pollicis longus muscle
Flexor pollicis longus muscle
Pronator quadratus muscle
Radial artery
Palmar carpal branch of radial artery
Tendon of flexor carpi radialis muscle

Palmar branch of median nerve
Flexor retinaculum
Superficial palmar branch of radial artery

Brachial artery
Median nerve
Ulnar nerve
Superior ulnar collateral artery
Inferior ulnar collateral artery

Triceps brachii muscle
Medial intermuscular septum of arm
Brachialis muscle

Medial epicondyle of humerus
⟨Common flexor tendon⟩
Muscular branches of median nerve
Ulnar recurrent artery
– Anterior branch
– Posterior branch
Ulnar artery
Common interosseous artery
Ulnar nerve
Ulnar artery
Anterior interosseous artery

Median nerve

Median artery

Anterior interosseous nerve
(of median nerve)

Flexor digitorum profundus muscle
Flexor carpi ulnaris muscle

Ulnar artery

Palmar branch of ulnar nerve

Dorsal branch of ulnar nerve

Tendon of flexor carpi ulnaris muscle

Pisiform
Tendons of flexor digitorum superficialis muscle
Deep palmar branch of ulnar artery
Ulnar nerve
– Deep branch
– Superficial branch

74 Arteries and nerves of the right forearm (50%)
The superficial muscles were removed.
Ventral aspect

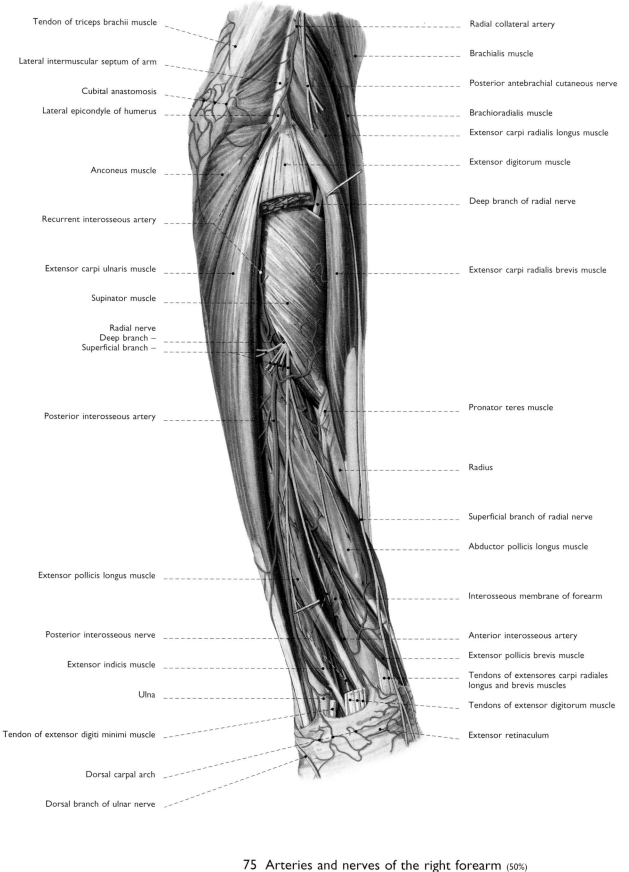

Tendon of triceps brachii muscle

Lateral intermuscular septum of arm

Cubital anastomosis

Lateral epicondyle of humerus

Anconeus muscle

Recurrent interosseous artery

Extensor carpi ulnaris muscle

Supinator muscle

Radial nerve
Deep branch –
Superficial branch –

Posterior interosseous artery

Extensor pollicis longus muscle

Posterior interosseous nerve

Extensor indicis muscle

Ulna

Tendon of extensor digiti minimi muscle

Dorsal carpal arch

Dorsal branch of ulnar nerve

Radial collateral artery

Brachialis muscle

Posterior antebrachial cutaneous nerve

Brachioradialis muscle

Extensor carpi radialis longus muscle

Extensor digitorum muscle

Deep branch of radial nerve

Extensor carpi radialis brevis muscle

Pronator teres muscle

Radius

Superficial branch of radial nerve

Abductor pollicis longus muscle

Interosseous membrane of forearm

Anterior interosseous artery

Extensor pollicis brevis muscle

Tendons of extensores carpi radiales
longus and brevis muscles

Tendons of extensor digitorum muscle

Extensor retinaculum

75 Arteries and nerves of the right forearm (50%)
The superficial muscles were partially removed.
Dorsolateral aspect

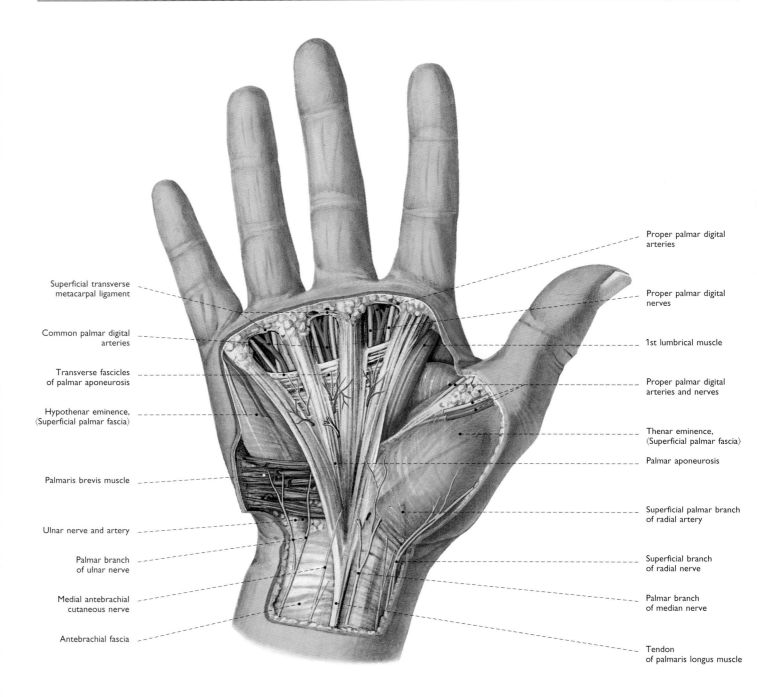

Superficial transverse metacarpal ligament

Common palmar digital arteries

Transverse fascicles of palmar aponeurosis

Hypothenar eminence, ⟨Superficial palmar fascia⟩

Palmaris brevis muscle

Ulnar nerve and artery

Palmar branch of ulnar nerve

Medial antebrachial cutaneous nerve

Antebrachial fascia

Proper palmar digital arteries

Proper palmar digital nerves

1st lumbrical muscle

Proper palmar digital arteries and nerves

Thenar eminence, ⟨Superficial palmar fascia⟩

Palmar aponeurosis

Superficial palmar branch of radial artery

Superficial branch of radial nerve

Palmar branch of median nerve

Tendon of palmaris longus muscle

76 Arteries and nerves of the palm of the right hand (75%)
Palmar aspect

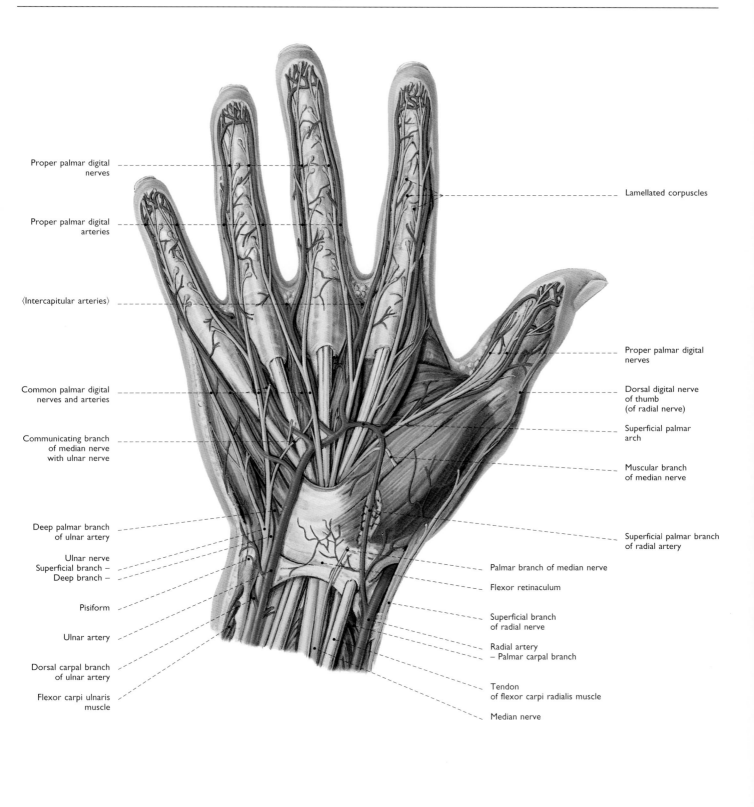

Proper palmar digital nerves

Proper palmar digital arteries

⟨Intercapitular arteries⟩

Common palmar digital nerves and arteries

Communicating branch of median nerve with ulnar nerve

Deep palmar branch of ulnar artery

Ulnar nerve
Superficial branch –
Deep branch –

Pisiform

Ulnar artery

Dorsal carpal branch of ulnar artery

Flexor carpi ulnaris muscle

Lamellated corpuscles

Proper palmar digital nerves

Dorsal digital nerve of thumb (of radial nerve)

Superficial palmar arch

Muscular branch of median nerve

Superficial palmar branch of radial artery

Palmar branch of median nerve

Flexor retinaculum

Superficial branch of radial nerve

Radial artery
– Palmar carpal branch

Tendon of flexor carpi radialis muscle

Median nerve

77 Arteries and nerves of the palm of the right hand (75%)
The palmar aponeurosis was removed.
Palmar aspect

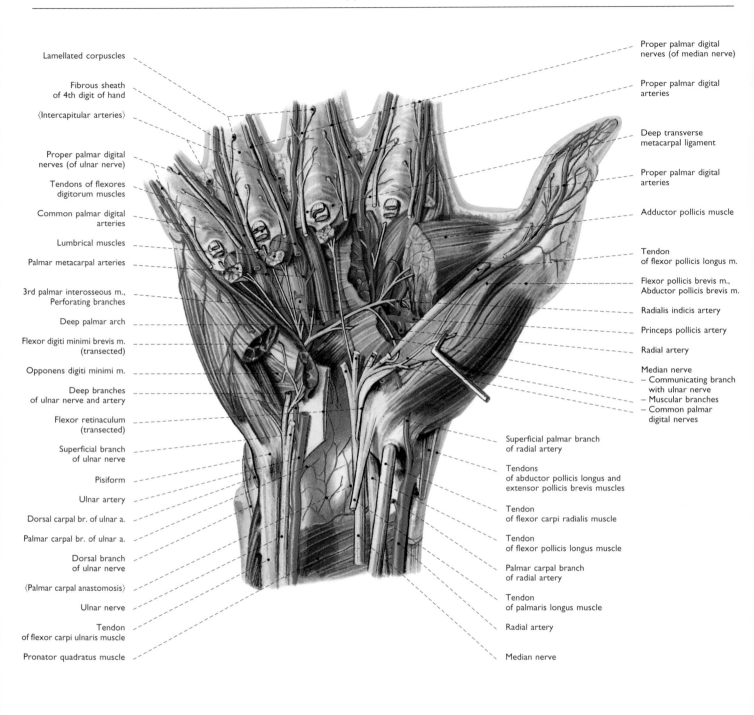

Lamellated corpuscles

Fibrous sheath
of 4th digit of hand

〈Intercapitular arteries〉

Proper palmar digital
nerves (of ulnar nerve)

Tendons of flexores
digitorum muscles

Common palmar digital
arteries

Lumbrical muscles

Palmar metacarpal arteries

3rd palmar interosseous m.,
Perforating branches

Deep palmar arch

Flexor digiti minimi brevis m.
(transected)

Opponens digiti minimi m.

Deep branches
of ulnar nerve and artery

Flexor retinaculum
(transected)

Superficial branch
of ulnar nerve

Pisiform

Ulnar artery

Dorsal carpal br. of ulnar a.

Palmar carpal br. of ulnar a.

Dorsal branch
of ulnar nerve

〈Palmar carpal anastomosis〉

Ulnar nerve

Tendon
of flexor carpi ulnaris muscle

Pronator quadratus muscle

Proper palmar digital
nerves (of median nerve)

Proper palmar digital
arteries

Deep transverse
metacarpal ligament

Proper palmar digital
arteries

Adductor pollicis muscle

Tendon
of flexor pollicis longus m.

Flexor pollicis brevis m.,
Abductor pollicis brevis m.

Radialis indicis artery

Princeps pollicis artery

Radial artery

Median nerve
– Communicating branch
  with ulnar nerve
– Muscular branches
– Common palmar
  digital nerves

Superficial palmar branch
of radial artery

Tendons
of abductor pollicis longus and
extensor pollicis brevis muscles

Tendon
of flexor carpi radialis muscle

Tendon
of flexor pollicis longus muscle

Palmar carpal branch
of radial artery

Tendon
of palmaris longus muscle

Radial artery

Median nerve

**78 Arteries and nerves
of the palm of the right hand** (75%)
The palmar aponeurosis and the flexor muscles
of the fingers were removed. Palmar aspect

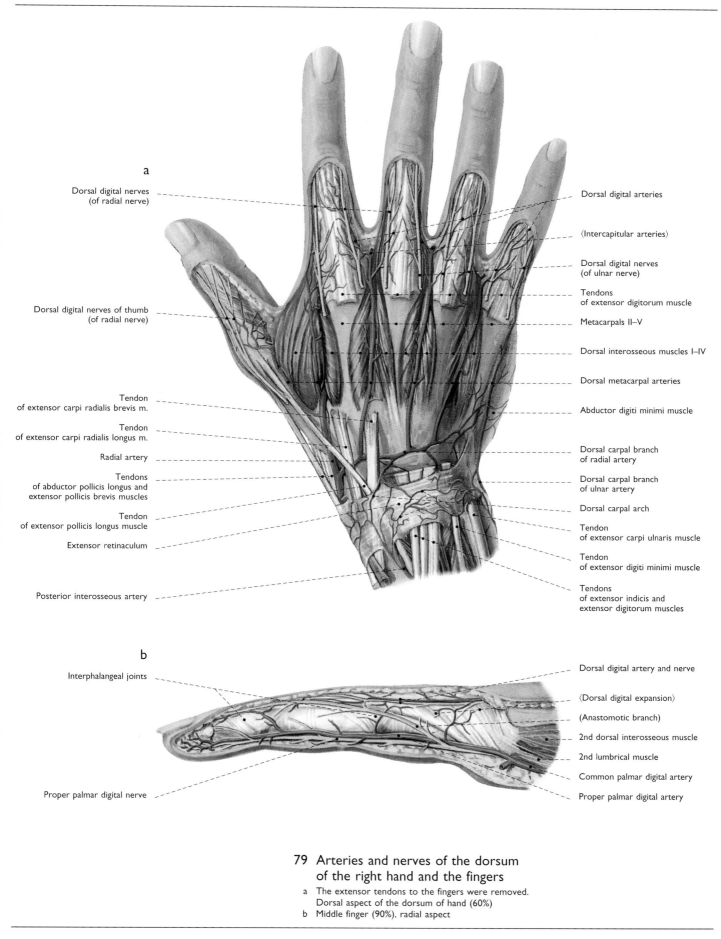

a

Dorsal digital nerves
(of radial nerve)

Dorsal digital nerves of thumb
(of radial nerve)

Tendon
of extensor carpi radialis brevis m.

Tendon
of extensor carpi radialis longus m.

Radial artery

Tendons
of abductor pollicis longus and
extensor pollicis brevis muscles

Tendon
of extensor pollicis longus muscle

Extensor retinaculum

Posterior interosseous artery

Dorsal digital arteries

⟨Intercapitular arteries⟩

Dorsal digital nerves
(of ulnar nerve)

Tendons
of extensor digitorum muscle

Metacarpals II–V

Dorsal interosseous muscles I–IV

Dorsal metacarpal arteries

Abductor digiti minimi muscle

Dorsal carpal branch
of radial artery

Dorsal carpal branch
of ulnar artery

Dorsal carpal arch

Tendon
of extensor carpi ulnaris muscle

Tendon
of extensor digiti minimi muscle

Tendons
of extensor indicis and
extensor digitorum muscles

b

Interphalangeal joints

Proper palmar digital nerve

Dorsal digital artery and nerve

⟨Dorsal digital expansion⟩

(Anastomotic branch)

2nd dorsal interosseous muscle

2nd lumbrical muscle

Common palmar digital artery

Proper palmar digital artery

79   Arteries and nerves of the dorsum of the right hand and the fingers

a   The extensor tendons to the fingers were removed.
Dorsal aspect of the dorsum of hand (60%)
b   Middle finger (90%), radial aspect

Lower Limb

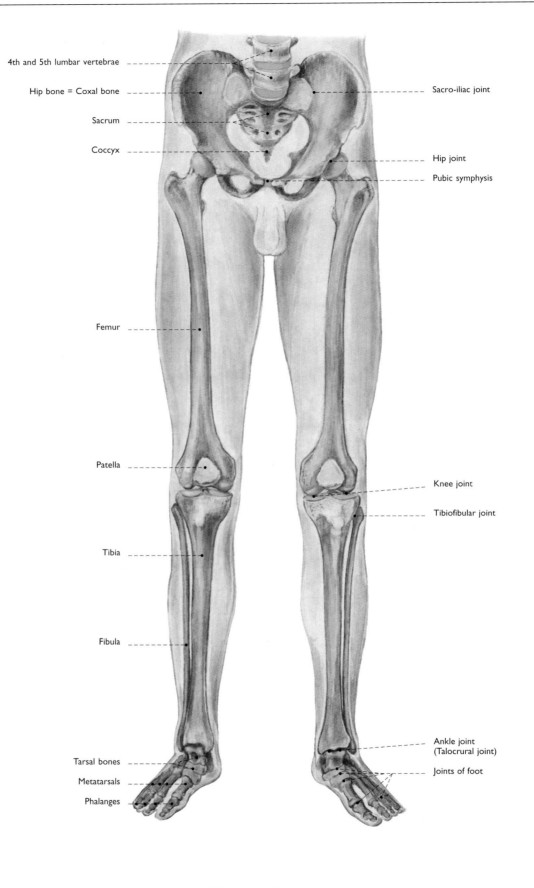

4th and 5th lumbar vertebrae

Hip bone = Coxal bone

Sacrum

Coccyx

Femur

Patella

Tibia

Fibula

Tarsal bones

Metatarsals

Phalanges

Sacro-iliac joint

Hip joint

Pubic symphysis

Knee joint

Tibiofibular joint

Ankle joint
(Talocrural joint)

Joints of foot

82 Lower limb (20%)
Ventral aspect

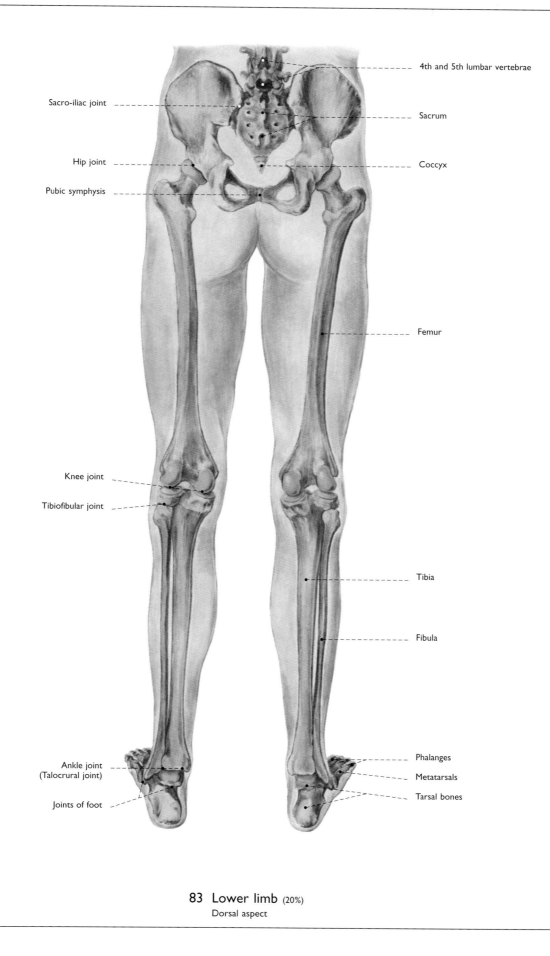

Sacro-iliac joint

Hip joint

Pubic symphysis

4th and 5th lumbar vertebrae

Sacrum

Coccyx

Femur

Knee joint

Tibiofibular joint

Tibia

Fibula

Ankle joint
(Talocrural joint)

Joints of foot

Phalanges

Metatarsals

Tarsal bones

83 Lower limb (20%)
Dorsal aspect

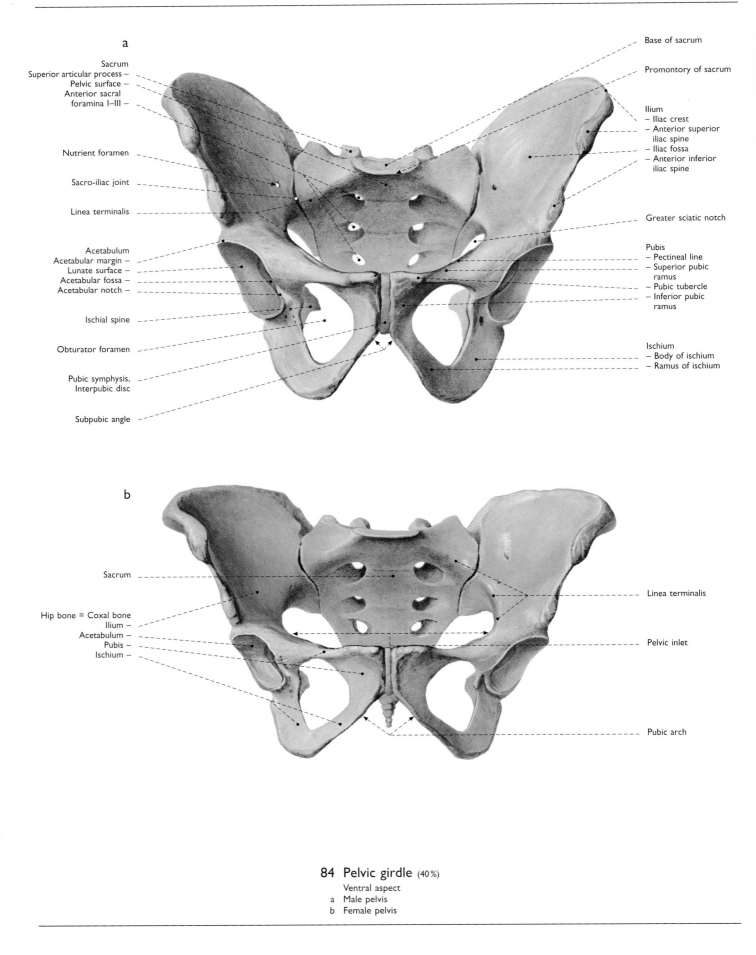

a

Sacrum
Superior articular process —
Pelvic surface —
Anterior sacral
foramina I–III —

Nutrient foramen

Sacro-iliac joint

Linea terminalis

Acetabulum
Acetabular margin —
Lunate surface —
Acetabular fossa —
Acetabular notch —

Ischial spine

Obturator foramen

Pubic symphysis,
Interpubic disc

Subpubic angle

Base of sacrum

Promontory of sacrum

Ilium
— Iliac crest
— Anterior superior
  iliac spine
— Iliac fossa
— Anterior inferior
  iliac spine

Greater sciatic notch

Pubis
— Pectineal line
— Superior pubic
  ramus
— Pubic tubercle
— Inferior pubic
  ramus

Ischium
— Body of ischium
— Ramus of ischium

b

Sacrum

Hip bone = Coxal bone
Ilium —
Acetabulum —
Pubis —
Ischium —

Linea terminalis

Pelvic inlet

Pubic arch

84 Pelvic girdle (40%)
Ventral aspect
a Male pelvis
b Female pelvis

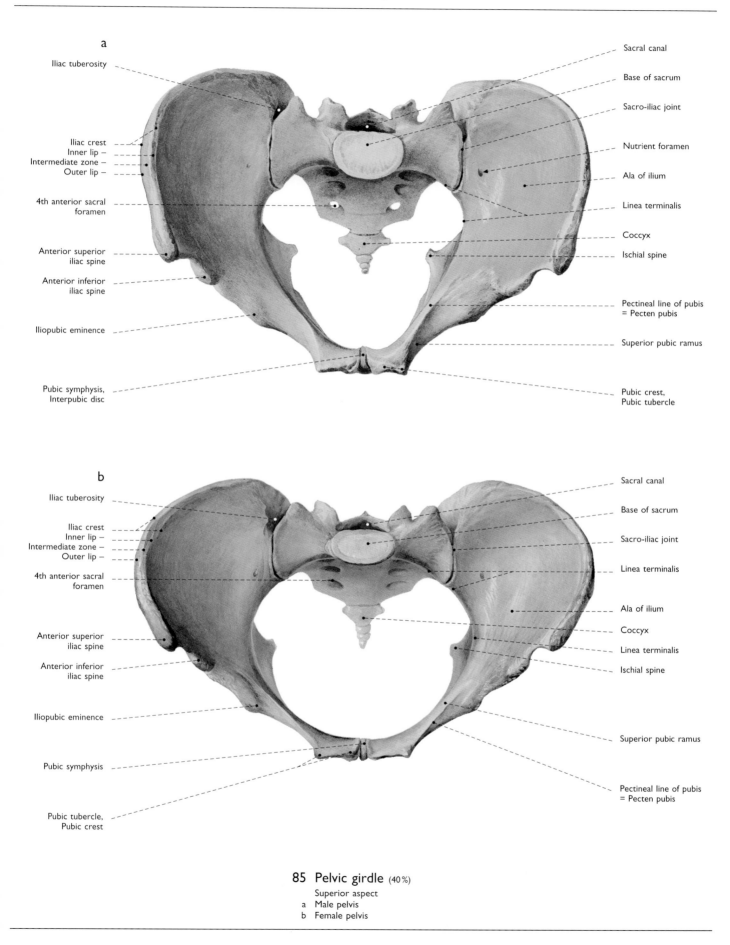

a

Iliac tuberosity

Iliac crest
Inner lip –
Intermediate zone –
Outer lip –

4th anterior sacral
foramen

Anterior superior
iliac spine

Anterior inferior
iliac spine

Iliopubic eminence

Pubic symphysis,
Interpubic disc

Sacral canal

Base of sacrum

Sacro-iliac joint

Nutrient foramen

Ala of ilium

Linea terminalis

Coccyx

Ischial spine

Pectineal line of pubis
= Pecten pubis

Superior pubic ramus

Pubic crest,
Pubic tubercle

b

Iliac tuberosity

Iliac crest
Inner lip –
Intermediate zone –
Outer lip –

4th anterior sacral
foramen

Anterior superior
iliac spine

Anterior inferior
iliac spine

Iliopubic eminence

Pubic symphysis

Pubic tubercle,
Pubic crest

Sacral canal

Base of sacrum

Sacro-iliac joint

Linea terminalis

Ala of ilium

Coccyx

Linea terminalis

Ischial spine

Superior pubic ramus

Pectineal line of pubis
= Pecten pubis

85 Pelvic girdle (40%)

Superior aspect
a Male pelvis
b Female pelvis

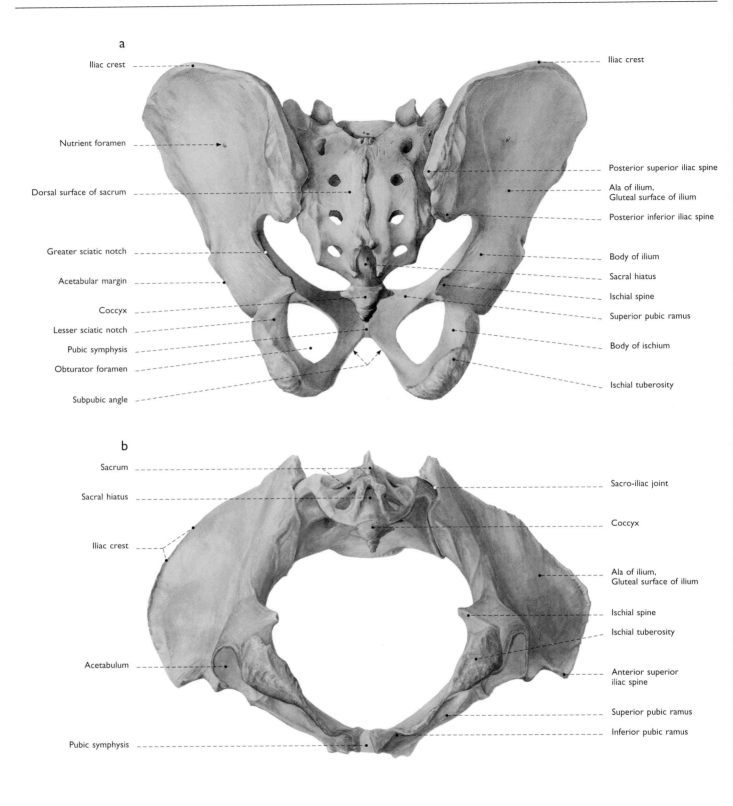

a

Iliac crest

Nutrient foramen

Dorsal surface of sacrum

Greater sciatic notch

Acetabular margin

Coccyx

Lesser sciatic notch

Pubic symphysis

Obturator foramen

Subpubic angle

Iliac crest

Posterior superior iliac spine

Ala of ilium,
Gluteal surface of ilium

Posterior inferior iliac spine

Body of ilium

Sacral hiatus

Ischial spine

Superior pubic ramus

Body of ischium

Ischial tuberosity

b

Sacrum

Sacral hiatus

Iliac crest

Acetabulum

Pubic symphysis

Sacro-iliac joint

Coccyx

Ala of ilium,
Gluteal surface of ilium

Ischial spine

Ischial tuberosity

Anterior superior
iliac spine

Superior pubic ramus

Inferior pubic ramus

86 Pelvic girdle of a female (40%)
a Dorsal aspect
b Inferior aspect

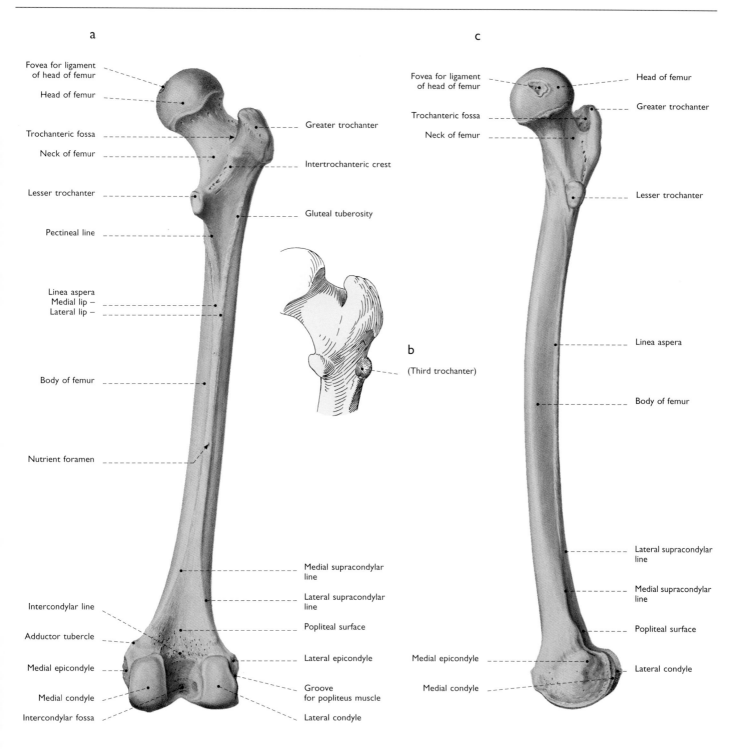

a

Fovea for ligament
of head of femur

Head of femur

Trochanteric fossa

Neck of femur

Lesser trochanter

Pectineal line

Linea aspera
Medial lip –
Lateral lip –

Body of femur

Nutrient foramen

Intercondylar line

Adductor tubercle

Medial epicondyle

Medial condyle

Intercondylar fossa

Greater trochanter

Intertrochanteric crest

Gluteal tuberosity

b

(Third trochanter)

Medial supracondylar
line

Lateral supracondylar
line

Popliteal surface

Lateral epicondyle

Groove
for popliteus muscle

Lateral condyle

c

Fovea for ligament
of head of femur

Trochanteric fossa

Neck of femur

Head of femur

Greater trochanter

Lesser trochanter

Linea aspera

Body of femur

Lateral supracondylar
line

Medial supracondylar
line

Popliteal surface

Medial epicondyle

Medial condyle

Lateral condyle

87 Right thigh bone (= femur) (40%)
   a Dorsal aspect
   b Proximal end of the thigh bone
      with a third trochanter, dorsal aspect
   c Medial aspect

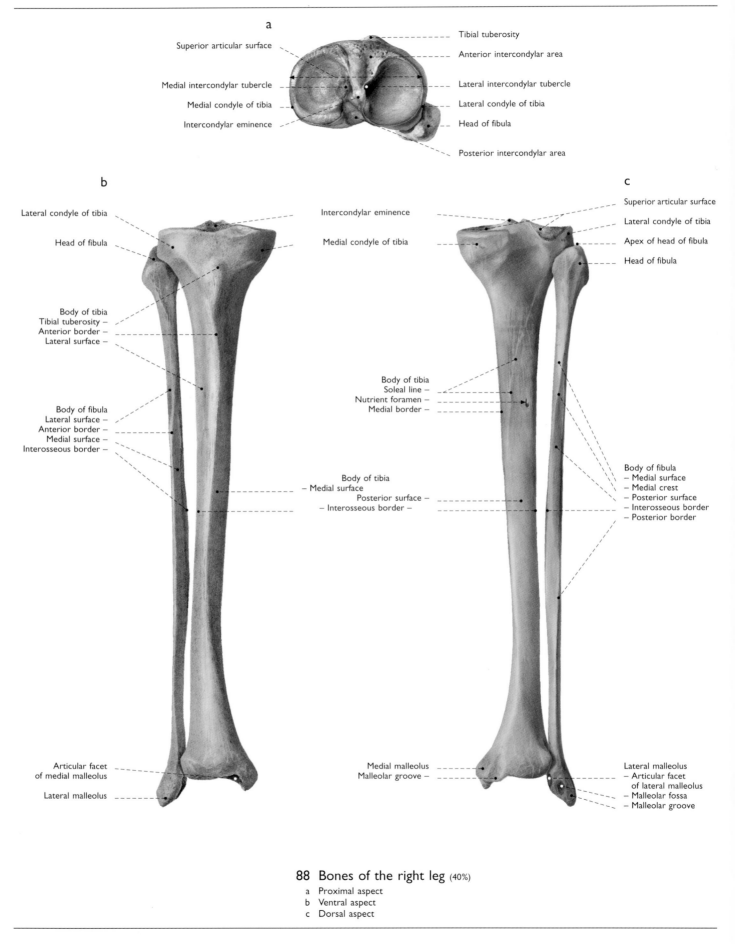

a

Superior articular surface

Tibial tuberosity
Anterior intercondylar area

Medial intercondylar tubercle
Medial condyle of tibia
Intercondylar eminence

Lateral intercondylar tubercle
Lateral condyle of tibia
Head of fibula

Posterior intercondylar area

b

Lateral condyle of tibia
Head of fibula

Intercondylar eminence
Medial condyle of tibia

Body of tibia
Tibial tuberosity −
Anterior border −
Lateral surface −

Body of fibula
Lateral surface −
Anterior border −
Medial surface −
Interosseous border −

Body of tibia
− Medial surface
Posterior surface −
− Interosseous border −

Articular facet
of medial malleolus

Lateral malleolus

c

Superior articular surface
Lateral condyle of tibia
Apex of head of fibula
Head of fibula

Body of tibia
Soleal line −
Nutrient foramen −
Medial border −

Body of fibula
− Medial surface
− Medial crest
− Posterior surface
− Interosseous border
− Posterior border

Medial malleolus
Malleolar groove −

Lateral malleolus
− Articular facet
of lateral malleolus
− Malleolar fossa
− Malleolar groove

88 Bones of the right leg (40%)
a Proximal aspect
b Ventral aspect
c Dorsal aspect

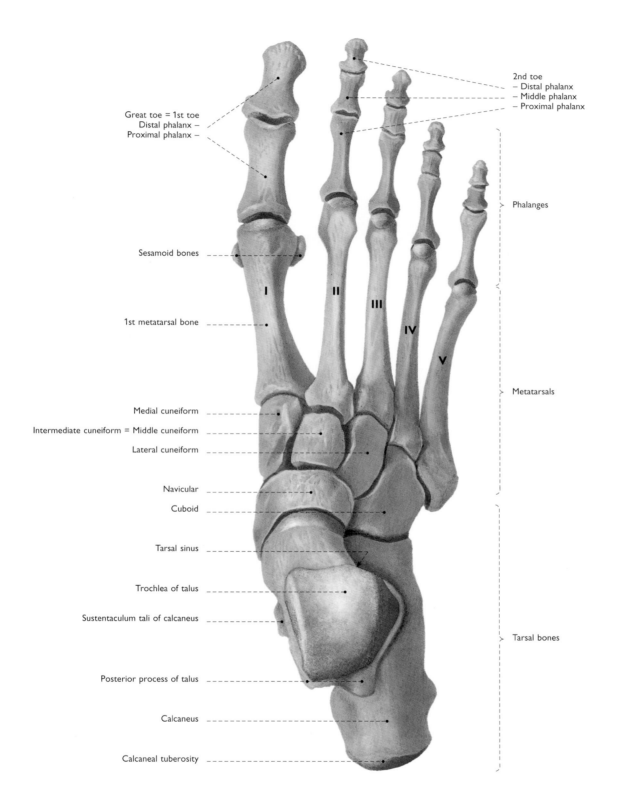

2nd toe
– Distal phalanx
– Middle phalanx
– Proximal phalanx

Great toe = 1st toe
Distal phalanx –
Proximal phalanx –

Phalanges

Sesamoid bones

I

II

III

IV

V

1st metatarsal bone

Metatarsals

Medial cuneiform
Intermediate cuneiform = Middle cuneiform
Lateral cuneiform

Navicular

Cuboid

Tarsal sinus

Trochlea of talus

Sustentaculum tali of calcaneus

Tarsal bones

Posterior process of talus

Calcaneus

Calcaneal tuberosity

89 Bones of the right foot (80%)
Dorsal aspect

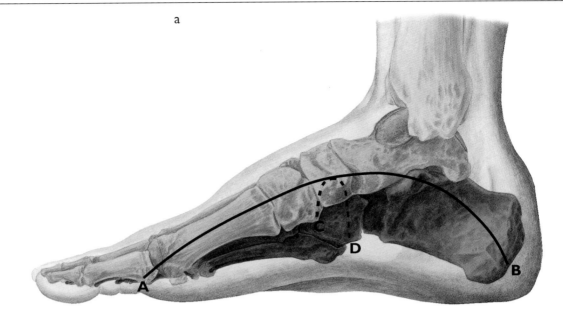

a

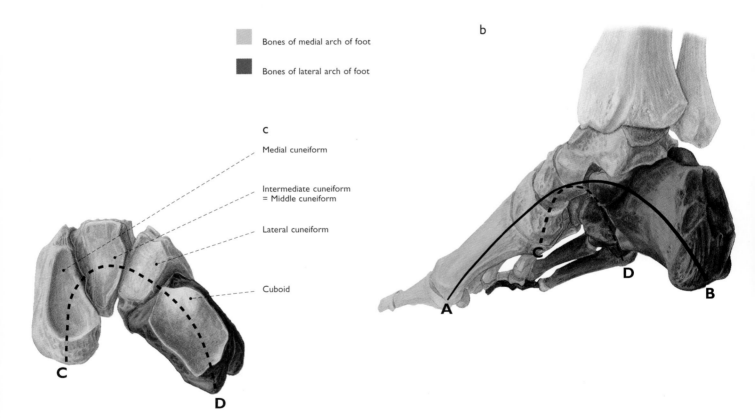

Bones of medial arch of foot

Bones of lateral arch of foot

b

c

Medial cuneiform

Intermediate cuneiform
= Middle cuneiform

Lateral cuneiform

Cuboid

**90 Longitudinal and transverse arches
of the skeleton of the right foot**

The bones of the medial arch are illustrated in clear brown,
those of the lateral arch in dark brown color.
The longitudinal arch is shown by a continuous line (A–B),
the transverse arch by a broken line (C–D).
a   Medial aspect (55%)
b   Mediodorsal aspect (55%)
c   Proximal aspect of the cuneiform bones and the cuboid bone (85%)

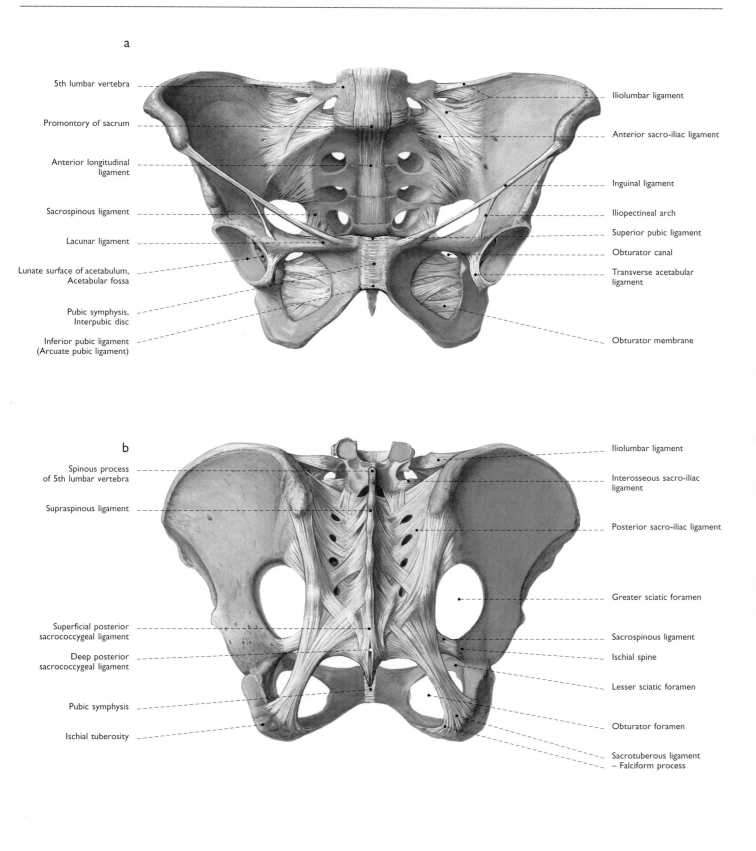

a

5th lumbar vertebra

Promontory of sacrum

Anterior longitudinal
ligament

Sacrospinous ligament

Lacunar ligament

Lunate surface of acetabulum,
Acetabular fossa

Pubic symphysis,
Interpubic disc

Inferior pubic ligament
(Arcuate pubic ligament)

Iliolumbar ligament

Anterior sacro-iliac ligament

Inguinal ligament

Iliopectineal arch

Superior pubic ligament

Obturator canal

Transverse acetabular
ligament

Obturator membrane

b

Spinous process
of 5th lumbar vertebra

Supraspinous ligament

Superficial posterior
sacrococcygeal ligament

Deep posterior
sacrococcygeal ligament

Pubic symphysis

Ischial tuberosity

Iliolumbar ligament

Interosseous sacro-iliac
ligament

Posterior sacro-iliac ligament

Greater sciatic foramen

Sacrospinous ligament

Ischial spine

Lesser sciatic foramen

Obturator foramen

Sacrotuberous ligament
– Falciform process

**91 Joints and ligaments of the pelvic girdle
of a female** (40%)

a   Ventral aspect
b   The obturator membrane was removed. Dorsal aspect

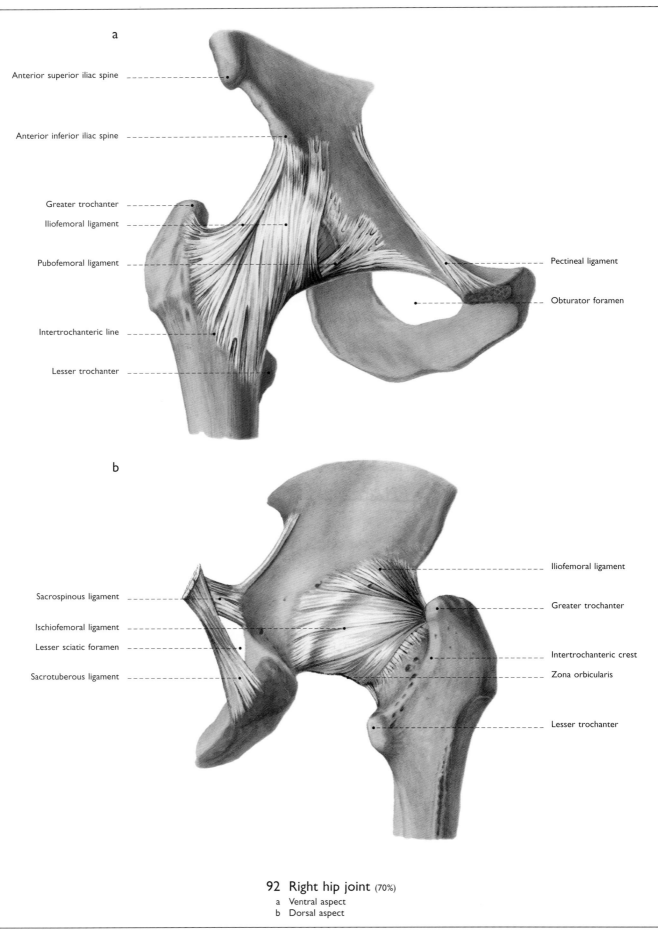

a

Anterior superior iliac spine

Anterior inferior iliac spine

Greater trochanter
Iliofemoral ligament

Pubofemoral ligament

Pectineal ligament

Obturator foramen

Intertrochanteric line

Lesser trochanter

b

Sacrospinous ligament

Ischiofemoral ligament
Lesser sciatic foramen

Sacrotuberous ligament

Iliofemoral ligament

Greater trochanter

Intertrochanteric crest
Zona orbicularis

Lesser trochanter

92 Right hip joint (70%)
  a Ventral aspect
  b Dorsal aspect

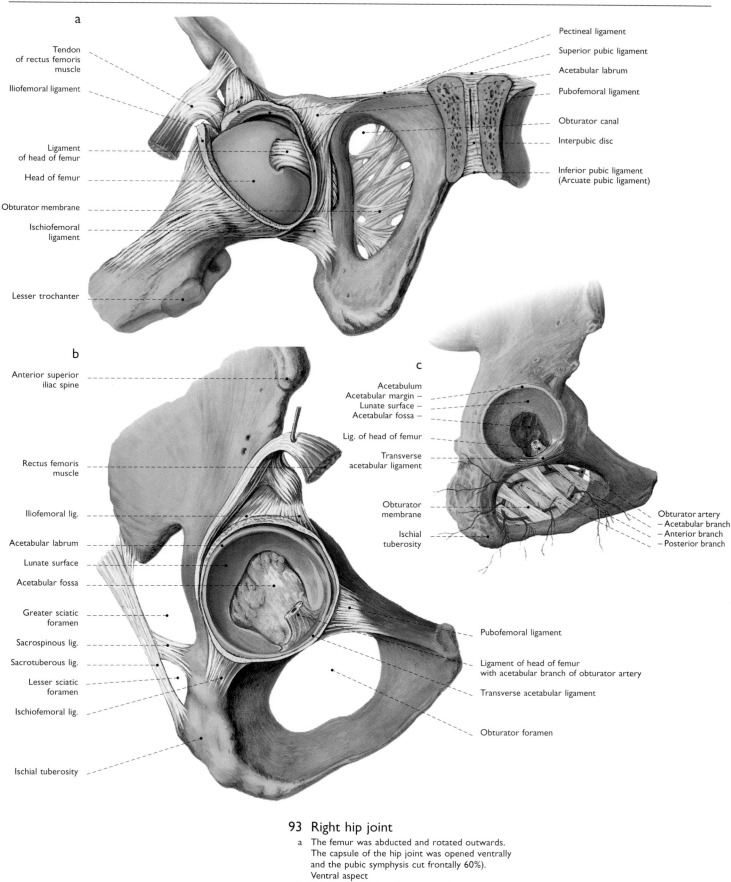

a

Tendon
of rectus femoris
muscle

Iliofemoral ligament

Ligament
of head of femur

Head of femur

Obturator membrane

Ischiofemoral
ligament

Lesser trochanter

Pectineal ligament

Superior pubic ligament

Acetabular labrum

Pubofemoral ligament

Obturator canal

Interpubic disc

Inferior pubic ligament
(Arcuate pubic ligament)

b

Anterior superior
iliac spine

Rectus femoris
muscle

Iliofemoral lig.

Acetabular labrum

Lunate surface

Acetabular fossa

Greater sciatic
foramen

Sacrospinous lig.

Sacrotuberous lig.

Lesser sciatic
foramen

Ischiofemoral lig.

Ischial tuberosity

c

Acetabulum
Acetabular margin –
Lunate surface –
Acetabular fossa –

Lig. of head of femur

Transverse
acetabular ligament

Obturator
membrane

Ischial
tuberosity

Obturator artery
– Acetabular branch
– Anterior branch
– Posterior branch

Pubofemoral ligament

Ligament of head of femur
with acetabular branch of obturator artery

Transverse acetabular ligament

Obturator foramen

93 Right hip joint

a The femur was abducted and rotated outwards.
   The capsule of the hip joint was opened ventrally
   and the pubic symphysis cut frontally 60%).
   Ventral aspect
b View of the socket of hip joint (60%), ventrolateral aspect
c Ligament of the head of femur with acetabular artery (35%),
   ventrolateral aspect

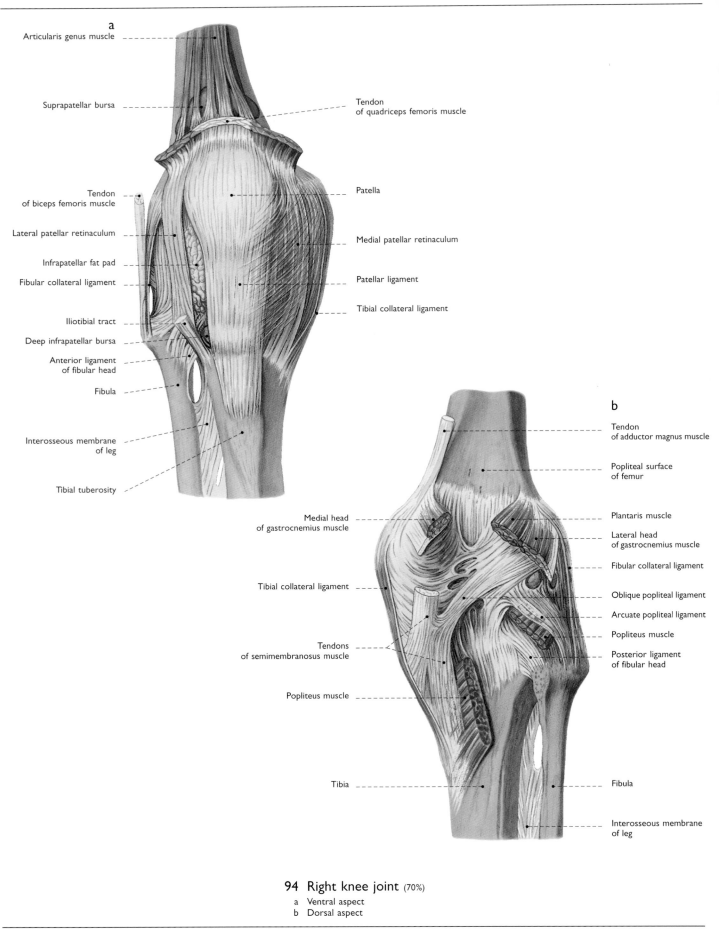

a
Articularis genus muscle

Suprapatellar bursa

Tendon
of biceps femoris muscle

Lateral patellar retinaculum

Infrapatellar fat pad

Fibular collateral ligament

Iliotibial tract

Deep infrapatellar bursa

Anterior ligament
of fibular head

Fibula

Interosseous membrane
of leg

Tibial tuberosity

Tendon
of quadriceps femoris muscle

Patella

Medial patellar retinaculum

Patellar ligament

Tibial collateral ligament

b

Tendon
of adductor magnus muscle

Popliteal surface
of femur

Plantaris muscle

Lateral head
of gastrocnemius muscle

Fibular collateral ligament

Oblique popliteal ligament

Arcuate popliteal ligament

Popliteus muscle

Posterior ligament
of fibular head

Medial head
of gastrocnemius muscle

Tibial collateral ligament

Tendons
of semimembranosus muscle

Popliteus muscle

Tibia

Fibula

Interosseous membrane
of leg

94 Right knee joint (70%)
a Ventral aspect
b Dorsal aspect

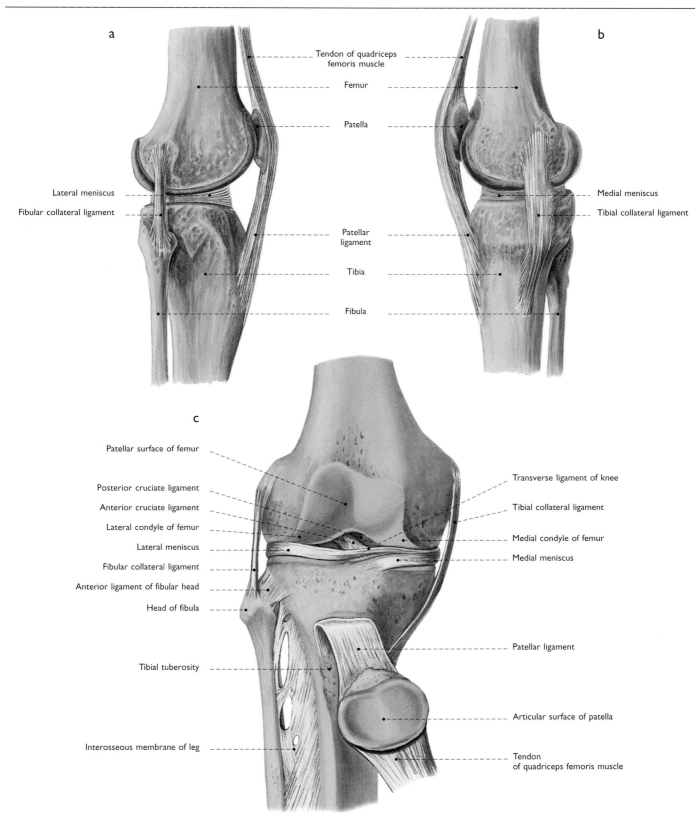

a

Tendon of quadriceps
femoris muscle

Femur

Patella

Lateral meniscus

Fibular collateral ligament

Patellar
ligament

Tibia

Fibula

b

Medial meniscus

Tibial collateral ligament

c

Patellar surface of femur

Posterior cruciate ligament

Anterior cruciate ligament

Lateral condyle of femur

Lateral meniscus

Fibular collateral ligament

Anterior ligament of fibular head

Head of fibula

Tibial tuberosity

Interosseous membrane of leg

Transverse ligament of knee

Tibial collateral ligament

Medial condyle of femur

Medial meniscus

Patellar ligament

Articular surface of patella

Tendon
of quadriceps femoris muscle

95 Right knee joint
The capsule was removed.
a Lateral aspect (50%)
b Medial aspect (50%)
c The patella was turned downwards (70%). Ventral aspect

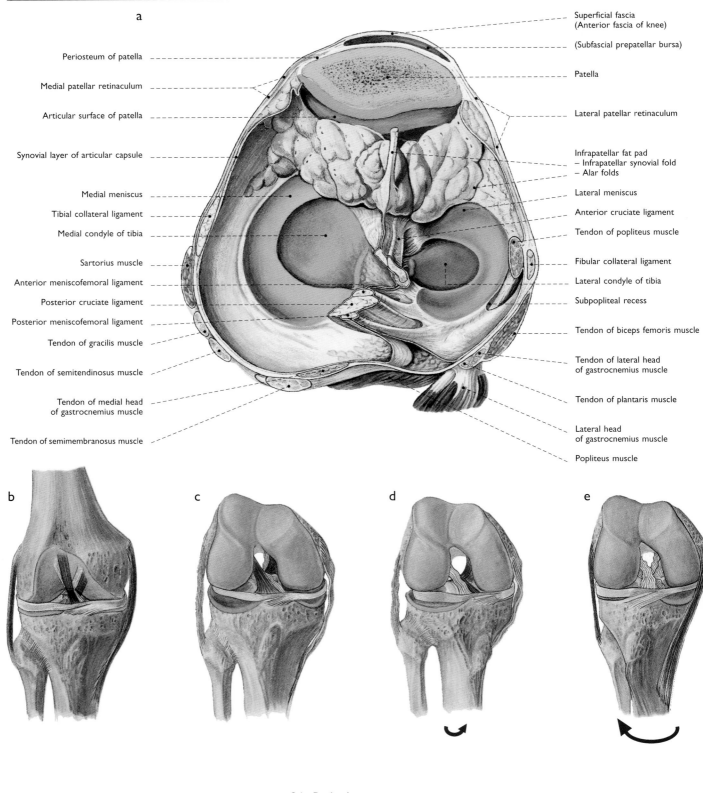

a

Periosteum of patella

Medial patellar retinaculum

Articular surface of patella

Synovial layer of articular capsule

Medial meniscus

Tibial collateral ligament

Medial condyle of tibia

Sartorius muscle

Anterior meniscofemoral ligament

Posterior cruciate ligament

Posterior meniscofemoral ligament

Tendon of gracilis muscle

Tendon of semitendinosus muscle

Tendon of medial head
of gastrocnemius muscle

Tendon of semimembranosus muscle

Superficial fascia
(Anterior fascia of knee)

(Subfascial prepatellar bursa)

Patella

Lateral patellar retinaculum

Infrapatellar fat pad
– Infrapatellar synovial fold
– Alar folds

Lateral meniscus

Anterior cruciate ligament

Tendon of popliteus muscle

Fibular collateral ligament

Lateral condyle of tibia

Subpopliteal recess

Tendon of biceps femoris muscle

Tendon of lateral head
of gastrocnemius muscle

Tendon of plantaris muscle

Lateral head
of gastrocnemius muscle

Popliteus muscle

b   c   d   e

96 Right knee joint

a   The joint was cut transversally through the middle of the patella (100%).
    Cranial aspect of the distal part

b–e State of tautening of the cruciate and collateral ligaments (50%)
    (according to von Lanz and Wachsmuth, 1972)

b   knee in extension
c   knee in flexion
d   knee in flexion and medial rotation
e   knee in flexion and lateral rotation
    The taut parts of ligaments are dark-colored. Ventral aspect

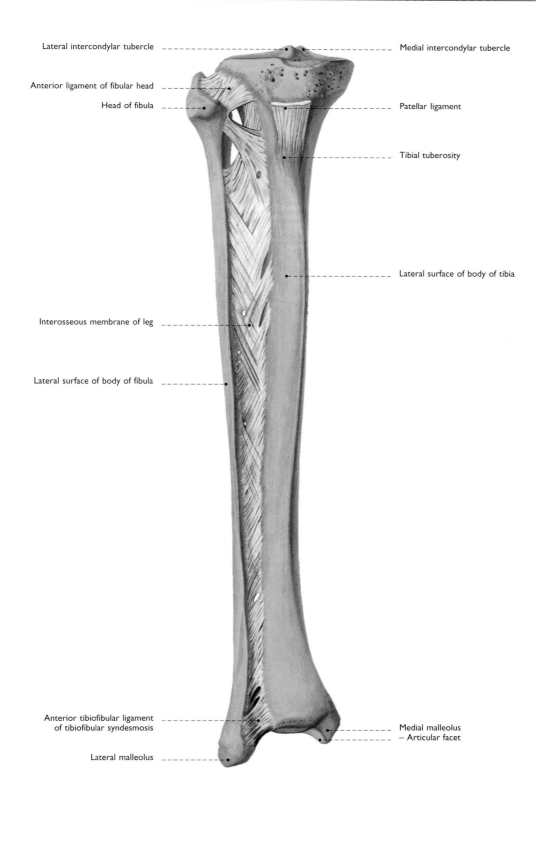

Lateral intercondylar tubercle

Medial intercondylar tubercle

Anterior ligament of fibular head

Head of fibula

Patellar ligament

Tibial tuberosity

Lateral surface of body of tibia

Interosseous membrane of leg

Lateral surface of body of fibula

Anterior tibiofibular ligament
of tibiofibular syndesmosis

Medial malleolus
– Articular facet

Lateral malleolus

**97 Tibiofibular joints and syndesmoses
of the right leg** (50%)
Ventral aspect

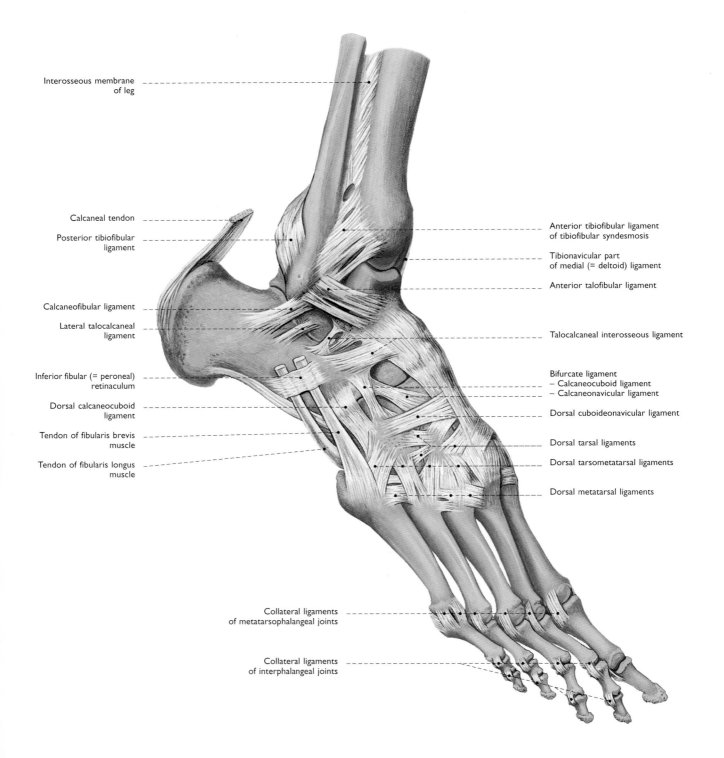

Interosseous membrane
of leg

Calcaneal tendon

Posterior tibiofibular
ligament

Calcaneofibular ligament

Lateral talocalcaneal
ligament

Inferior fibular (= peroneal)
retinaculum

Dorsal calcaneocuboid
ligament

Tendon of fibularis brevis
muscle

Tendon of fibularis longus
muscle

Collateral ligaments
of metatarsophalangeal joints

Collateral ligaments
of interphalangeal joints

Anterior tibiofibular ligament
of tibiofibular syndesmosis

Tibionavicular part
of medial (= deltoid) ligament

Anterior talofibular ligament

Talocalcaneal interosseous ligament

Bifurcate ligament
− Calcaneocuboid ligament
− Calcaneonavicular ligament

Dorsal cuboideonavicular ligament

Dorsal tarsal ligaments

Dorsal tarsometatarsal ligaments

Dorsal metatarsal ligaments

98 Joints and ligaments of the right foot (70%)
Lateral aspect

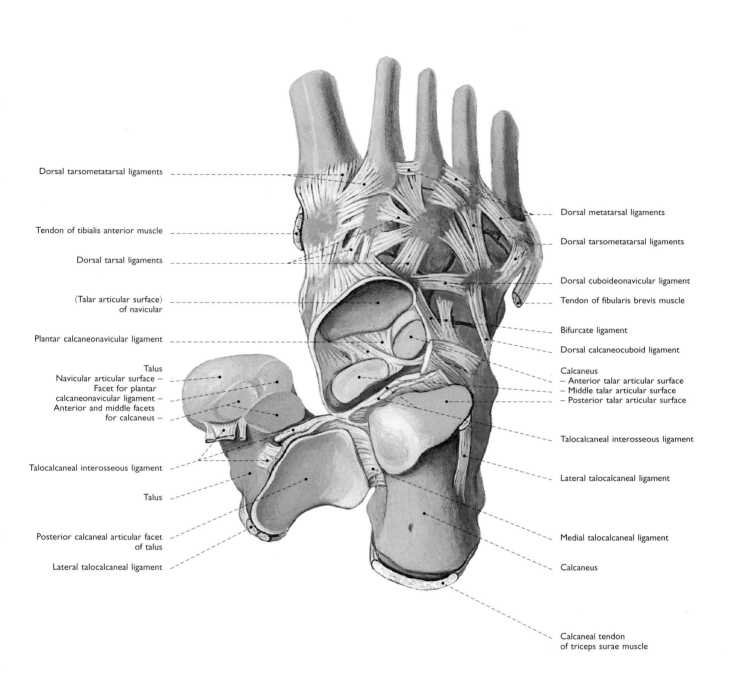

Dorsal tarsometatarsal ligaments

Tendon of tibialis anterior muscle

Dorsal tarsal ligaments

⟨Talar articular surface⟩
of navicular

Plantar calcaneonavicular ligament

Talus
Navicular articular surface –
Facet for plantar
calcaneonavicular ligament –
Anterior and middle facets
for calcaneus –

Talocalcaneal interosseous ligament

Talus

Posterior calcaneal articular facet
of talus

Lateral talocalcaneal ligament

Dorsal metatarsal ligaments

Dorsal tarsometatarsal ligaments

Dorsal cuboideonavicular ligament

Tendon of fibularis brevis muscle

Bifurcate ligament

Dorsal calcaneocuboid ligament

Calcaneus
– Anterior talar articular surface
– Middle talar articular surface
– Posterior talar articular surface

Talocalcaneal interosseous ligament

Lateral talocalcaneal ligament

Medial talocalcaneal ligament

Calcaneus

Calcaneal tendon
of triceps surae muscle

**99 Subtalar (= talocalcaneal), talocalcaneonavicular,
and tarsometatarsal joint of the right foot** (90%)
The talus was turned medially. Dorsal aspect

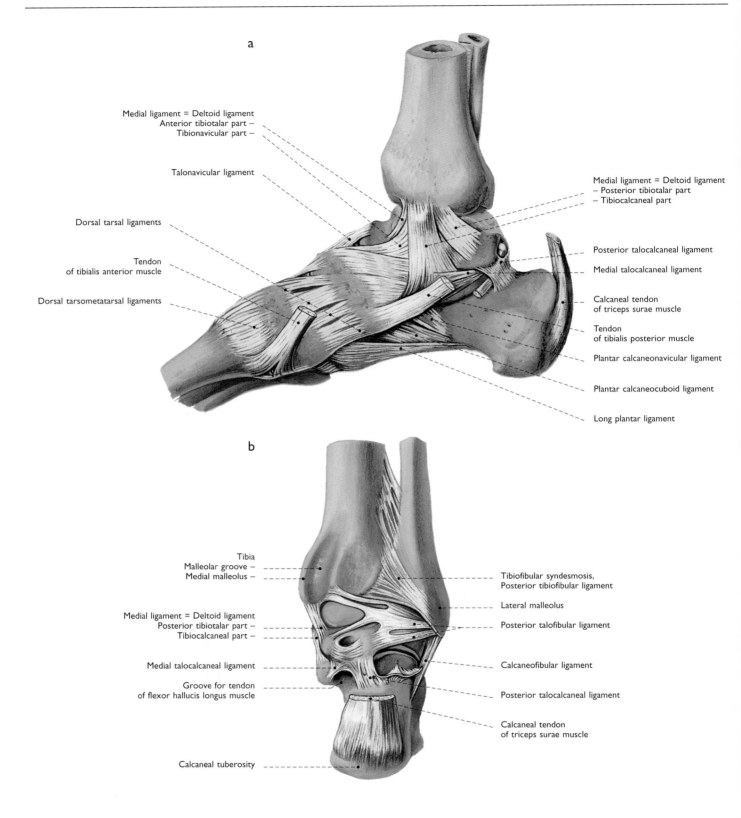

a

Medial ligament = Deltoid ligament
Anterior tibiotalar part –
Tibionavicular part –

Talonavicular ligament

Dorsal tarsal ligaments

Tendon
of tibialis anterior muscle

Dorsal tarsometatarsal ligaments

Medial ligament = Deltoid ligament
– Posterior tibiotalar part
– Tibiocalcaneal part

Posterior talocalcaneal ligament

Medial talocalcaneal ligament

Calcaneal tendon
of triceps surae muscle

Tendon
of tibialis posterior muscle

Plantar calcaneonavicular ligament

Plantar calcaneocuboid ligament

Long plantar ligament

b

Tibia
Malleolar groove –
Medial malleolus –

Medial ligament = Deltoid ligament
Posterior tibiotalar part –
Tibiocalcaneal part –

Medial talocalcaneal ligament

Groove for tendon
of flexor hallucis longus muscle

Calcaneal tuberosity

Tibiofibular syndesmosis,
Posterior tibiofibular ligament

Lateral malleolus

Posterior talofibular ligament

Calcaneofibular ligament

Posterior talocalcaneal ligament

Calcaneal tendon
of triceps surae muscle

100 Joints and ligaments of the right foot (60%)
a   Medial aspect
b   Dorsal aspect

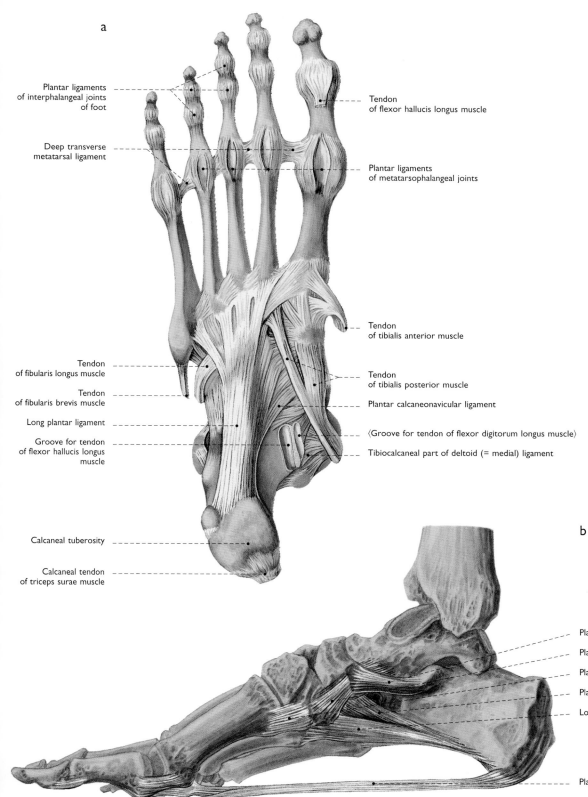

a

Plantar ligaments of interphalangeal joints of foot

Deep transverse metatarsal ligament

Tendon of flexor hallucis longus muscle

Plantar ligaments of metatarsophalangeal joints

Tendon of tibialis anterior muscle

Tendon of fibularis longus muscle

Tendon of tibialis posterior muscle

Tendon of fibularis brevis muscle

Plantar calcaneonavicular ligament

Long plantar ligament

⟨Groove for tendon of flexor digitorum longus muscle⟩

Groove for tendon of flexor hallucis longus muscle

Tibiocalcaneal part of deltoid (= medial) ligament

Calcaneal tuberosity

Calcaneal tendon of triceps surae muscle

b

Plantar calcaneonavicular ligament

Plantar cuneonavicular ligaments

Plantar tarsometatarsal ligaments

Plantar calcaneocuboid ligament

Long plantar ligament

Plantar aponeurosis

101 Joints and ligaments of the right foot (60%)
a Plantar aspect
b Ligaments stabilizing the subtalar (= talocalcaneal) and talocalcaneonavicular joints, medial aspect

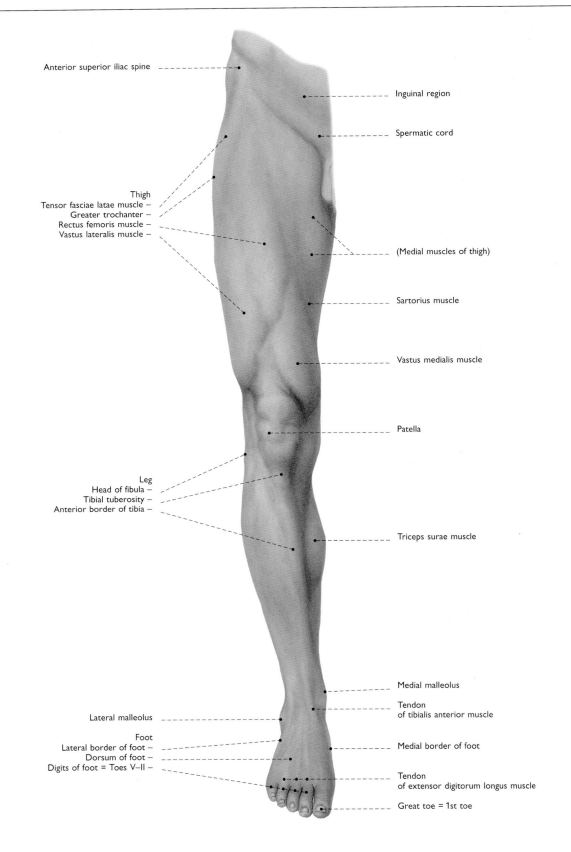

Anterior superior iliac spine

Inguinal region

Spermatic cord

Thigh
Tensor fasciae latae muscle –
Greater trochanter –
Rectus femoris muscle –
Vastus lateralis muscle –

(Medial muscles of thigh)

Sartorius muscle

Vastus medialis muscle

Patella

Leg
Head of fibula –
Tibial tuberosity –
Anterior border of tibia –

Triceps surae muscle

Medial malleolus

Tendon
of tibialis anterior muscle

Lateral malleolus

Foot
Lateral border of foot –
Dorsum of foot –
Digits of foot = Toes V–II –

Medial border of foot

Tendon
of extensor digitorum longus muscle

Great toe = 1st toe

102 Surface anatomy of the right lower limb (20%)
Ventral aspect

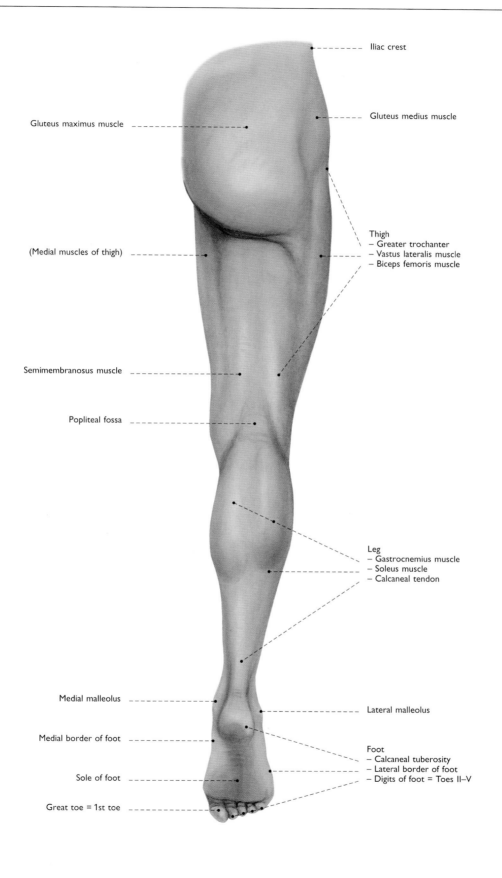

Iliac crest

Gluteus medius muscle

Gluteus maximus muscle

Thigh
– Greater trochanter
– Vastus lateralis muscle
– Biceps femoris muscle

(Medial muscles of thigh)

Semimembranosus muscle

Popliteal fossa

Leg
– Gastrocnemius muscle
– Soleus muscle
– Calcaneal tendon

Medial malleolus

Lateral malleolus

Medial border of foot

Foot
– Calcaneal tuberosity
– Lateral border of foot
– Digits of foot = Toes II–V

Sole of foot

Great toe = 1st toe

103 Surface anatomy of the right lower limb (20%)
Dorsal aspect

a

b

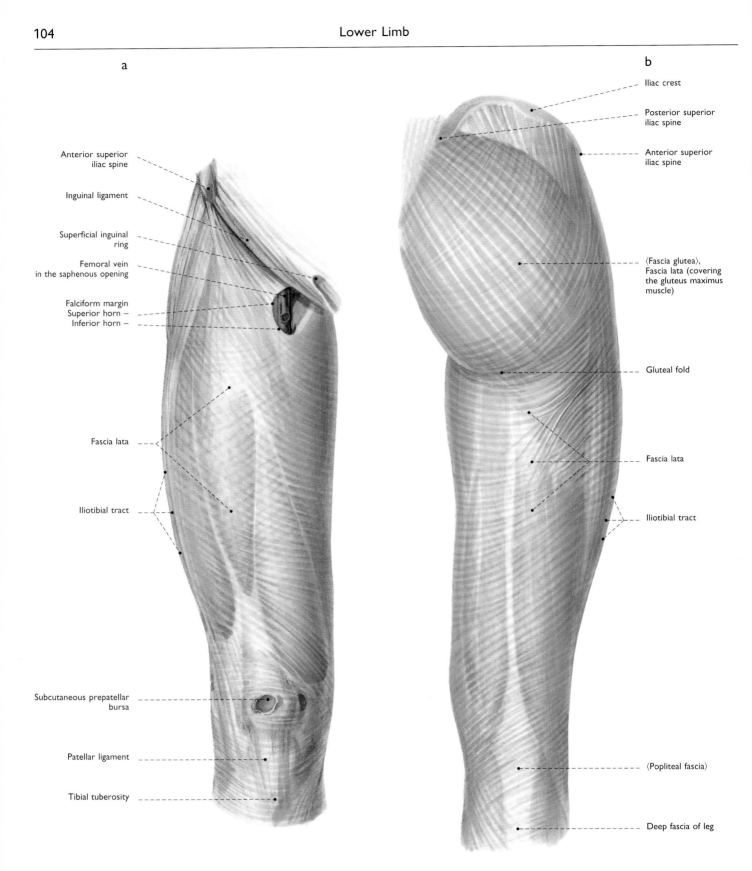

Anterior superior
iliac spine

Inguinal ligament

Superficial inguinal
ring

Femoral vein
in the saphenous opening

Falciform margin
Superior horn –
Inferior horn –

Fascia lata

Iliotibial tract

Subcutaneous prepatellar
bursa

Patellar ligament

Tibial tuberosity

Iliac crest

Posterior superior
iliac spine

Anterior superior
iliac spine

⟨Fascia glutea⟩,
Fascia lata (covering
the gluteus maximus
muscle)

Gluteal fold

Fascia lata

Iliotibial tract

⟨Popliteal fascia⟩

Deep fascia of leg

104 Fascia lata of the right thigh (30%)
a The cribriform fascia in the saphenous opening
   was removed. Ventral aspect
b Dorsal aspect

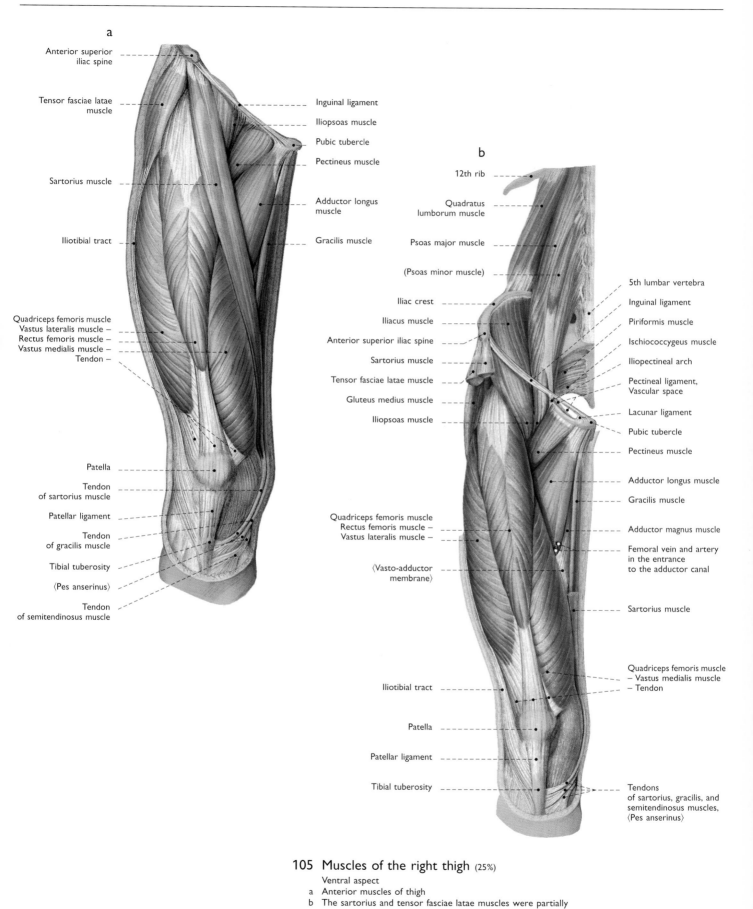

a

Anterior superior iliac spine

Tensor fasciae latae muscle

Sartorius muscle

Iliotibial tract

Quadriceps femoris muscle
Vastus lateralis muscle –
Rectus femoris muscle –
Vastus medialis muscle –
Tendon –

Patella

Tendon of sartorius muscle

Patellar ligament

Tendon of gracilis muscle

Tibial tuberosity

⟨Pes anserinus⟩

Tendon of semitendinosus muscle

Inguinal ligament

Iliopsoas muscle

Pubic tubercle

Pectineus muscle

Adductor longus muscle

Gracilis muscle

b

12th rib

Quadratus lumborum muscle

Psoas major muscle

(Psoas minor muscle)

Iliac crest

Iliacus muscle

Anterior superior iliac spine

Sartorius muscle

Tensor fasciae latae muscle

Gluteus medius muscle

Iliopsoas muscle

Quadriceps femoris muscle
Rectus femoris muscle –
Vastus lateralis muscle –

⟨Vasto-adductor membrane⟩

Iliotibial tract

Patella

Patellar ligament

Tibial tuberosity

5th lumbar vertebra

Inguinal ligament

Piriformis muscle

Ischiococcygeus muscle

Iliopectineal arch

Pectineal ligament, Vascular space

Lacunar ligament

Pubic tubercle

Pectineus muscle

Adductor longus muscle

Gracilis muscle

Adductor magnus muscle

Femoral vein and artery in the entrance to the adductor canal

Sartorius muscle

Quadriceps femoris muscle
– Vastus medialis muscle
– Tendon

Tendons of sartorius, gracilis, and semitendinosus muscles, ⟨Pes anserinus⟩

105   Muscles of the right thigh (25%)

Ventral aspect
a   Anterior muscles of thigh
b   The sartorius and tensor fasciae latae muscles were partially removed. Some muscles of the pelvis are additionally shown.

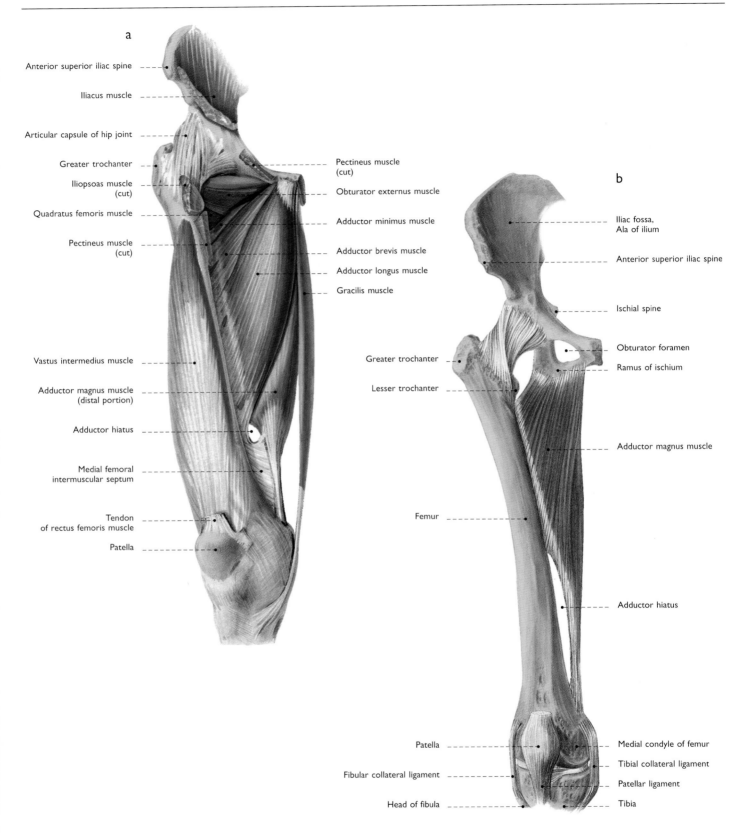

a

Anterior superior iliac spine

Iliacus muscle

Articular capsule of hip joint

Greater trochanter

Iliopsoas muscle
(cut)

Quadratus femoris muscle

Pectineus muscle
(cut)

Vastus intermedius muscle

Adductor magnus muscle
(distal portion)

Adductor hiatus

Medial femoral
intermuscular septum

Tendon
of rectus femoris muscle

Patella

Pectineus muscle
(cut)

Obturator externus muscle

Adductor minimus muscle

Adductor brevis muscle

Adductor longus muscle

Gracilis muscle

b

Iliac fossa,
Ala of ilium

Anterior superior iliac spine

Ischial spine

Obturator foramen

Ramus of ischium

Greater trochanter

Lesser trochanter

Adductor magnus muscle

Femur

Adductor hiatus

Patella

Fibular collateral ligament

Head of fibula

Medial condyle of femur

Tibial collateral ligament

Patellar ligament

Tibia

106 Muscles of the right thigh (25%)

Ventral aspect
a Medial muscles (adductors) of thigh
    and deep part of the quadriceps femoris muscle
b Adductor magnus muscle

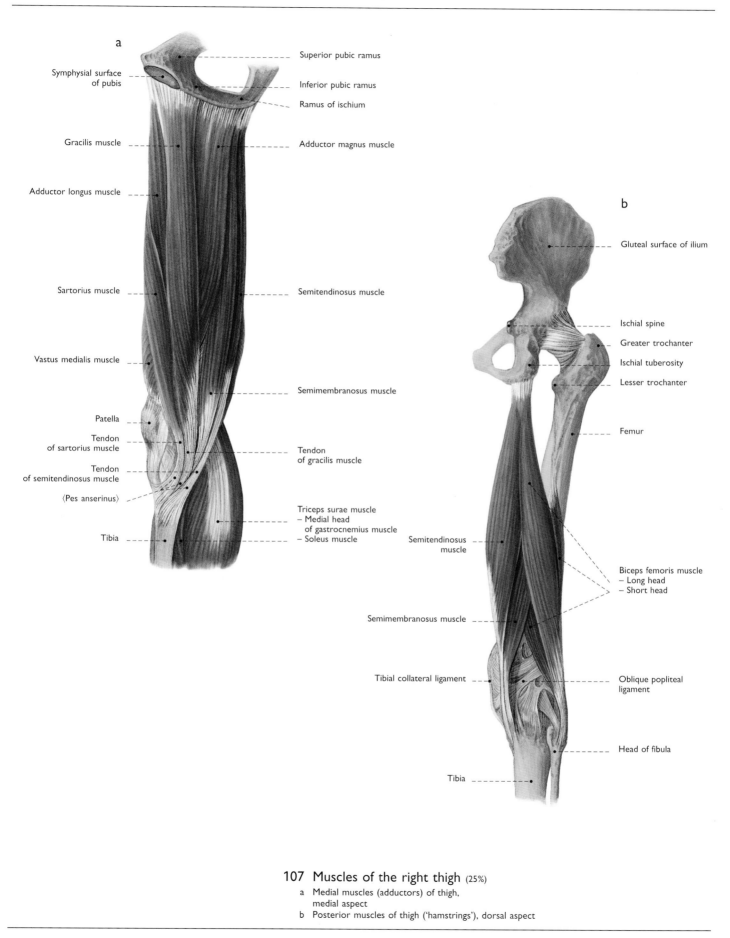

a

Symphysial surface of pubis

Gracilis muscle

Adductor longus muscle

Sartorius muscle

Vastus medialis muscle

Patella

Tendon of sartorius muscle

Tendon of semitendinosus muscle

⟨Pes anserinus⟩

Tibia

Superior pubic ramus

Inferior pubic ramus

Ramus of ischium

Adductor magnus muscle

Semitendinosus muscle

Semimembranosus muscle

Tendon of gracilis muscle

Triceps surae muscle
– Medial head of gastrocnemius muscle
– Soleus muscle

b

Gluteal surface of ilium

Ischial spine

Greater trochanter

Ischial tuberosity

Lesser trochanter

Femur

Semitendinosus muscle

Biceps femoris muscle
– Long head
– Short head

Semimembranosus muscle

Tibial collateral ligament

Oblique popliteal ligament

Head of fibula

Tibia

107 Muscles of the right thigh (25%)
　a Medial muscles (adductors) of thigh, medial aspect
　b Posterior muscles of thigh ('hamstrings'), dorsal aspect

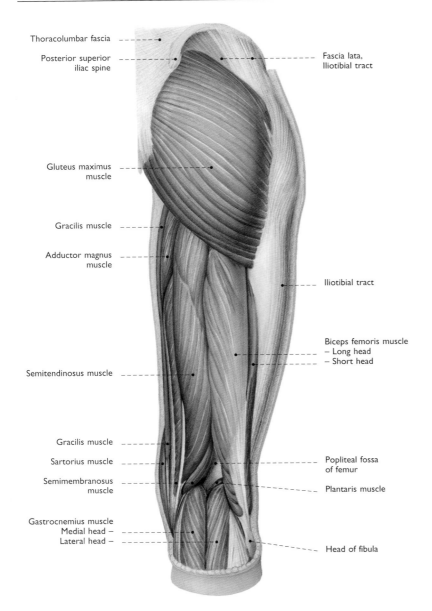

Thoracolumbar fascia

Posterior superior
iliac spine

Fascia lata,
Iliotibial tract

Gluteus maximus
muscle

Gracilis muscle

Adductor magnus
muscle

Iliotibial tract

Biceps femoris muscle
– Long head
– Short head

Semitendinosus muscle

Gracilis muscle

Sartorius muscle

Popliteal fossa
of femur

Semimembranosus
muscle

Plantaris muscle

Gastrocnemius muscle
Medial head –
Lateral head –

Head of fibula

**108 Muscles of the right thigh and
superficial layer of the muscles of the hip**
Dorsal aspect (20%)

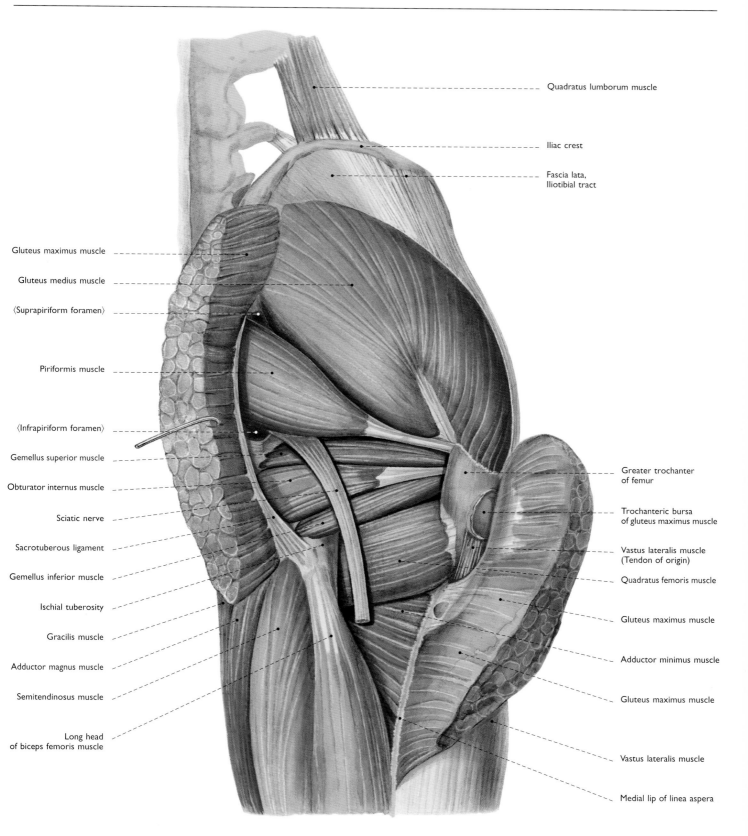

Quadratus lumborum muscle

Iliac crest

Fascia lata,
Iliotibial tract

Gluteus maximus muscle

Gluteus medius muscle

⟨Suprapiriform foramen⟩

Piriformis muscle

⟨Infrapiriform foramen⟩

Gemellus superior muscle

Obturator internus muscle

Sciatic nerve

Sacrotuberous ligament

Gemellus inferior muscle

Ischial tuberosity

Gracilis muscle

Adductor magnus muscle

Semitendinosus muscle

Long head
of biceps femoris muscle

Greater trochanter
of femur

Trochanteric bursa
of gluteus maximus muscle

Vastus lateralis muscle
(Tendon of origin)

Quadratus femoris muscle

Gluteus maximus muscle

Adductor minimus muscle

Gluteus maximus muscle

Vastus lateralis muscle

Medial lip of linea aspera

109 Muscles of the right hip (50%)
Deep layer. The gluteus maximus muscle was divided.
Dorsal aspect

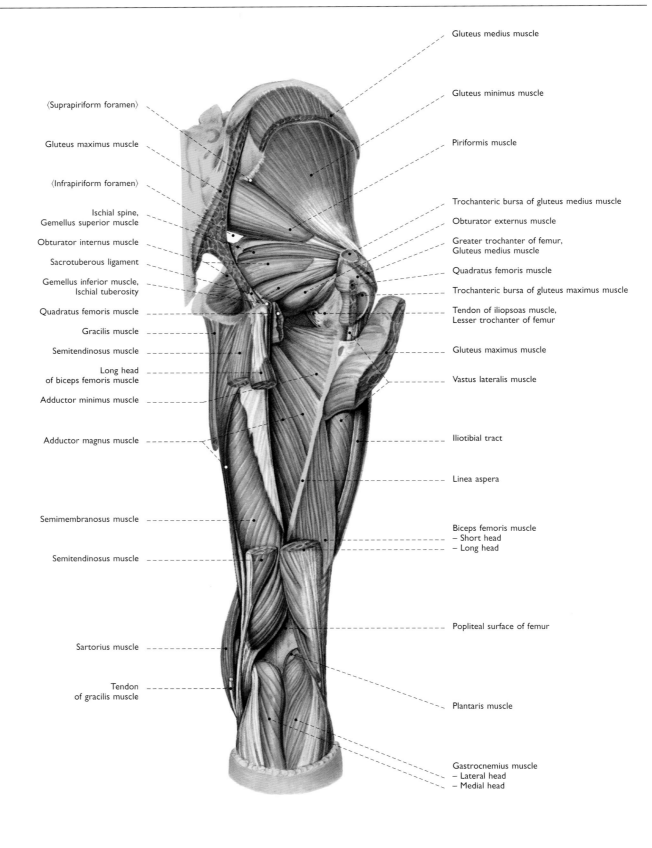

⟨Suprapiriform foramen⟩

Gluteus maximus muscle

⟨Infrapiriform foramen⟩

Ischial spine,
Gemellus superior muscle

Obturator internus muscle

Sacrotuberous ligament

Gemellus inferior muscle,
Ischial tuberosity

Quadratus femoris muscle

Gracilis muscle

Semitendinosus muscle

Long head
of biceps femoris muscle

Adductor minimus muscle

Adductor magnus muscle

Semimembranosus muscle

Semitendinosus muscle

Sartorius muscle

Tendon
of gracilis muscle

Gluteus medius muscle

Gluteus minimus muscle

Piriformis muscle

Trochanteric bursa of gluteus medius muscle

Obturator externus muscle

Greater trochanter of femur,
Gluteus medius muscle

Quadratus femoris muscle

Trochanteric bursa of gluteus maximus muscle

Tendon of iliopsoas muscle,
Lesser trochanter of femur

Gluteus maximus muscle

Vastus lateralis muscle

Iliotibial tract

Linea aspera

Biceps femoris muscle
– Short head
– Long head

Popliteal surface of femur

Plantaris muscle

Gastrocnemius muscle
– Lateral head
– Medial head

110 Muscles of the right thigh and hip (30%)
Deep muscular layer. The superficial muscles were partially removed.
Dorsal aspect

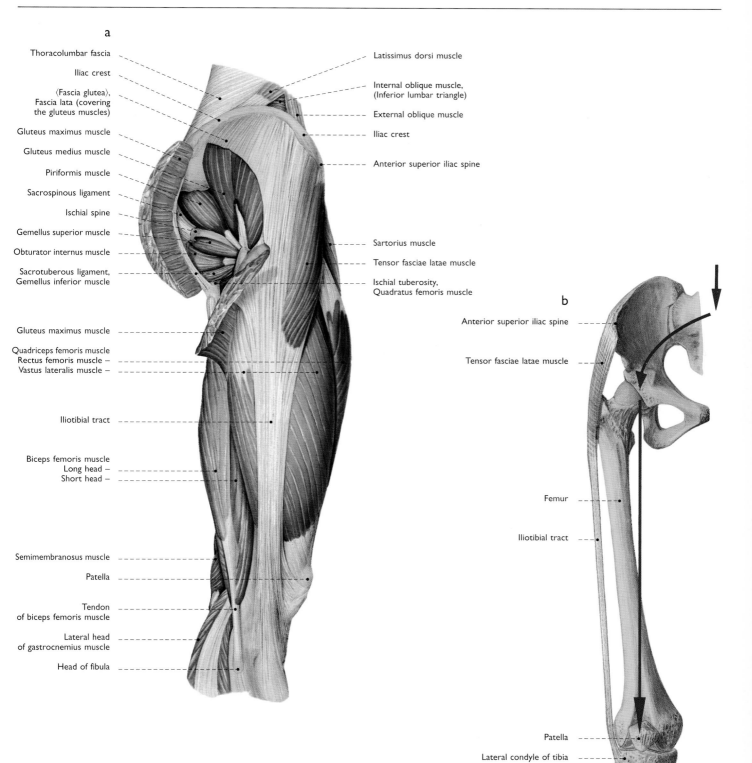

a

Thoracolumbar fascia

Iliac crest

⟨Fascia glutea⟩,
Fascia lata (covering
the gluteus muscles)

Gluteus maximus muscle

Gluteus medius muscle

Piriformis muscle

Sacrospinous ligament

Ischial spine

Gemellus superior muscle

Obturator internus muscle

Sacrotuberous ligament,
Gemellus inferior muscle

Gluteus maximus muscle

Quadriceps femoris muscle
Rectus femoris muscle –
Vastus lateralis muscle –

Iliotibial tract

Biceps femoris muscle
Long head –
Short head –

Semimembranosus muscle

Patella

Tendon
of biceps femoris muscle

Lateral head
of gastrocnemius muscle

Head of fibula

Latissimus dorsi muscle

Internal oblique muscle,
(Inferior lumbar triangle)

External oblique muscle

Iliac crest

Anterior superior iliac spine

Sartorius muscle

Tensor fasciae latae muscle

Ischial tuberosity,
Quadratus femoris muscle

b

Anterior superior iliac spine

Tensor fasciae latae muscle

Femur

Iliotibial tract

Patella

Lateral condyle of tibia

111 Muscles of the right thigh and hip (20%)

    a The gluteus maximus muscle was divided and turned up. Lateral aspect
    b Tensor fasciae latae muscle and iliotibial tract, ventral aspect.
       The arrow indicates the weight line in erect posture.

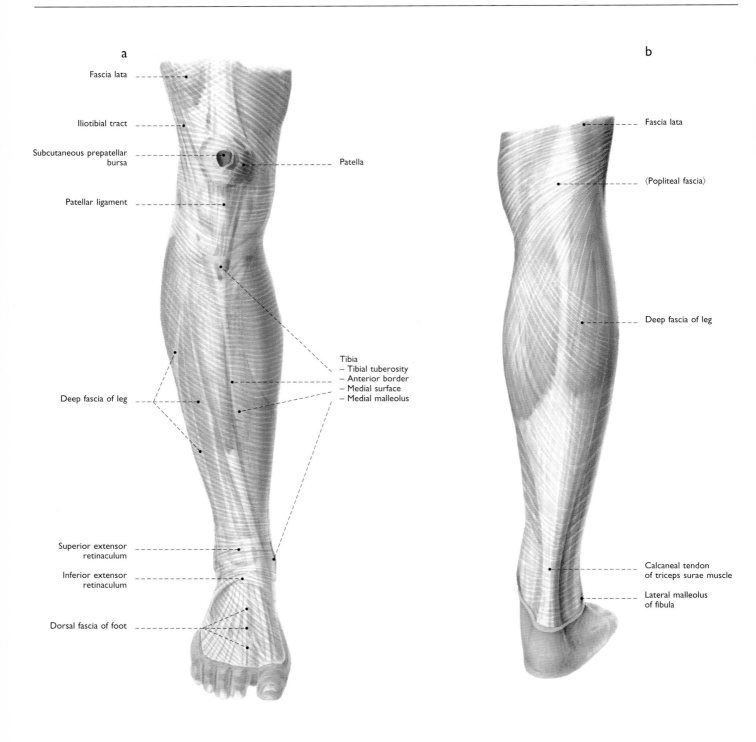

a

Fascia lata

Iliotibial tract

Subcutaneous prepatellar
bursa

Patellar ligament

Patella

Deep fascia of leg

Tibia
– Tibial tuberosity
– Anterior border
– Medial surface
– Medial malleolus

Superior extensor
retinaculum

Inferior extensor
retinaculum

Dorsal fascia of foot

b

Fascia lata

⟨Popliteal fascia⟩

Deep fascia of leg

Calcaneal tendon
of triceps surae muscle

Lateral malleolus
of fibula

112 Fasciae of the right leg
and the dorsum of foot (25%)
a Ventral aspect
b Dorsal aspect

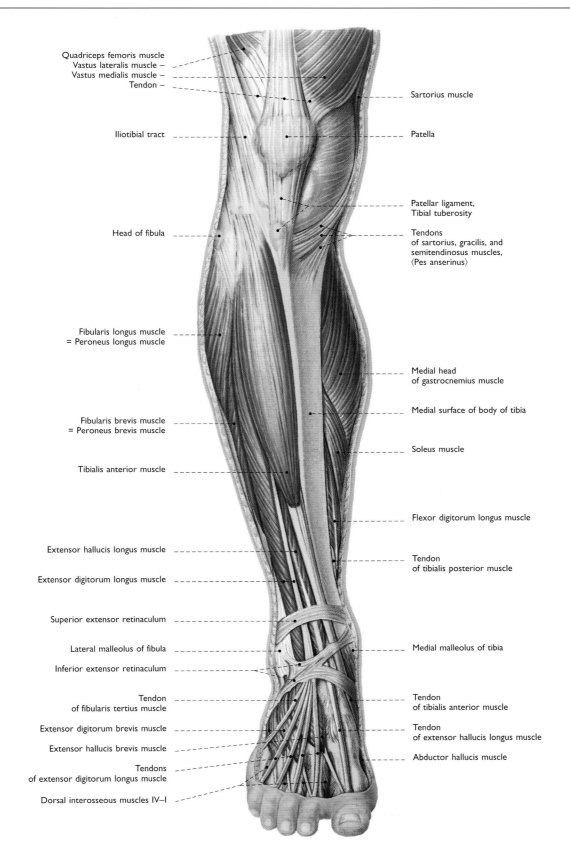

Quadriceps femoris muscle
Vastus lateralis muscle –
Vastus medialis muscle –
Tendon –

Sartorius muscle

Iliotibial tract

Patella

Patellar ligament,
Tibial tuberosity

Head of fibula

Tendons
of sartorius, gracilis, and
semitendinosus muscles,
⟨Pes anserinus⟩

Fibularis longus muscle
= Peroneus longus muscle

Medial head
of gastrocnemius muscle

Medial surface of body of tibia

Fibularis brevis muscle
= Peroneus brevis muscle

Soleus muscle

Tibialis anterior muscle

Flexor digitorum longus muscle

Extensor hallucis longus muscle

Tendon
of tibialis posterior muscle

Extensor digitorum longus muscle

Superior extensor retinaculum

Lateral malleolus of fibula

Medial malleolus of tibia

Inferior extensor retinaculum

Tendon
of fibularis tertius muscle

Tendon
of tibialis anterior muscle

Extensor digitorum brevis muscle

Tendon
of extensor hallucis longus muscle

Extensor hallucis brevis muscle

Abductor hallucis muscle

Tendons
of extensor digitorum longus muscle

Dorsal interosseous muscles IV–I

**113 Muscles of the right leg
and the dorsum of foot** (30%)
Ventral aspect

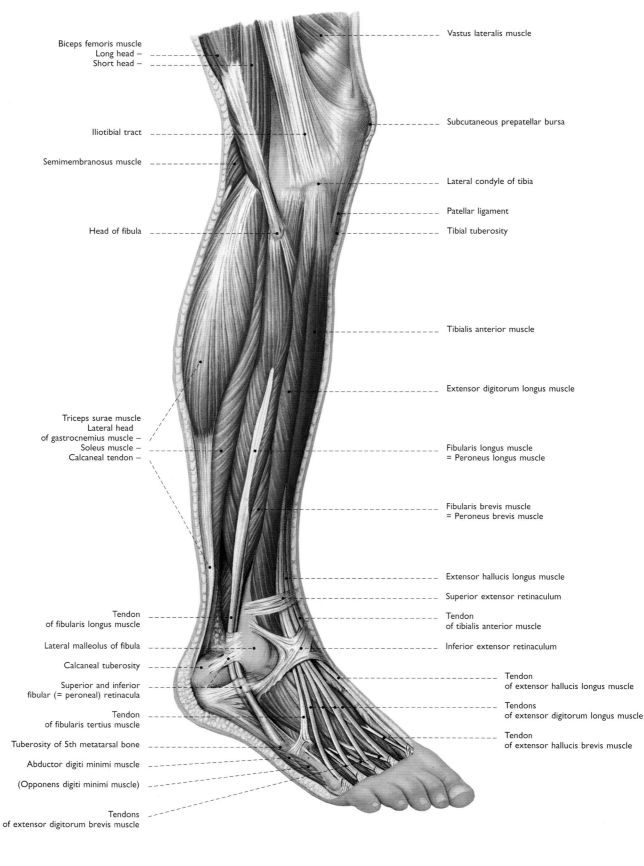

Biceps femoris muscle
Long head –
Short head –

Vastus lateralis muscle

Subcutaneous prepatellar bursa

Iliotibial tract

Semimembranosus muscle

Lateral condyle of tibia

Patellar ligament

Head of fibula

Tibial tuberosity

Tibialis anterior muscle

Extensor digitorum longus muscle

Triceps surae muscle
Lateral head
of gastrocnemius muscle –
Soleus muscle –
Calcaneal tendon –

Fibularis longus muscle
= Peroneus longus muscle

Fibularis brevis muscle
= Peroneus brevis muscle

Extensor hallucis longus muscle

Superior extensor retinaculum

Tendon
of fibularis longus muscle

Tendon
of tibialis anterior muscle

Lateral malleolus of fibula

Inferior extensor retinaculum

Calcaneal tuberosity

Superior and inferior
fibular (= peroneal) retinacula

Tendon
of extensor hallucis longus muscle

Tendon
of fibularis tertius muscle

Tendons
of extensor digitorum longus muscle

Tuberosity of 5th metatarsal bone

Tendon
of extensor hallucis brevis muscle

Abductor digiti minimi muscle

(Opponens digiti minimi muscle)

Tendons
of extensor digitorum brevis muscle

**114 Muscles of the right leg and
the dorsum of foot** (30%)
Lateral aspect

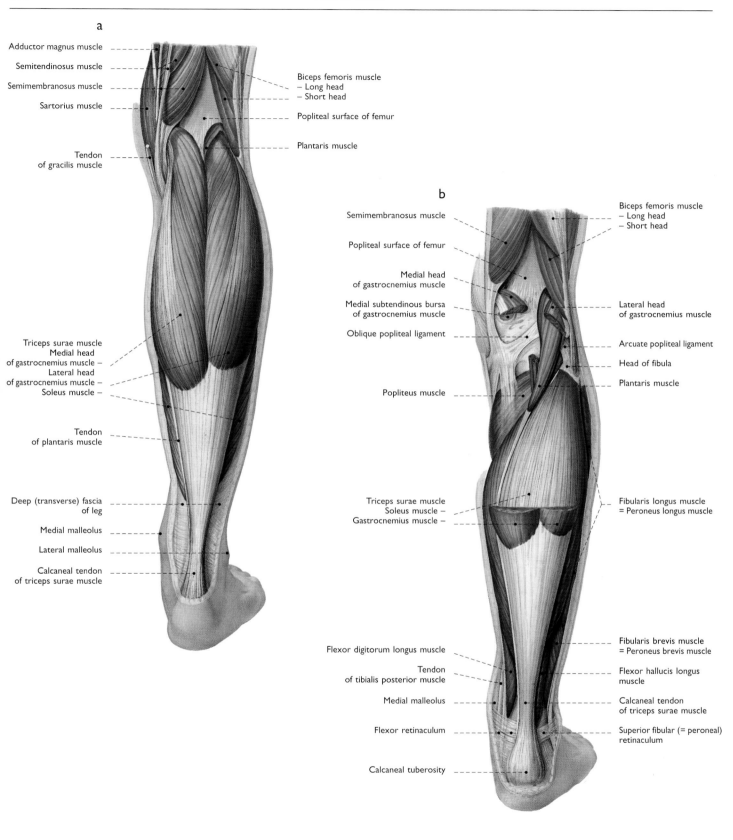

a

Adductor magnus muscle

Semitendinosus muscle

Semimembranosus muscle

Sartorius muscle

Tendon
of gracilis muscle

Biceps femoris muscle
– Long head
– Short head

Popliteal surface of femur

Plantaris muscle

Triceps surae muscle
Medial head
of gastrocnemius muscle –
Lateral head
of gastrocnemius muscle –
Soleus muscle –

Tendon
of plantaris muscle

Deep (transverse) fascia
of leg

Medial malleolus

Lateral malleolus

Calcaneal tendon
of triceps surae muscle

b

Semimembranosus muscle

Popliteal surface of femur

Medial head
of gastrocnemius muscle

Medial subtendinous bursa
of gastrocnemius muscle

Oblique popliteal ligament

Popliteus muscle

Triceps surae muscle
Soleus muscle –
Gastrocnemius muscle –

Biceps femoris muscle
– Long head
– Short head

Lateral head
of gastrocnemius muscle

Arcuate popliteal ligament

Head of fibula

Plantaris muscle

Fibularis longus muscle
= Peroneus longus muscle

Flexor digitorum longus muscle

Tendon
of tibialis posterior muscle

Medial malleolus

Flexor retinaculum

Calcaneal tuberosity

Fibularis brevis muscle
= Peroneus brevis muscle

Flexor hallucis longus
muscle

Calcaneal tendon
of triceps surae muscle

Superior fibular (= peroneal)
retinaculum

115 Muscles of the right leg (25%)
      Dorsal aspect
   a Most superficial layer
   b Superficial layer after partial removal
      of the gastrocnemius muscle

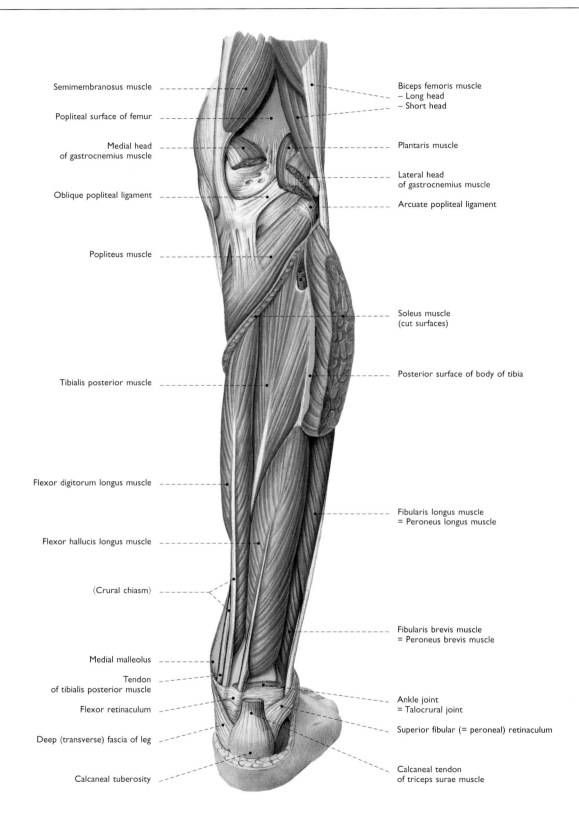

Semimembranosus muscle

Popliteal surface of femur

Medial head
of gastrocnemius muscle

Oblique popliteal ligament

Popliteus muscle

Tibialis posterior muscle

Flexor digitorum longus muscle

Flexor hallucis longus muscle

⟨Crural chiasm⟩

Medial malleolus

Tendon
of tibialis posterior muscle

Flexor retinaculum

Deep (transverse) fascia of leg

Calcaneal tuberosity

Biceps femoris muscle
– Long head
– Short head

Plantaris muscle

Lateral head
of gastrocnemius muscle

Arcuate popliteal ligament

Soleus muscle
(cut surfaces)

Posterior surface of body of tibia

Fibularis longus muscle
= Peroneus longus muscle

Fibularis brevis muscle
= Peroneus brevis muscle

Ankle joint
= Talocrural joint

Superior fibular (= peroneal) retinaculum

Calcaneal tendon
of triceps surae muscle

116 Muscles of the right leg (30%)
Deep layer, dorsal aspect

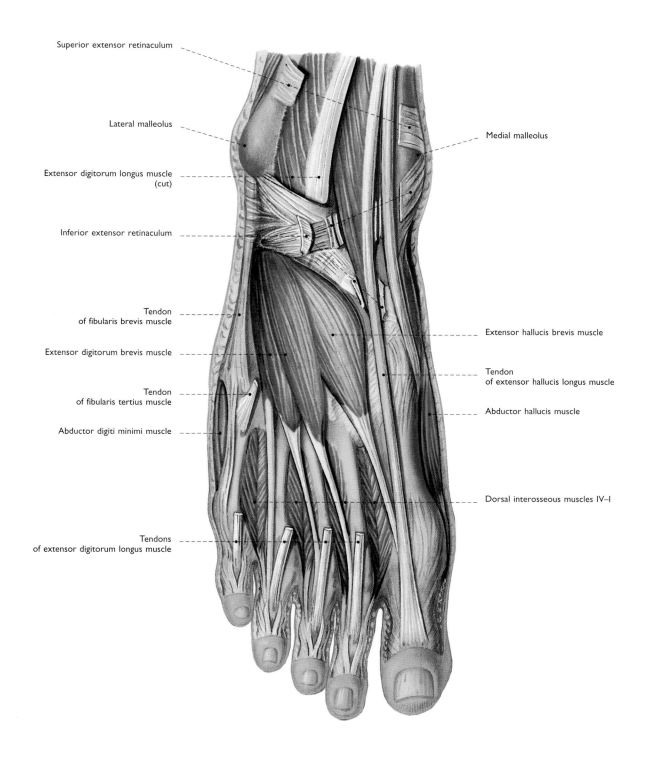

Superior extensor retinaculum

Lateral malleolus

Extensor digitorum longus muscle
(cut)

Inferior extensor retinaculum

Tendon
of fibularis brevis muscle

Extensor digitorum brevis muscle

Tendon
of fibularis tertius muscle

Abductor digiti minimi muscle

Tendons
of extensor digitorum longus muscle

Medial malleolus

Extensor hallucis brevis muscle

Tendon
of extensor hallucis longus muscle

Abductor hallucis muscle

Dorsal interosseous muscles IV–I

117 Muscles of the dorsum of the right foot (75%)
The extensor digitorum longus muscle and the retinacula
were partially removed. Ventral aspect

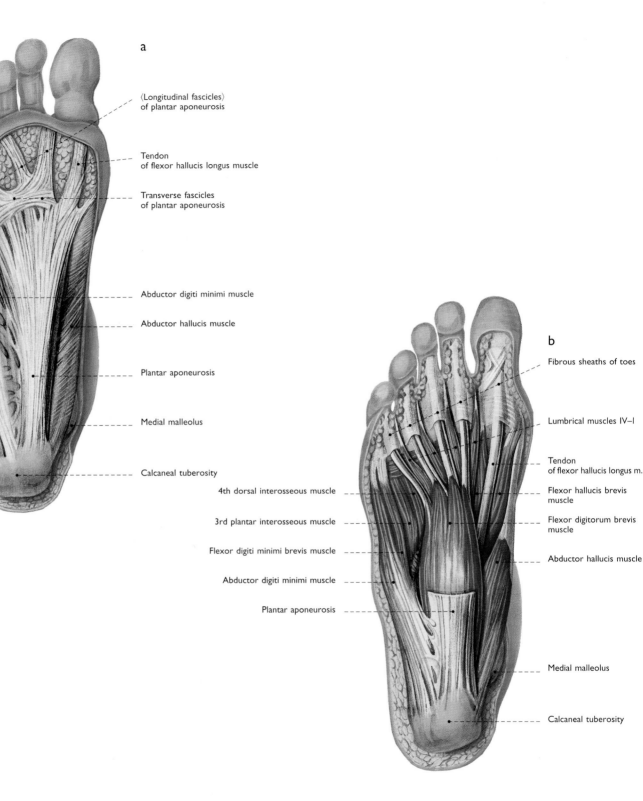

a

⟨Longitudinal fascicles⟩
of plantar aponeurosis

Tendon
of flexor hallucis longus muscle

Transverse fascicles
of plantar aponeurosis

Abductor digiti minimi muscle

Abductor hallucis muscle

Plantar aponeurosis

Medial malleolus

Calcaneal tuberosity

b

Fibrous sheaths of toes

Lumbrical muscles IV–I

Tendon
of flexor hallucis longus m.

Flexor hallucis brevis
muscle

Flexor digitorum brevis
muscle

Abductor hallucis muscle

Medial malleolus

Calcaneal tuberosity

4th dorsal interosseous muscle

3rd plantar interosseous muscle

Flexor digiti minimi brevis muscle

Abductor digiti minimi muscle

Plantar aponeurosis

118 Muscles of the sole of the right foot (50%)

Plantar aspect
a Plantar aponeurosis and superficial muscles
b Superficial layer after partial removal of the plantar aponeurosis

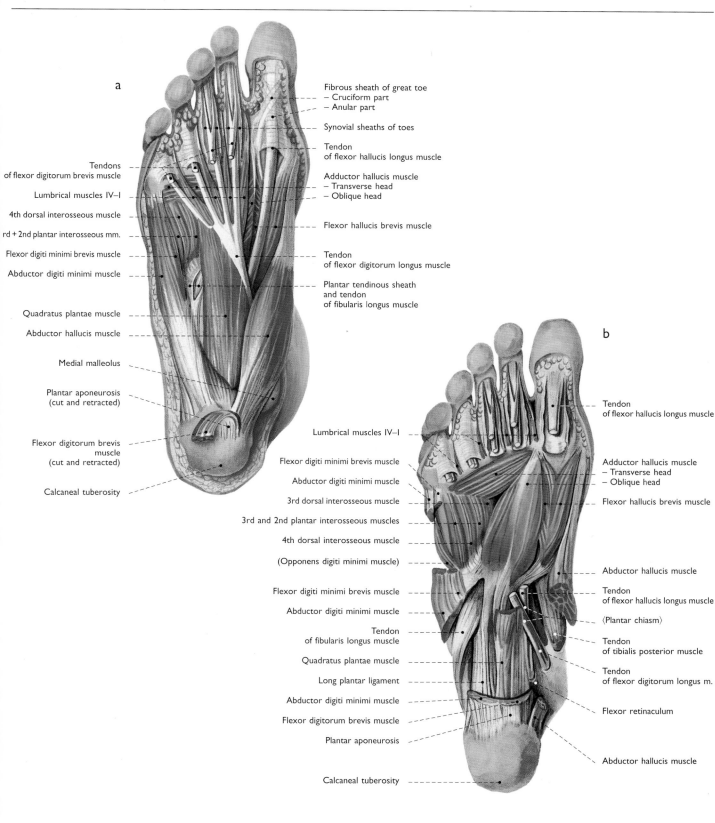

a

Tendons
of flexor digitorum brevis muscle

Lumbrical muscles IV–I

4th dorsal interosseous muscle

rd + 2nd plantar interosseous mm.

Flexor digiti minimi brevis muscle

Abductor digiti minimi muscle

Quadratus plantae muscle

Abductor hallucis muscle

Medial malleolus

Plantar aponeurosis
(cut and retracted)

Flexor digitorum brevis
muscle
(cut and retracted)

Calcaneal tuberosity

Fibrous sheath of great toe
– Cruciform part
– Anular part

Synovial sheaths of toes

Tendon
of flexor hallucis longus muscle

Adductor hallucis muscle
– Transverse head
– Oblique head

Flexor hallucis brevis muscle

Tendon
of flexor digitorum longus muscle

Plantar tendinous sheath
and tendon
of fibularis longus muscle

b

Lumbrical muscles IV–I

Flexor digiti minimi brevis muscle

Abductor digiti minimi muscle

3rd dorsal interosseous muscle

3rd and 2nd plantar interosseous muscles

4th dorsal interosseous muscle

(Opponens digiti minimi muscle)

Flexor digiti minimi brevis muscle

Abductor digiti minimi muscle

Tendon
of fibularis longus muscle

Quadratus plantae muscle

Long plantar ligament

Abductor digiti minimi muscle

Flexor digitorum brevis muscle

Plantar aponeurosis

Calcaneal tuberosity

Tendon
of flexor hallucis longus muscle

Adductor hallucis muscle
– Transverse head
– Oblique head

Flexor hallucis brevis muscle

Abductor hallucis muscle

Tendon
of flexor hallucis longus muscle

〈Plantar chiasm〉

Tendon
of tibialis posterior muscle

Tendon
of flexor digitorum longus m.

Flexor retinaculum

Abductor hallucis muscle

119 Muscles of the sole of the right foot (50%)
Plantar aspect
a Deep layer after partial removal of the plantar aponeurosis
and the flexor digitorum brevis muscle
b Deepest layer after extensive removal of the muscles
of the superficial and deep layers

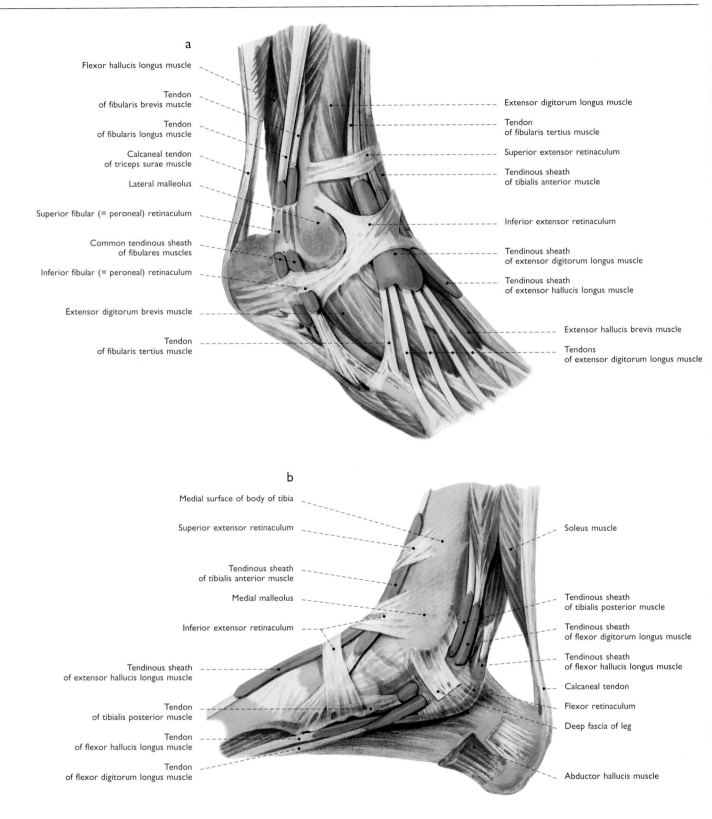

a

Flexor hallucis longus muscle

Tendon
of fibularis brevis muscle

Tendon
of fibularis longus muscle

Calcaneal tendon
of triceps surae muscle

Lateral malleolus

Superior fibular (= peroneal) retinaculum

Common tendinous sheath
of fibulares muscles

Inferior fibular (= peroneal) retinaculum

Extensor digitorum brevis muscle

Tendon
of fibularis tertius muscle

Extensor digitorum longus muscle

Tendon
of fibularis tertius muscle

Superior extensor retinaculum

Tendinous sheath
of tibialis anterior muscle

Inferior extensor retinaculum

Tendinous sheath
of extensor digitorum longus muscle

Tendinous sheath
of extensor hallucis longus muscle

Extensor hallucis brevis muscle

Tendons
of extensor digitorum longus muscle

b

Medial surface of body of tibia

Superior extensor retinaculum

Tendinous sheath
of tibialis anterior muscle

Medial malleolus

Inferior extensor retinaculum

Tendinous sheath
of extensor hallucis longus muscle

Tendon
of tibialis posterior muscle

Tendon
of flexor hallucis longus muscle

Tendon
of flexor digitorum longus muscle

Soleus muscle

Tendinous sheath
of tibialis posterior muscle

Tendinous sheath
of flexor digitorum longus muscle

Tendinous sheath
of flexor hallucis longus muscle

Calcaneal tendon

Flexor retinaculum

Deep fascia of leg

Abductor hallucis muscle

120 Tarsal tendinous sheaths of the right foot (50%)
a Lateral aspect
b Medial aspect

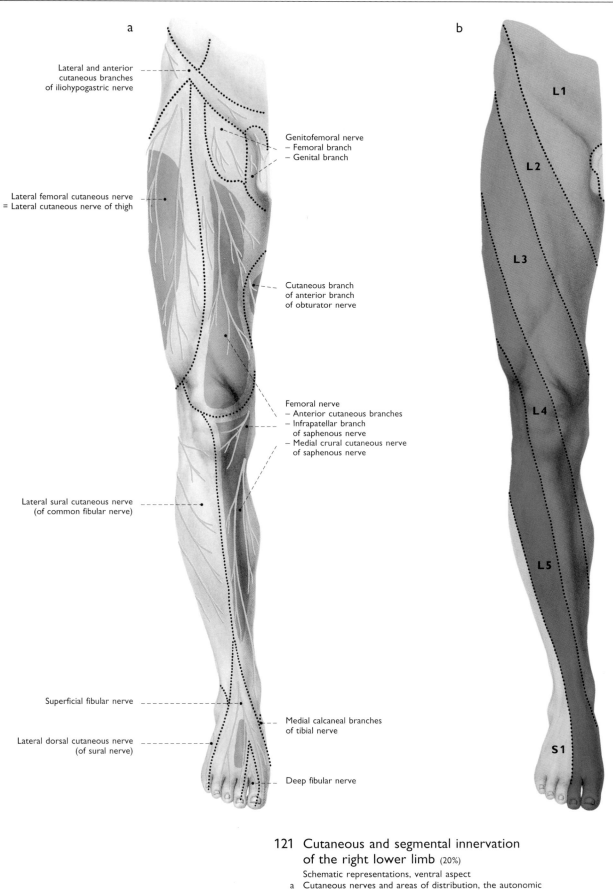

a

Lateral and anterior
cutaneous branches
of iliohypogastric nerve

Genitofemoral nerve
– Femoral branch
– Genital branch

Lateral femoral cutaneous nerve
= Lateral cutaneous nerve of thigh

Cutaneous branch
of anterior branch
of obturator nerve

Femoral nerve
– Anterior cutaneous branches
– Infrapatellar branch
of saphenous nerve
– Medial crural cutaneous nerve
of saphenous nerve

Lateral sural cutaneous nerve
(of common fibular nerve)

Superficial fibular nerve

Lateral dorsal cutaneous nerve
(of sural nerve)

Medial calcaneal branches
of tibial nerve

Deep fibular nerve

b

L1

L2

L3

L4

L5

S1

**121 Cutaneous and segmental innervation
of the right lower limb** (20%)

Schematic representations, ventral aspect
a Cutaneous nerves and areas of distribution, the autonomic
areas of the different nerves are given in a darker gray.
b Segmental innervation (dermatomes)

a

b

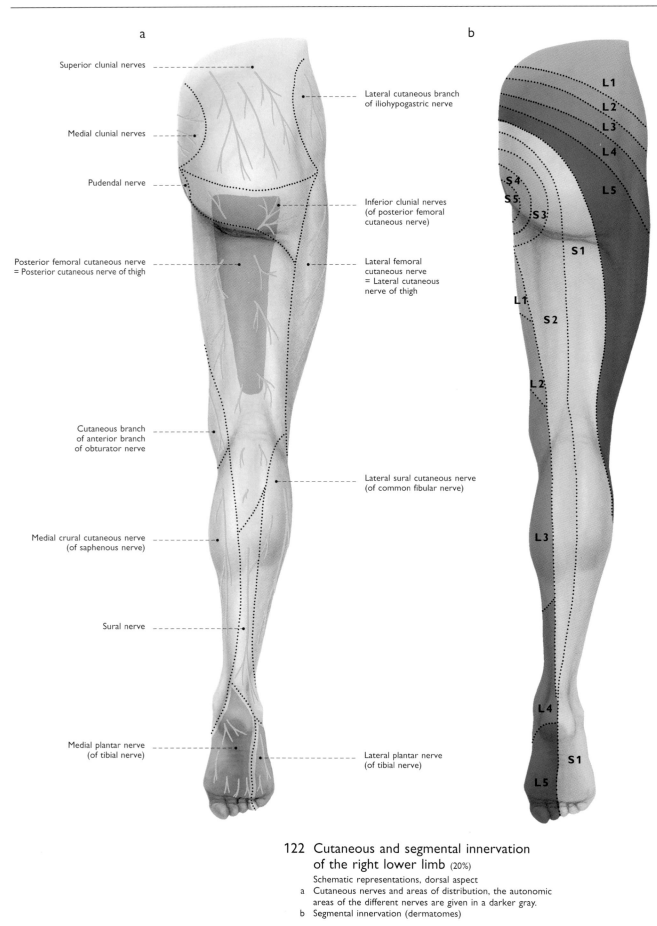

Superior clunial nerves

Lateral cutaneous branch
of iliohypogastric nerve

Medial clunial nerves

Pudendal nerve

Inferior clunial nerves
(of posterior femoral
cutaneous nerve)

Posterior femoral cutaneous nerve
= Posterior cutaneous nerve of thigh

Lateral femoral
cutaneous nerve
= Lateral cutaneous
nerve of thigh

Cutaneous branch
of anterior branch
of obturator nerve

Lateral sural cutaneous nerve
(of common fibular nerve)

Medial crural cutaneous nerve
(of saphenous nerve)

Sural nerve

Medial plantar nerve
(of tibial nerve)

Lateral plantar nerve
(of tibial nerve)

L1
L2
L3
L4
L5
S4
S5
S3
S1
L1
S2
L2
L3
L4
S1
L5

122 Cutaneous and segmental innervation
of the right lower limb (20%)

Schematic representations, dorsal aspect
a Cutaneous nerves and areas of distribution, the autonomic
    areas of the different nerves are given in a darker gray.
b Segmental innervation (dermatomes)

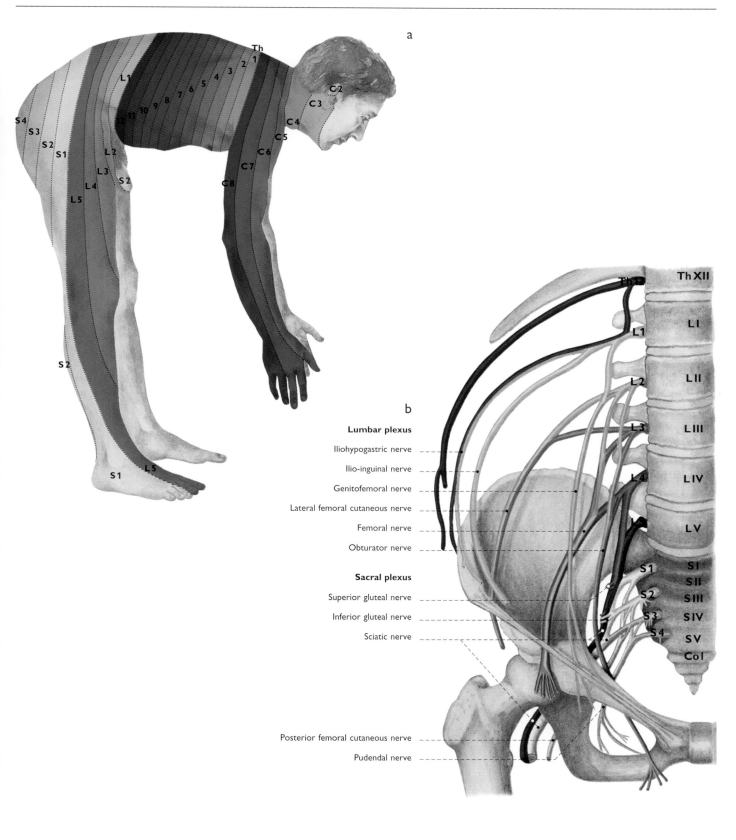

a

b

Lumbar plexus

Iliohypogastric nerve

Ilio-inguinal nerve

Genitofemoral nerve

Lateral femoral cutaneous nerve

Femoral nerve

Obturator nerve

Sacral plexus

Superior gluteal nerve

Inferior gluteal nerve

Sciatic nerve

Posterior femoral cutaneous nerve

Pudendal nerve

123 Segmental innervation and lumbosacral plexus

   a Segmental innervation (dermatomes) of the upper limb, trunk, and lower limb (according to von Lanz and Wachsmuth, 1972)

   b Plan of the lumbosacral plexus

a

b

Superficial epigastric vein

Superolateral and
superomedial superficial
inguinal lymph nodes

Inferior superficial
inguinal lymph nodes

Accessory saphenous vein
(medial branch)

Great saphenous vein

Great saphenous vein

Small saphenous vein

c

Tibial nerve,
Common fibular n.

Popliteal vein

Popliteal artery

Great saphenous vein

Superficial and deep
popliteal lymph nodes

(Anastomosing vein)

Small saphenous vein

Dorsal venous arch
of foot

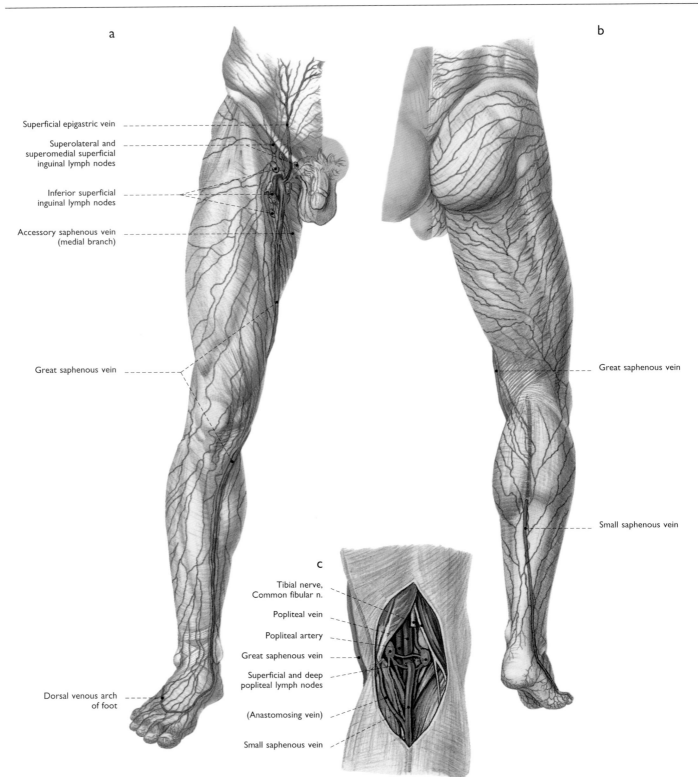

**124 Lymphatic vessels and lymph nodes
of the right lower limb**

  a   Ventral aspect (20%)
  b   Dorsal aspect (20%)
  c   Lymphatic vessels of the popliteal fossa (30%), dorsal aspect

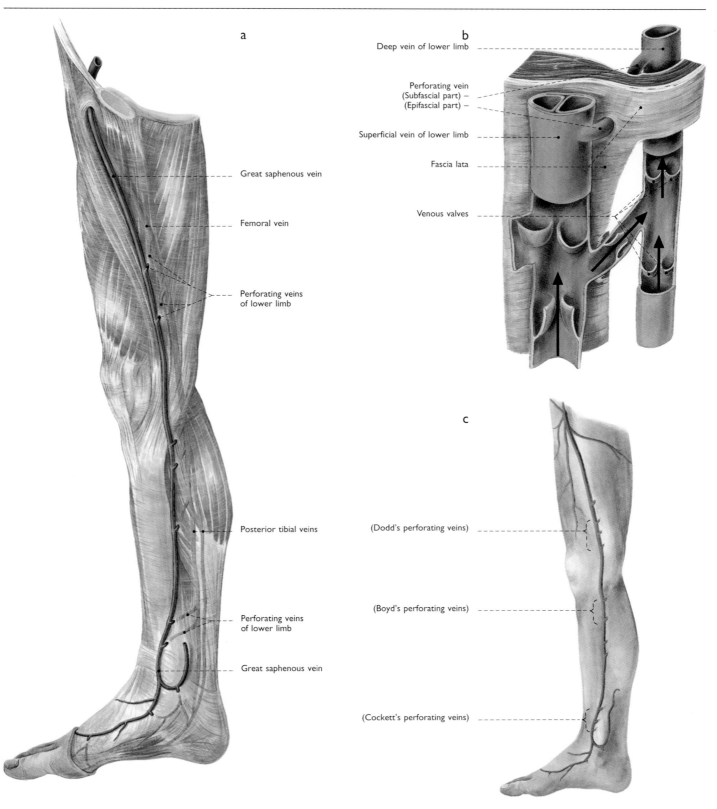

a

b

Deep vein of lower limb

Perforating vein
(Subfascial part) –
(Epifascial part) –

Superficial vein of lower limb

Fascia lata

Venous valves

Great saphenous vein

Femoral vein

Perforating veins
of lower limb

c

Posterior tibial veins

(Dodd's perforating veins)

Perforating veins
of lower limb

(Boyd's perforating veins)

Great saphenous vein

(Cockett's perforating veins)

125 Veins of the right lower limb
   a Superficial and deep veins of the lower limb (20%), medial aspect
   b Connection between superficial and deep veins of the lower limb
      by perforating veins (200%), schematic representation
   c Main localization of perforating veins over the lower limb (10%),
      medial aspect

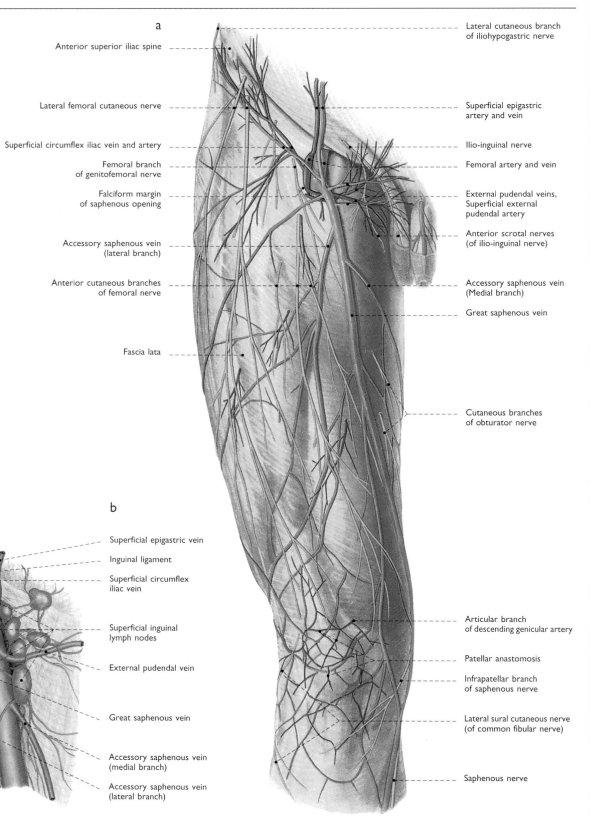

a

Anterior superior iliac spine

Lateral femoral cutaneous nerve

Superficial circumflex iliac vein and artery

Femoral branch
of genitofemoral nerve

Falciform margin
of saphenous opening

Accessory saphenous vein
(lateral branch)

Anterior cutaneous branches
of femoral nerve

Fascia lata

Lateral cutaneous branch
of iliohypogastric nerve

Superficial epigastric
artery and vein

Ilio-inguinal nerve

Femoral artery and vein

External pudendal veins,
Superficial external
pudendal artery

Anterior scrotal nerves
(of ilio-inguinal nerve)

Accessory saphenous vein
(Medial branch)

Great saphenous vein

Cutaneous branches
of obturator nerve

Articular branch
of descending genicular artery

Patellar anastomosis

Infrapatellar branch
of saphenous nerve

Lateral sural cutaneous nerve
(of common fibular nerve)

Saphenous nerve

b

Superficial epigastric vein

Inguinal ligament

Superficial circumflex
iliac vein

Superficial inguinal
lymph nodes

External pudendal vein

Great saphenous vein

Accessory saphenous vein
(medial branch)

Accessory saphenous vein
(lateral branch)

126 Subcutaneous blood vessels, nerves,
and lymph nodes of the right thigh

a Ventral aspect (30%)
b Superficial veins and lymph nodes in and around
the saphenous opening (70%)

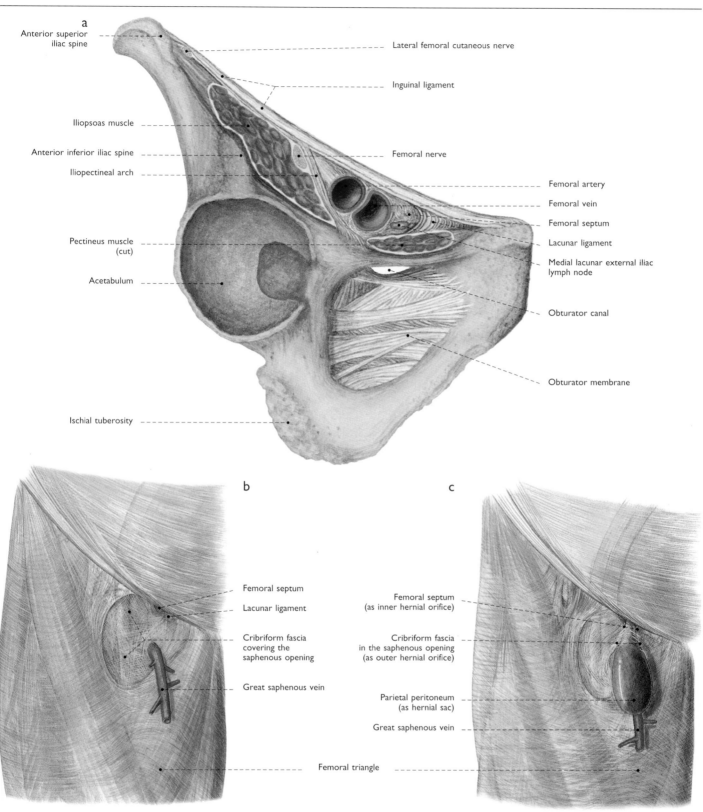

a

Anterior superior iliac spine

Iliopsoas muscle

Anterior inferior iliac spine

Iliopectineal arch

Pectineus muscle (cut)

Acetabulum

Ischial tuberosity

Lateral femoral cutaneous nerve

Inguinal ligament

Femoral nerve

Femoral artery

Femoral vein

Femoral septum

Lacunar ligament

Medial lacunar external iliac lymph node

Obturator canal

Obturator membrane

b

Femoral septum

Lacunar ligament

Cribriform fascia covering the saphenous opening

Great saphenous vein

Femoral triangle

c

Femoral septum (as inner hernial orifice)

Cribriform fascia in the saphenous opening (as outer hernial orifice)

Parietal peritoneum (as hernial sac)

Great saphenous vein

127 Inguinal region and femoral triangle

a Structures passing posterior to the inguinal ligament (60%) (according to von Lanz and Wachsmuth, 1972), inferior (distal) aspect

b, c Femoral triangle, saphenous opening, and femoral hernia (30%), ventral aspect

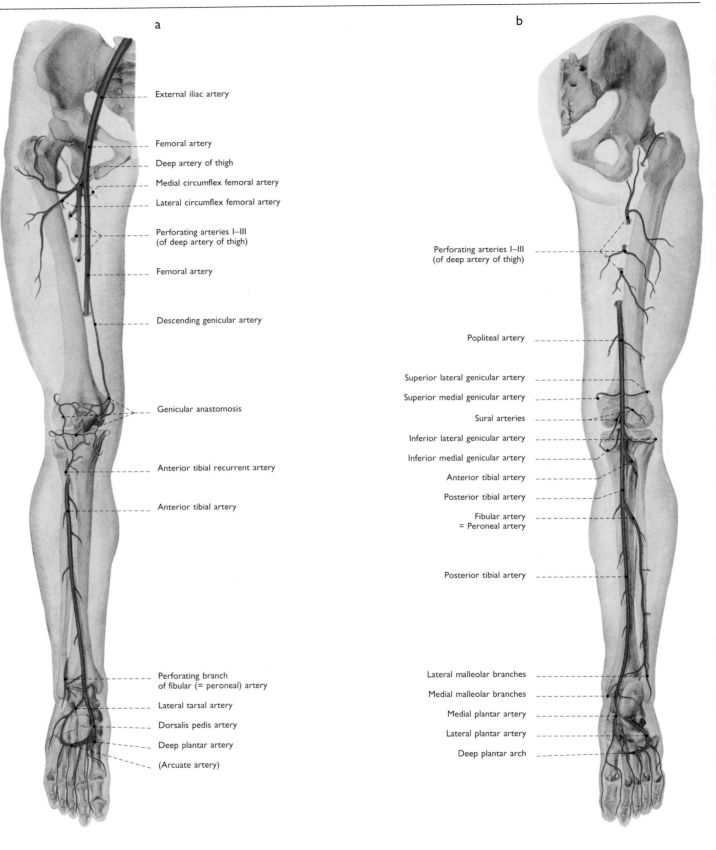

a

External iliac artery

Femoral artery

Deep artery of thigh

Medial circumflex femoral artery

Lateral circumflex femoral artery

Perforating arteries I–III
(of deep artery of thigh)

Femoral artery

Descending genicular artery

Genicular anastomosis

Anterior tibial recurrent artery

Anterior tibial artery

Perforating branch
of fibular (= peroneal) artery

Lateral tarsal artery

Dorsalis pedis artery

Deep plantar artery

(Arcuate artery)

b

Perforating arteries I–III
(of deep artery of thigh)

Popliteal artery

Superior lateral genicular artery

Superior medial genicular artery

Sural arteries

Inferior lateral genicular artery

Inferior medial genicular artery

Anterior tibial artery

Posterior tibial artery

Fibular artery
= Peroneal artery

Posterior tibial artery

Lateral malleolar branches

Medial malleolar branches

Medial plantar artery

Lateral plantar artery

Deep plantar arch

128 Arteries of the right lower limb (20%)
Schematic representations
a Ventral aspect
b Dorsal aspect

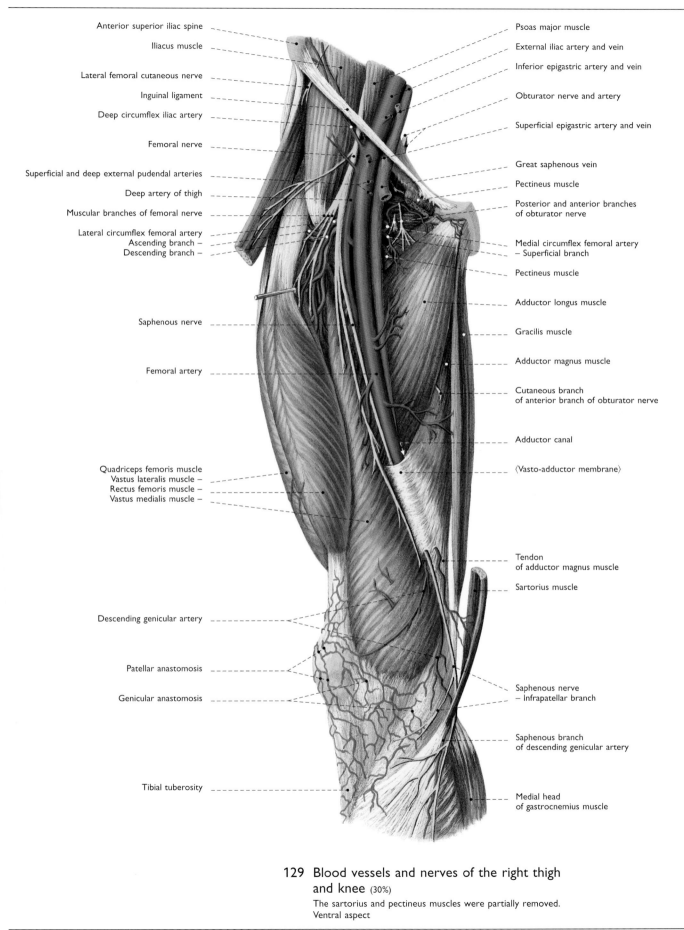

Anterior superior iliac spine

Iliacus muscle

Lateral femoral cutaneous nerve

Inguinal ligament

Deep circumflex iliac artery

Femoral nerve

Superficial and deep external pudendal arteries

Deep artery of thigh

Muscular branches of femoral nerve

Lateral circumflex femoral artery
Ascending branch –
Descending branch –

Saphenous nerve

Femoral artery

Quadriceps femoris muscle
Vastus lateralis muscle –
Rectus femoris muscle –
Vastus medialis muscle –

Descending genicular artery

Patellar anastomosis

Genicular anastomosis

Tibial tuberosity

Psoas major muscle

External iliac artery and vein

Inferior epigastric artery and vein

Obturator nerve and artery

Superficial epigastric artery and vein

Great saphenous vein

Pectineus muscle

Posterior and anterior branches
of obturator nerve

Medial circumflex femoral artery
– Superficial branch

Pectineus muscle

Adductor longus muscle

Gracilis muscle

Adductor magnus muscle

Cutaneous branch
of anterior branch of obturator nerve

Adductor canal

〈Vasto-adductor membrane〉

Tendon
of adductor magnus muscle

Sartorius muscle

Saphenous nerve
– Infrapatellar branch

Saphenous branch
of descending genicular artery

Medial head
of gastrocnemius muscle

**129 Blood vessels and nerves of the right thigh
and knee** (30%)
The sartorius and pectineus muscles were partially removed.
Ventral aspect

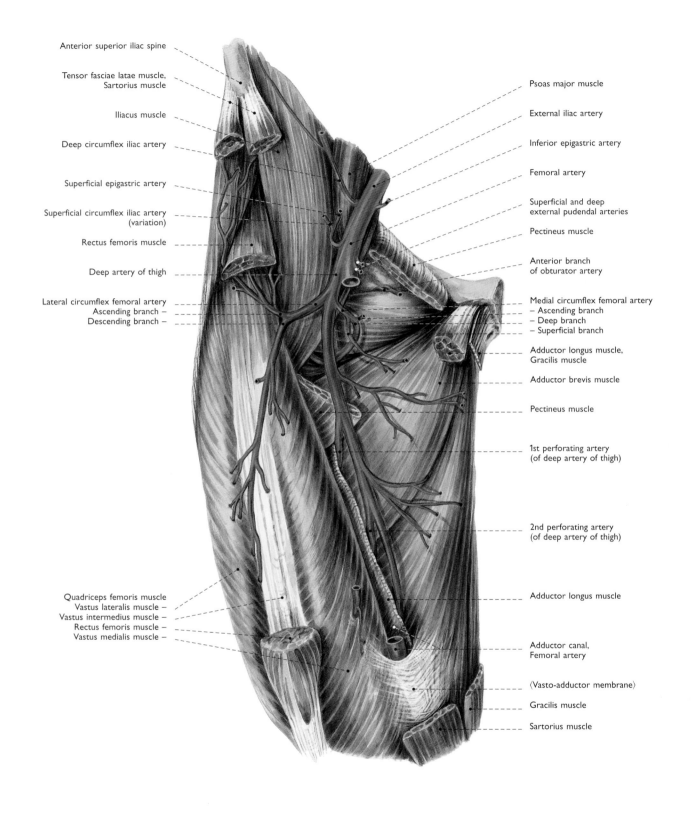

Anterior superior iliac spine

Tensor fasciae latae muscle, Sartorius muscle

Iliacus muscle

Deep circumflex iliac artery

Superficial epigastric artery

Superficial circumflex iliac artery (variation)

Rectus femoris muscle

Deep artery of thigh

Lateral circumflex femoral artery
Ascending branch –
Descending branch –

Quadriceps femoris muscle
Vastus lateralis muscle –
Vastus intermedius muscle –
Rectus femoris muscle –
Vastus medialis muscle –

Psoas major muscle

External iliac artery

Inferior epigastric artery

Femoral artery

Superficial and deep external pudendal arteries

Pectineus muscle

Anterior branch of obturator artery

Medial circumflex femoral artery
– Ascending branch
– Deep branch
– Superficial branch

Adductor longus muscle, Gracilis muscle

Adductor brevis muscle

Pectineus muscle

1st perforating artery (of deep artery of thigh)

2nd perforating artery (of deep artery of thigh)

Adductor longus muscle

Adductor canal, Femoral artery

⟨Vasto-adductor membrane⟩

Gracilis muscle

Sartorius muscle

130 Deep artery of thigh and its branches in the right thigh (40%)
The superficial muscles were partially removed.
Ventral aspect

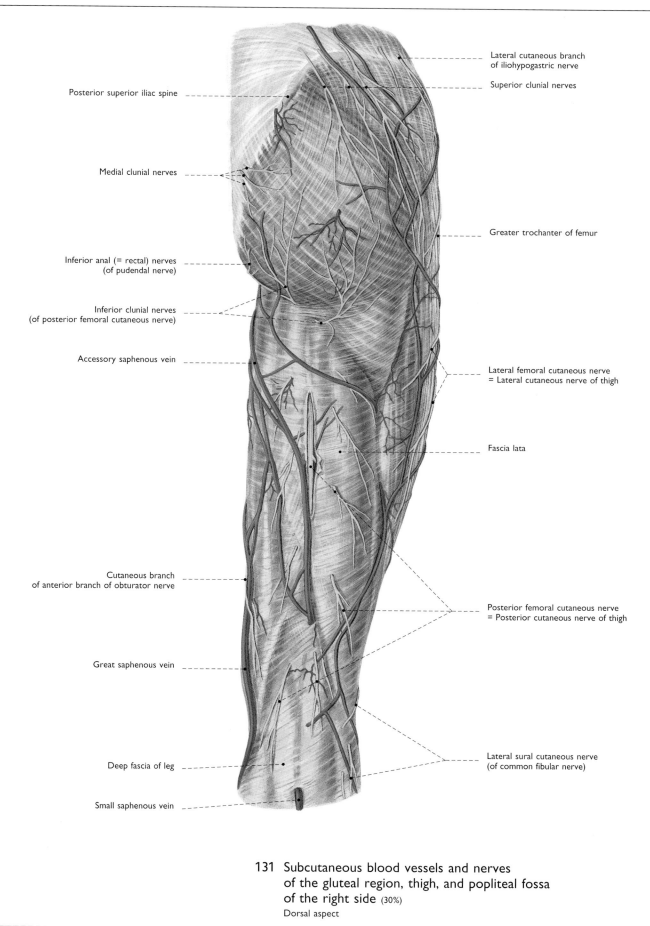

Posterior superior iliac spine

Medial clunial nerves

Inferior anal (= rectal) nerves
(of pudendal nerve)

Inferior clunial nerves
(of posterior femoral cutaneous nerve)

Accessory saphenous vein

Cutaneous branch
of anterior branch of obturator nerve

Great saphenous vein

Deep fascia of leg

Small saphenous vein

Lateral cutaneous branch
of iliohypogastric nerve

Superior clunial nerves

Greater trochanter of femur

Lateral femoral cutaneous nerve
= Lateral cutaneous nerve of thigh

Fascia lata

Posterior femoral cutaneous nerve
= Posterior cutaneous nerve of thigh

Lateral sural cutaneous nerve
(of common fibular nerve)

131 Subcutaneous blood vessels and nerves
of the gluteal region, thigh, and popliteal fossa
of the right side (30%)
Dorsal aspect

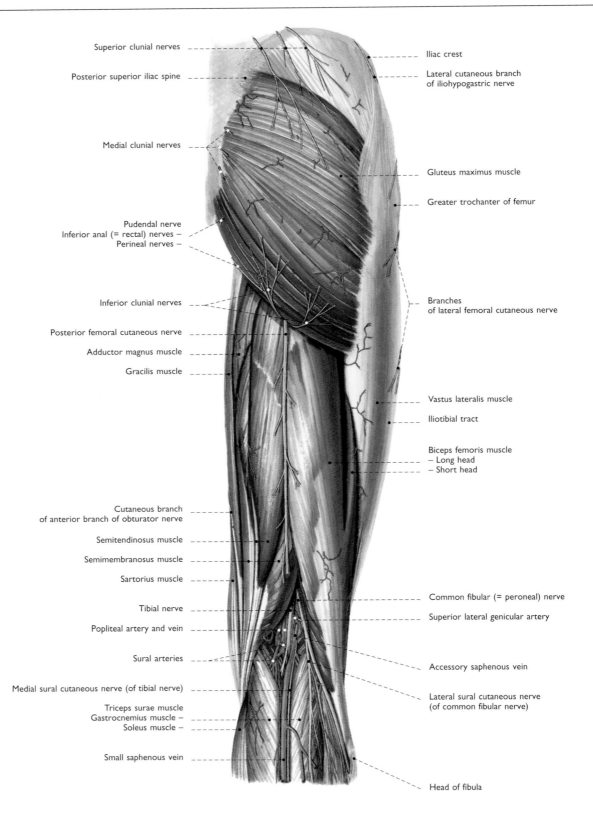

Superior clunial nerves

Posterior superior iliac spine

Medial clunial nerves

Pudendal nerve
Inferior anal (= rectal) nerves
Perineal nerves

Inferior clunial nerves

Posterior femoral cutaneous nerve

Adductor magnus muscle

Gracilis muscle

Cutaneous branch
of anterior branch of obturator nerve

Semitendinosus muscle

Semimembranosus muscle

Sartorius muscle

Tibial nerve

Popliteal artery and vein

Sural arteries

Medial sural cutaneous nerve (of tibial nerve)

Triceps surae muscle
Gastrocnemius muscle
Soleus muscle

Small saphenous vein

Iliac crest

Lateral cutaneous branch
of iliohypogastric nerve

Gluteus maximus muscle

Greater trochanter of femur

Branches
of lateral femoral cutaneous nerve

Vastus lateralis muscle

Iliotibial tract

Biceps femoris muscle
– Long head
– Short head

Common fibular (= peroneal) nerve

Superior lateral genicular artery

Accessory saphenous vein

Lateral sural cutaneous nerve
(of common fibular nerve)

Head of fibula

**132 Blood vessels and nerves of the gluteal region,
thigh, and popliteal fossa of the right side** (30%)
The fasciae of the lower limb were removed.

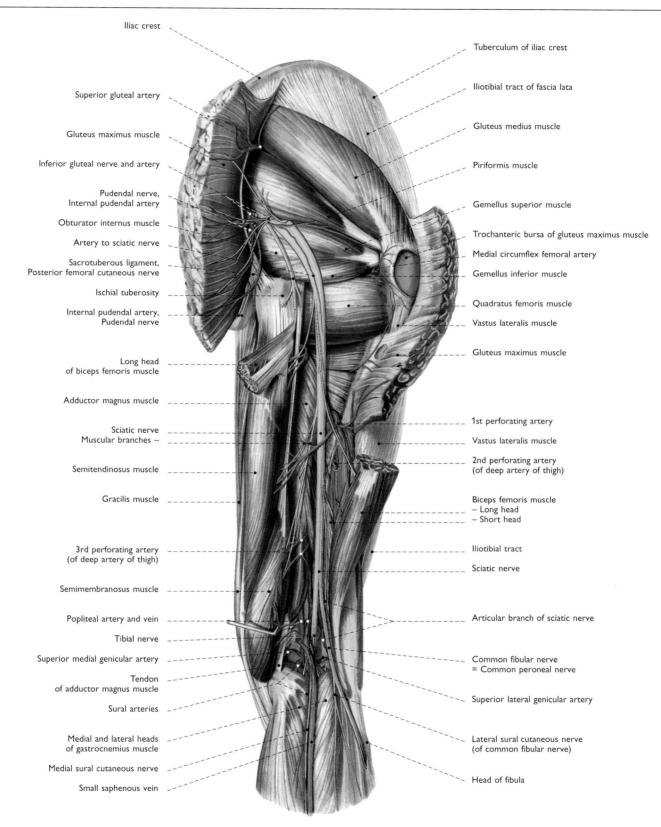

Iliac crest

Superior gluteal artery

Gluteus maximus muscle

Inferior gluteal nerve and artery

Pudendal nerve,
Internal pudendal artery

Obturator internus muscle

Artery to sciatic nerve

Sacrotuberous ligament,
Posterior femoral cutaneous nerve

Ischial tuberosity

Internal pudendal artery,
Pudendal nerve

Long head
of biceps femoris muscle

Adductor magnus muscle

Sciatic nerve
Muscular branches –

Semitendinosus muscle

Gracilis muscle

3rd perforating artery
(of deep artery of thigh)

Semimembranosus muscle

Popliteal artery and vein

Tibial nerve

Superior medial genicular artery

Tendon
of adductor magnus muscle

Sural arteries

Medial and lateral heads
of gastrocnemius muscle

Medial sural cutaneous nerve

Small saphenous vein

Tuberculum of iliac crest

Iliotibial tract of fascia lata

Gluteus medius muscle

Piriformis muscle

Gemellus superior muscle

Trochanteric bursa of gluteus maximus muscle

Medial circumflex femoral artery

Gemellus inferior muscle

Quadratus femoris muscle

Vastus lateralis muscle

Gluteus maximus muscle

1st perforating artery

Vastus lateralis muscle

2nd perforating artery
(of deep artery of thigh)

Biceps femoris muscle
– Long head
– Short head

Iliotibial tract

Sciatic nerve

Articular branch of sciatic nerve

Common fibular nerve
= Common peroneal nerve

Superior lateral genicular artery

Lateral sural cutaneous nerve
(of common fibular nerve)

Head of fibula

**133 Blood vessels and nerves of the gluteal region,
thigh, and popliteal fossa of the right side** (30%)
The gluteus maximus muscle and the long head
of the biceps femoris muscle were divided. Dorsal aspect

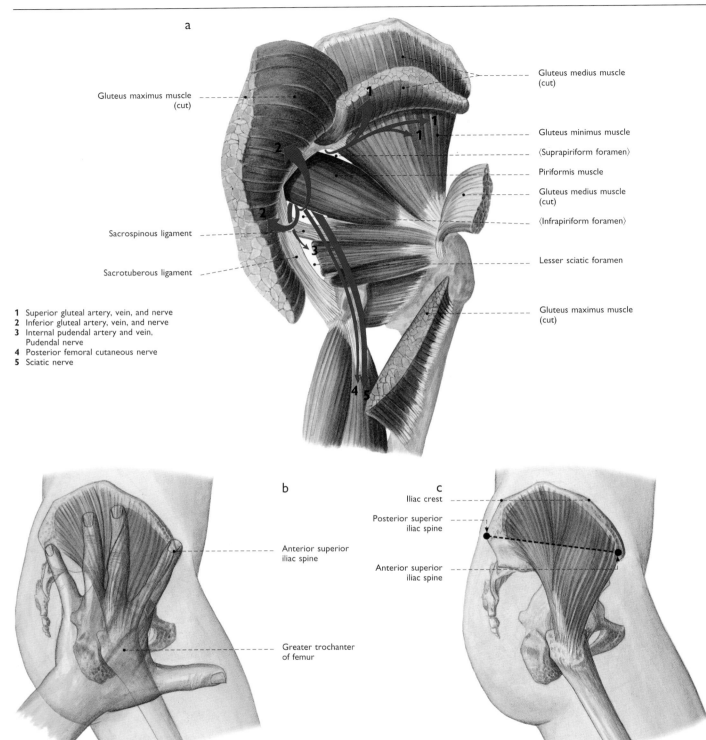

a

Gluteus maximus muscle
(cut)

Sacrospinous ligament

Sacrotuberous ligament

Gluteus medius muscle
(cut)

Gluteus minimus muscle

⟨Suprapiriform foramen⟩

Piriformis muscle

Gluteus medius muscle
(cut)

⟨Infrapiriform foramen⟩

Lesser sciatic foramen

Gluteus maximus muscle
(cut)

1 Superior gluteal artery, vein, and nerve
2 Inferior gluteal artery, vein, and nerve
3 Internal pudendal artery and vein,
   Pudendal nerve
4 Posterior femoral cutaneous nerve
5 Sciatic nerve

b

Anterior superior
iliac spine

Greater trochanter
of femur

c

Iliac crest

Posterior superior
iliac spine

Anterior superior
iliac spine

134　Gluteal region and intragluteal injection

a　The gluteus maximus and medius muscles were divided and retracted.
The arteries and nerves passing through the greater sciatic foramen
above and below the piriformis muscle are given by arrows (35%)
(according to von Lanz and Wachsmuth, 1972). Dorsal aspect

b, c　Intragluteal injection according to von Hochstetter (b) and
von Lanz and Wachsmuth (c). The injection areas are indicated
by red color (20%). Lateral aspect

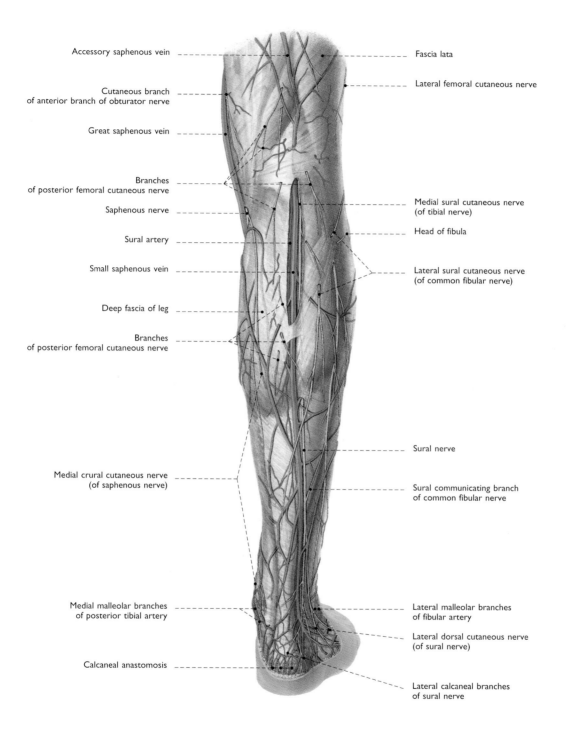

Accessory saphenous vein

Cutaneous branch
of anterior branch of obturator nerve

Great saphenous vein

Branches
of posterior femoral cutaneous nerve

Saphenous nerve

Sural artery

Small saphenous vein

Deep fascia of leg

Branches
of posterior femoral cutaneous nerve

Medial crural cutaneous nerve
(of saphenous nerve)

Medial malleolar branches
of posterior tibial artery

Calcaneal anastomosis

Fascia lata

Lateral femoral cutaneous nerve

Medial sural cutaneous nerve
(of tibial nerve)

Head of fibula

Lateral sural cutaneous nerve
(of common fibular nerve)

Sural nerve

Sural communicating branch
of common fibular nerve

Lateral malleolar branches
of fibular artery

Lateral dorsal cutaneous nerve
(of sural nerve)

Lateral calcaneal branches
of sural nerve

**135 Subcutaneous blood vessels and nerves
of the popliteal fossa and the leg
of the right side** (30%)
Dorsal aspect

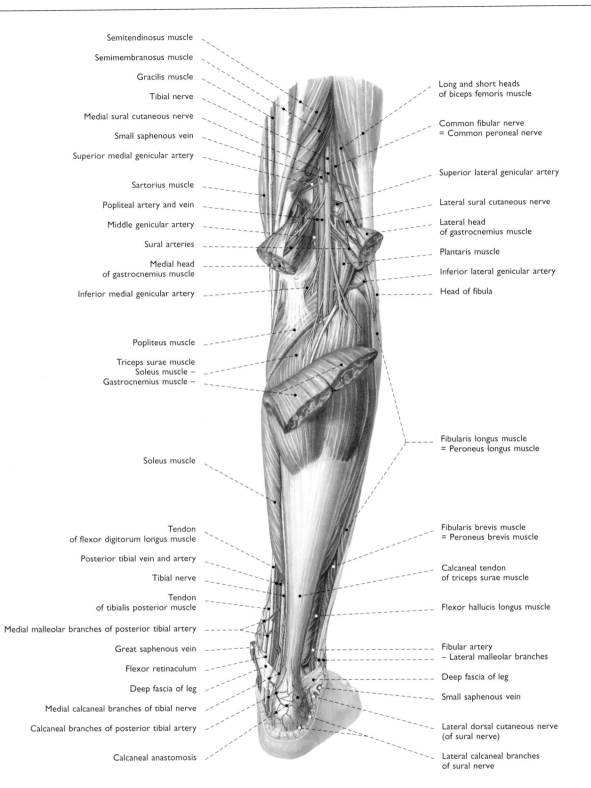

Semitendinosus muscle

Semimembranosus muscle

Gracilis muscle

Tibial nerve

Medial sural cutaneous nerve

Small saphenous vein

Superior medial genicular artery

Sartorius muscle

Popliteal artery and vein

Middle genicular artery

Sural arteries

Medial head
of gastrocnemius muscle

Inferior medial genicular artery

Popliteus muscle

Triceps surae muscle
Soleus muscle –
Gastrocnemius muscle –

Soleus muscle

Tendon
of flexor digitorum longus muscle

Posterior tibial vein and artery

Tibial nerve

Tendon
of tibialis posterior muscle

Medial malleolar branches of posterior tibial artery

Great saphenous vein

Flexor retinaculum

Deep fascia of leg

Medial calcaneal branches of tibial nerve

Calcaneal branches of posterior tibial artery

Calcaneal anastomosis

Long and short heads
of biceps femoris muscle

Common fibular nerve
= Common peroneal nerve

Superior lateral genicular artery

Lateral sural cutaneous nerve

Lateral head
of gastrocnemius muscle

Plantaris muscle

Inferior lateral genicular artery

Head of fibula

Fibularis longus muscle
= Peroneus longus muscle

Fibularis brevis muscle
= Peroneus brevis muscle

Calcaneal tendon
of triceps surae muscle

Flexor hallucis longus muscle

Fibular artery
– Lateral malleolar branches

Deep fascia of leg

Small saphenous vein

Lateral dorsal cutaneous nerve
(of sural nerve)

Lateral calcaneal branches
of sural nerve

136　Blood vessels and nerves
of the popliteal fossa and the leg
of the right side (30%)
The gastrocnemius muscle was divided. Dorsal aspect

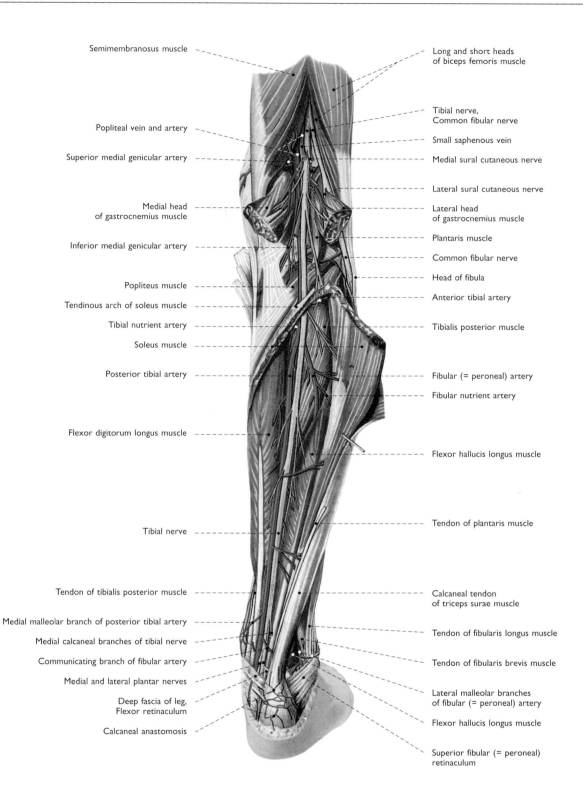

Semimembranosus muscle

Popliteal vein and artery

Superior medial genicular artery

Medial head
of gastrocnemius muscle

Inferior medial genicular artery

Popliteus muscle

Tendinous arch of soleus muscle

Tibial nutrient artery

Soleus muscle

Posterior tibial artery

Flexor digitorum longus muscle

Tibial nerve

Tendon of tibialis posterior muscle

Medial malleolar branch of posterior tibial artery

Medial calcaneal branches of tibial nerve

Communicating branch of fibular artery

Medial and lateral plantar nerves

Deep fascia of leg,
Flexor retinaculum

Calcaneal anastomosis

Long and short heads
of biceps femoris muscle

Tibial nerve,
Common fibular nerve

Small saphenous vein

Medial sural cutaneous nerve

Lateral sural cutaneous nerve

Lateral head
of gastrocnemius muscle

Plantaris muscle

Common fibular nerve

Head of fibula

Anterior tibial artery

Tibialis posterior muscle

Fibular (= peroneal) artery

Fibular nutrient artery

Flexor hallucis longus muscle

Tendon of plantaris muscle

Calcaneal tendon
of triceps surae muscle

Tendon of fibularis longus muscle

Tendon of fibularis brevis muscle

Lateral malleolar branches
of fibular (= peroneal) artery

Flexor hallucis longus muscle

Superior fibular (= peroneal)
retinaculum

**137 Blood vessels and nerves
of the popliteal fossa and the leg
of the right side** (30%)
The gastrocnemius and soleus muscles were divided,
the deep veins removed. Dorsal aspect

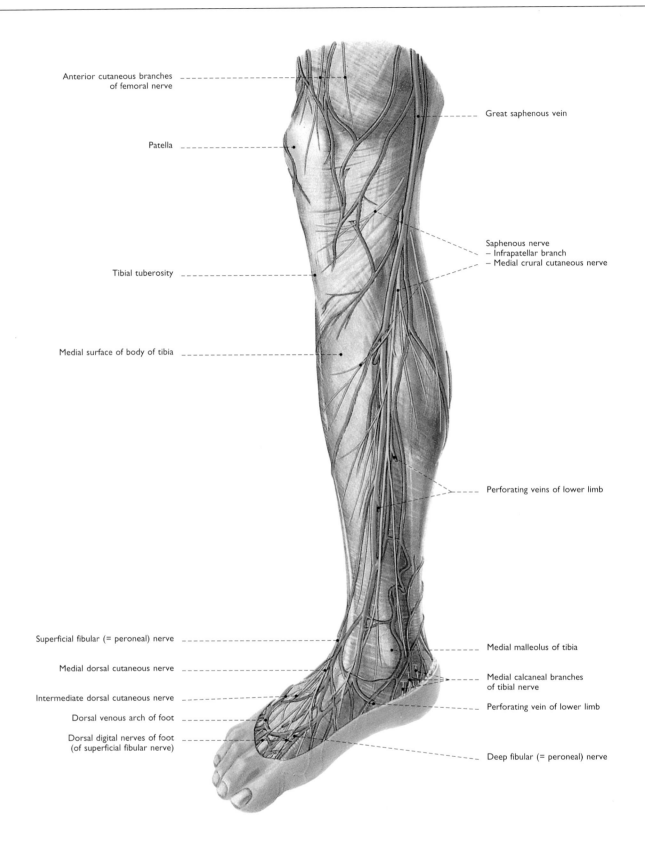

Anterior cutaneous branches
of femoral nerve

Patella

Tibial tuberosity

Medial surface of body of tibia

Great saphenous vein

Saphenous nerve
− Infrapatellar branch
− Medial crural cutaneous nerve

Perforating veins of lower limb

Superficial fibular (= peroneal) nerve

Medial dorsal cutaneous nerve

Intermediate dorsal cutaneous nerve

Dorsal venous arch of foot

Dorsal digital nerves of foot
(of superficial fibular nerve)

Medial malleolus of tibia

Medial calcaneal branches
of tibial nerve

Perforating vein of lower limb

Deep fibular (= peroneal) nerve

**138 Subcutaneous veins and nerves
of the right leg and foot** (30%)
Ventromedial aspect

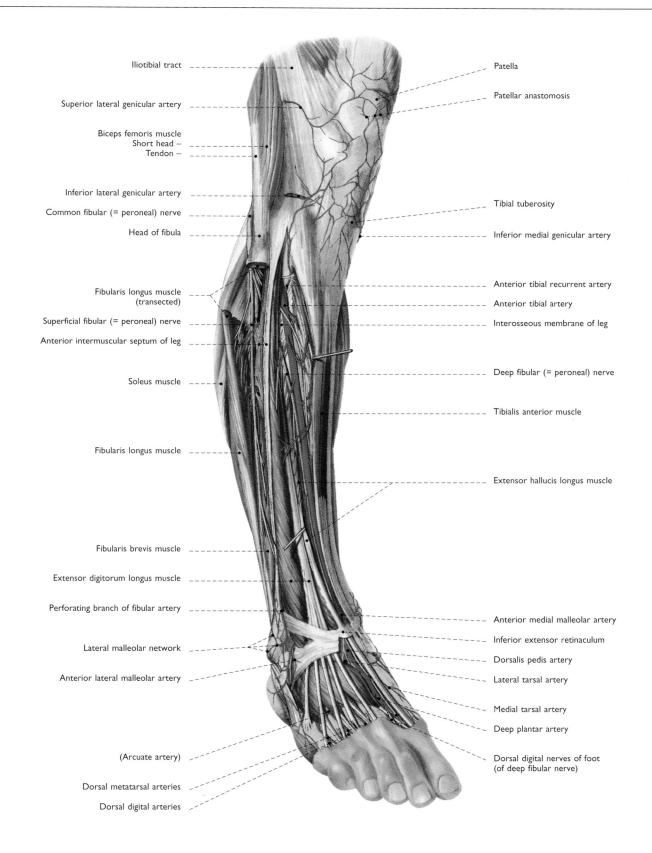

Iliotibial tract

Superior lateral genicular artery

Biceps femoris muscle
Short head –
Tendon –

Inferior lateral genicular artery

Common fibular (= peroneal) nerve

Head of fibula

Fibularis longus muscle
(transected)

Superficial fibular (= peroneal) nerve

Anterior intermuscular septum of leg

Soleus muscle

Fibularis longus muscle

Fibularis brevis muscle

Extensor digitorum longus muscle

Perforating branch of fibular artery

Lateral malleolar network

Anterior lateral malleolar artery

(Arcuate artery)

Dorsal metatarsal arteries

Dorsal digital arteries

Patella

Patellar anastomosis

Tibial tuberosity

Inferior medial genicular artery

Anterior tibial recurrent artery

Anterior tibial artery

Interosseous membrane of leg

Deep fibular (= peroneal) nerve

Tibialis anterior muscle

Extensor hallucis longus muscle

Anterior medial malleolar artery

Inferior extensor retinaculum

Dorsalis pedis artery

Lateral tarsal artery

Medial tarsal artery

Deep plantar artery

Dorsal digital nerves of foot
(of deep fibular nerve)

139 Arteries and nerves of the right leg and foot (30%)
The deep veins were removed. Ventrolateral aspect

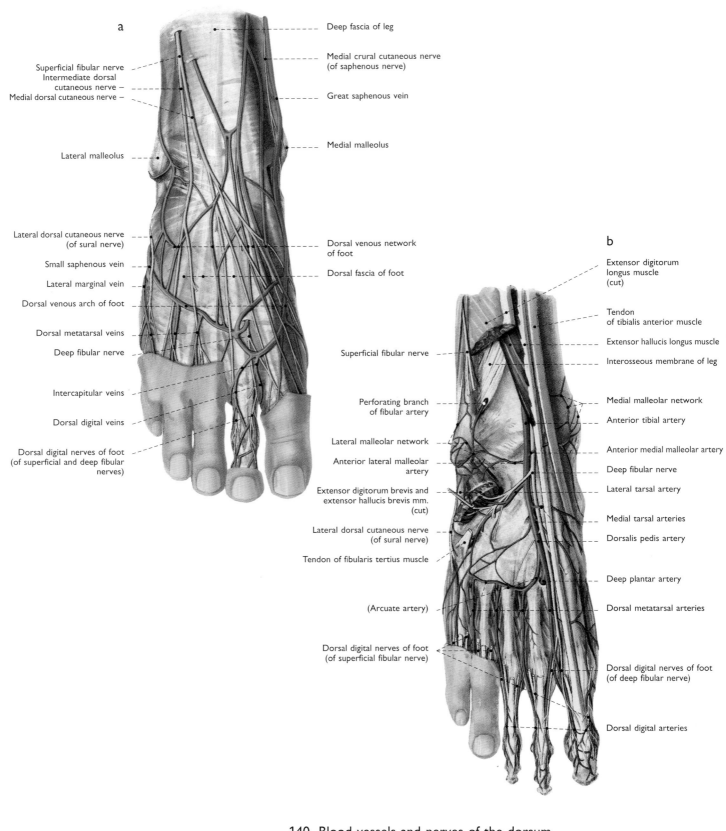

a

Superficial fibular nerve
Intermediate dorsal cutaneous nerve
Medial dorsal cutaneous nerve

Lateral malleolus

Lateral dorsal cutaneous nerve (of sural nerve)

Small saphenous vein

Lateral marginal vein

Dorsal venous arch of foot

Dorsal metatarsal veins

Deep fibular nerve

Intercapitular veins

Dorsal digital veins

Dorsal digital nerves of foot (of superficial and deep fibular nerves)

Deep fascia of leg

Medial crural cutaneous nerve (of saphenous nerve)

Great saphenous vein

Medial malleolus

Dorsal venous network of foot

Dorsal fascia of foot

Superficial fibular nerve

Perforating branch of fibular artery

Lateral malleolar network

Anterior lateral malleolar artery

Extensor digitorum brevis and extensor hallucis brevis mm. (cut)

Lateral dorsal cutaneous nerve (of sural nerve)

Tendon of fibularis tertius muscle

(Arcuate artery)

Dorsal digital nerves of foot (of superficial fibular nerve)

b

Extensor digitorum longus muscle (cut)

Tendon of tibialis anterior muscle

Extensor hallucis longus muscle

Interosseous membrane of leg

Medial malleolar network

Anterior tibial artery

Anterior medial malleolar artery

Deep fibular nerve

Lateral tarsal artery

Medial tarsal arteries

Dorsalis pedis artery

Deep plantar artery

Dorsal metatarsal arteries

Dorsal digital nerves of foot (of deep fibular nerve)

Dorsal digital arteries

140   Blood vessels and nerves of the dorsum of the right foot (50%)
Ventral aspect
a   Subcutaneous veins and nerves
b   Arteries and nerves after removal of the dorsal fascia of foot

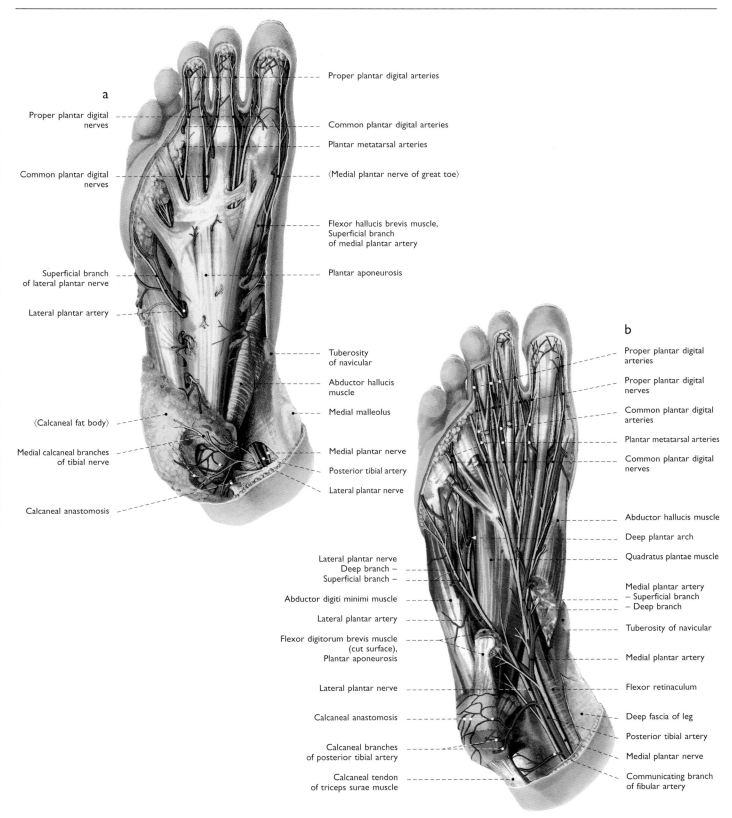

a

Proper plantar digital nerves

Common plantar digital nerves

Superficial branch of lateral plantar nerve

Lateral plantar artery

⟨Calcaneal fat body⟩

Medial calcaneal branches of tibial nerve

Calcaneal anastomosis

Proper plantar digital arteries

Common plantar digital arteries

Plantar metatarsal arteries

⟨Medial plantar nerve of great toe⟩

Flexor hallucis brevis muscle, Superficial branch of medial plantar artery

Plantar aponeurosis

Tuberosity of navicular

Abductor hallucis muscle

Medial malleolus

Medial plantar nerve

Posterior tibial artery

Lateral plantar nerve

b

Proper plantar digital arteries

Proper plantar digital nerves

Common plantar digital arteries

Plantar metatarsal arteries

Common plantar digital nerves

Abductor hallucis muscle

Deep plantar arch

Quadratus plantae muscle

Medial plantar artery
– Superficial branch
– Deep branch

Tuberosity of navicular

Medial plantar artery

Flexor retinaculum

Deep fascia of leg

Posterior tibial artery

Medial plantar nerve

Communicating branch of fibular artery

Lateral plantar nerve
Deep branch –
Superficial branch –

Abductor digiti minimi muscle

Lateral plantar artery

Flexor digitorum brevis muscle (cut surface), Plantar aponeurosis

Lateral plantar nerve

Calcaneal anastomosis

Calcaneal branches of posterior tibial artery

Calcaneal tendon of triceps surae muscle

141 Arteries and nerves of the sole of the right foot (50%)
Plantar aspect
a Superficial layer
b The abductor hallucis muscle and the short flexor muscle of toes were partially removed.

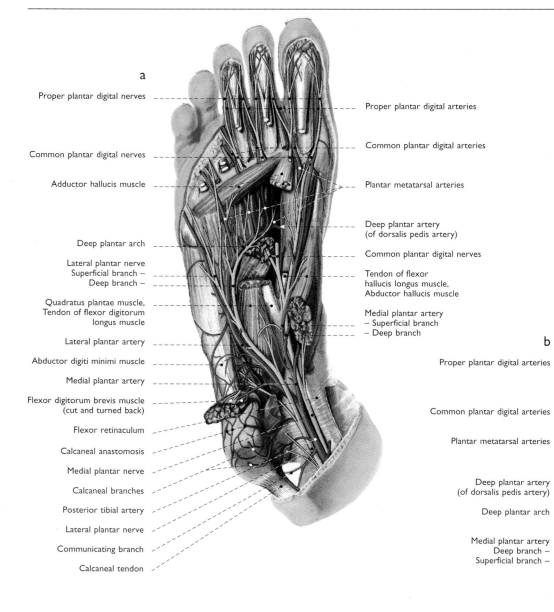

a

Proper plantar digital nerves

Common plantar digital nerves

Adductor hallucis muscle

Deep plantar arch

Lateral plantar nerve
Superficial branch –
Deep branch –

Quadratus plantae muscle,
Tendon of flexor digitorum
longus muscle

Lateral plantar artery

Abductor digiti minimi muscle

Medial plantar artery

Flexor digitorum brevis muscle
(cut and turned back)

Flexor retinaculum

Calcaneal anastomosis

Medial plantar nerve

Calcaneal branches

Posterior tibial artery

Lateral plantar nerve

Communicating branch

Calcaneal tendon

Proper plantar digital arteries

Common plantar digital arteries

Plantar metatarsal arteries

Deep plantar artery
(of dorsalis pedis artery)

Common plantar digital nerves

Tendon of flexor
hallucis longus muscle,
Abductor hallucis muscle

Medial plantar artery
– Superficial branch
– Deep branch

b

Proper plantar digital arteries

Common plantar digital arteries

Plantar metatarsal arteries

Deep plantar artery
(of dorsalis pedis artery)

Deep plantar arch

Medial plantar artery
Deep branch –
Superficial branch –

Lateral plantar artery

Medial plantar artery

Posterior tibial artery

Calcaneal anastomosis

**142 Arteries and nerves of the sole
of the right foot** (50%)
Plantar aspect
a The oblique head of the abductor hallucis muscle and
the short flexor muscle of toes were partially removed.
b Arteries of the sole of the right foot, schematic representation

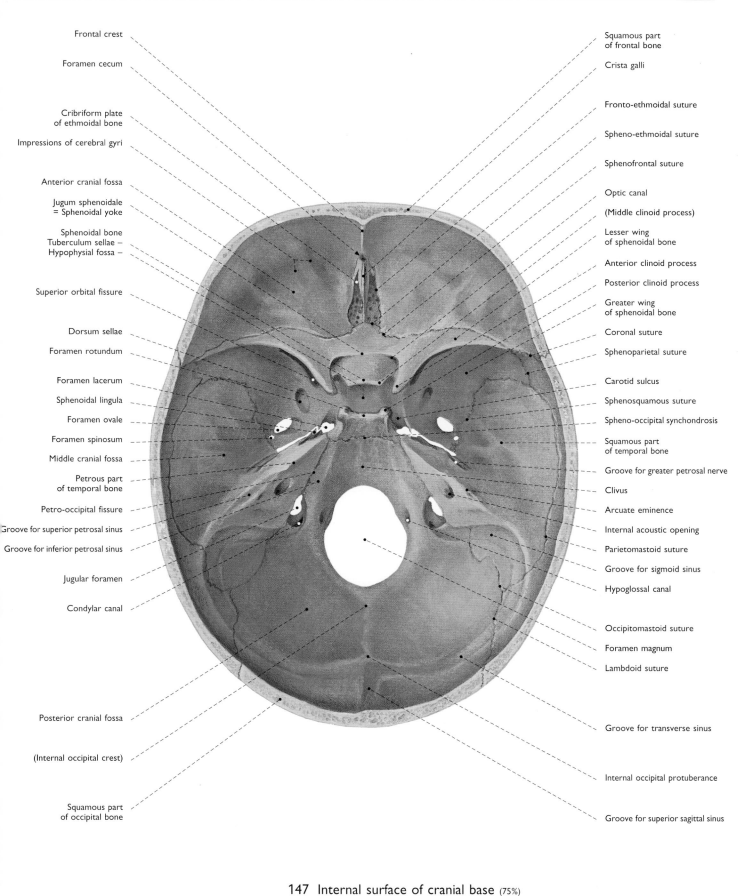

Frontal crest

Foramen cecum

Cribriform plate
of ethmoidal bone

Impressions of cerebral gyri

Anterior cranial fossa

Jugum sphenoidale
= Sphenoidal yoke

Sphenoidal bone
Tuberculum sellae –
Hypophysial fossa –

Superior orbital fissure

Dorsum sellae

Foramen rotundum

Foramen lacerum

Sphenoidal lingula

Foramen ovale

Foramen spinosum

Middle cranial fossa

Petrous part
of temporal bone

Petro-occipital fissure

Groove for superior petrosal sinus

Groove for inferior petrosal sinus

Jugular foramen

Condylar canal

Posterior cranial fossa

(Internal occipital crest)

Squamous part
of occipital bone

Squamous part
of frontal bone

Crista galli

Fronto-ethmoidal suture

Spheno-ethmoidal suture

Sphenofrontal suture

Optic canal

(Middle clinoid process)

Lesser wing
of sphenoidal bone

Anterior clinoid process

Posterior clinoid process

Greater wing
of sphenoidal bone

Coronal suture

Sphenoparietal suture

Carotid sulcus

Sphenosquamous suture

Spheno-occipital synchondrosis

Squamous part
of temporal bone

Groove for greater petrosal nerve

Clivus

Arcuate eminence

Internal acoustic opening

Parietomastoid suture

Groove for sigmoid sinus

Hypoglossal canal

Occipitomastoid suture

Foramen magnum

Lambdoid suture

Groove for transverse sinus

Internal occipital protuberance

Groove for superior sagittal sinus

147 Internal surface of cranial base (75%)
Superior aspect (vertical aspect)

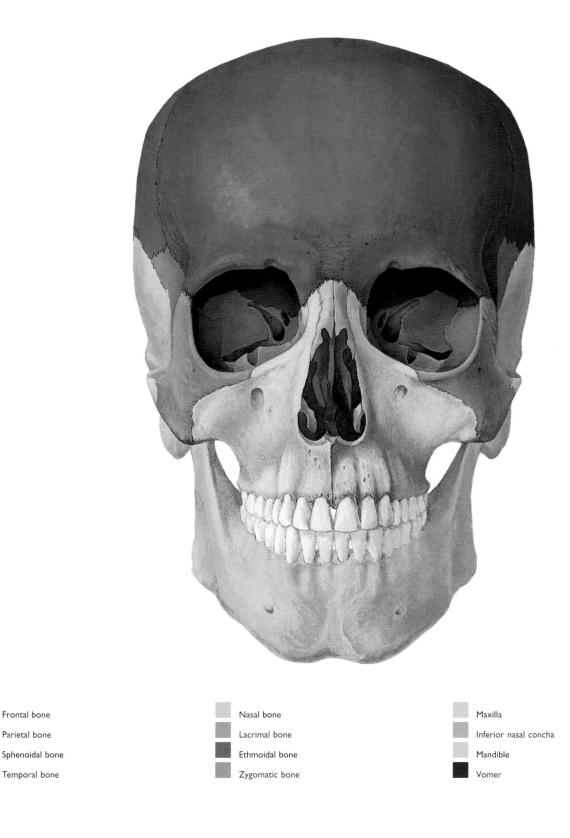

■ Frontal bone	■ Nasal bone	■ Maxilla
■ Parietal bone	■ Lacrimal bone	■ Inferior nasal concha
■ Sphenoidal bone	■ Ethmoidal bone	■ Mandible
■ Temporal bone	■ Zygomatic bone	■ Vomer

148 Skull = Cranium (80%)
The skull bones are marked by different colors.
Facial aspect (frontal aspect)

a

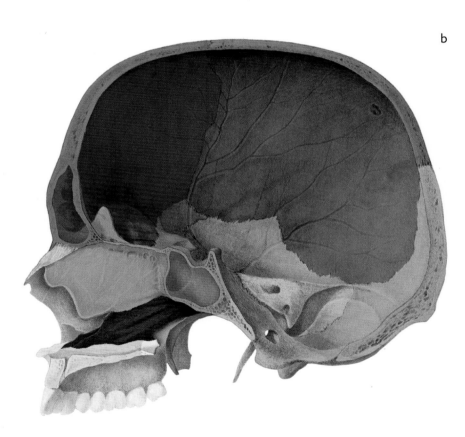

b

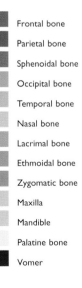

Frontal bone

Parietal bone

Sphenoidal bone

Occipital bone

Temporal bone

Nasal bone

Lacrimal bone

Ethmoidal bone

Zygomatic bone

Maxilla

Mandible

Palatine bone

Vomer

149 Skull = Cranium
        The skull bones are marked by different colors.
    a   Left lateral aspect
    b   Medial aspect of the right half of the skull

a

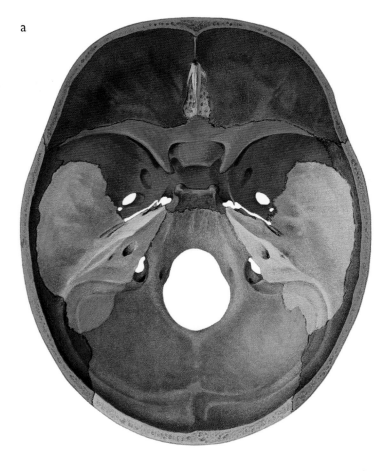

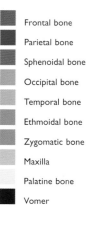

Frontal bone

Parietal bone

Sphenoidal bone

Occipital bone

Temporal bone

Ethmoidal bone

Zygomatic bone

Maxilla

Palatine bone

Vomer

b

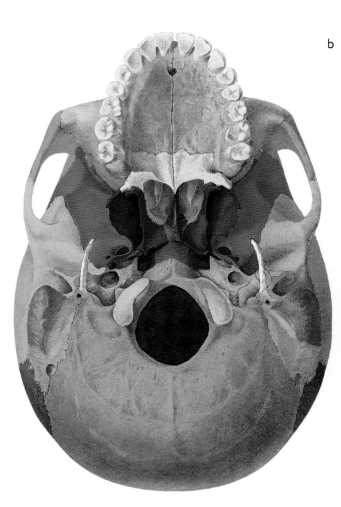

150 Cranial base (70%)

The skull bones are marked by different colors.
a   Internal surface
b   External surface

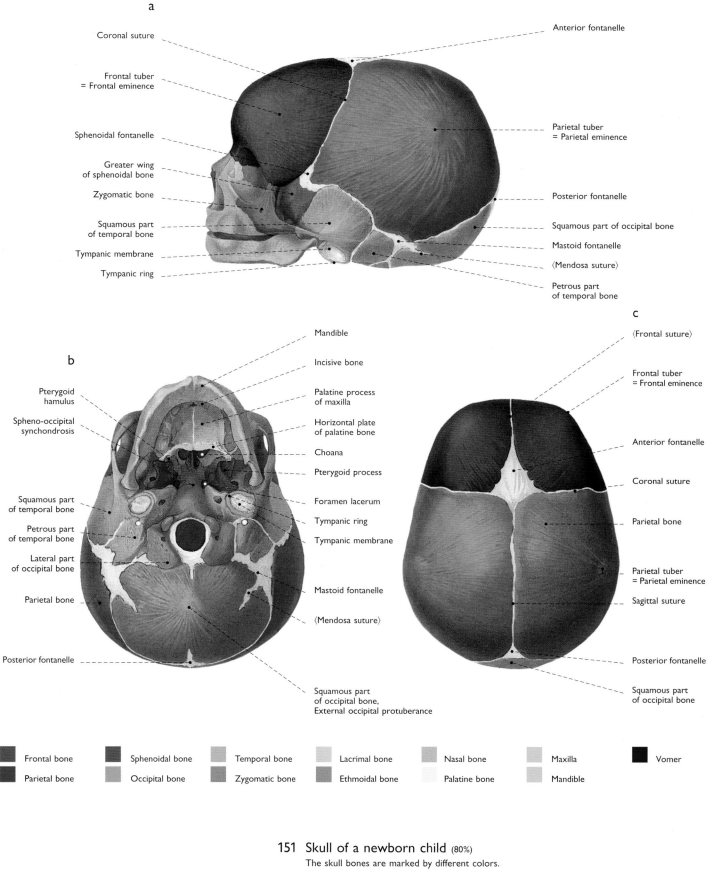

a

Coronal suture

Frontal tuber
= Frontal eminence

Sphenoidal fontanelle

Greater wing
of sphenoidal bone

Zygomatic bone

Squamous part
of temporal bone

Tympanic membrane

Tympanic ring

Anterior fontanelle

Parietal tuber
= Parietal eminence

Posterior fontanelle

Squamous part of occipital bone

Mastoid fontanelle

⟨Mendosa suture⟩

Petrous part
of temporal bone

b

Pterygoid
hamulus

Spheno-occipital
synchondrosis

Squamous part
of temporal bone

Petrous part
of temporal bone

Lateral part
of occipital bone

Parietal bone

Posterior fontanelle

Mandible

Incisive bone

Palatine process
of maxilla

Horizontal plate
of palatine bone

Choana

Pterygoid process

Foramen lacerum

Tympanic ring

Tympanic membrane

Mastoid fontanelle

⟨Mendosa suture⟩

Squamous part
of occipital bone,
External occipital protuberance

c

⟨Frontal suture⟩

Frontal tuber
= Frontal eminence

Anterior fontanelle

Coronal suture

Parietal bone

Parietal tuber
= Parietal eminence

Sagittal suture

Posterior fontanelle

Squamous part
of occipital bone

Frontal bone   Sphenoidal bone   Temporal bone   Lacrimal bone   Nasal bone   Maxilla   Vomer

Parietal bone   Occipital bone   Zygomatic bone   Ethmoidal bone   Palatine bone   Mandible

151 Skull of a newborn child (80%)
The skull bones are marked by different colors.
a Left lateral aspect
b External surface of cranial base, inferior aspect
c Skull cap = calvaria, superior aspect (vertical aspect)

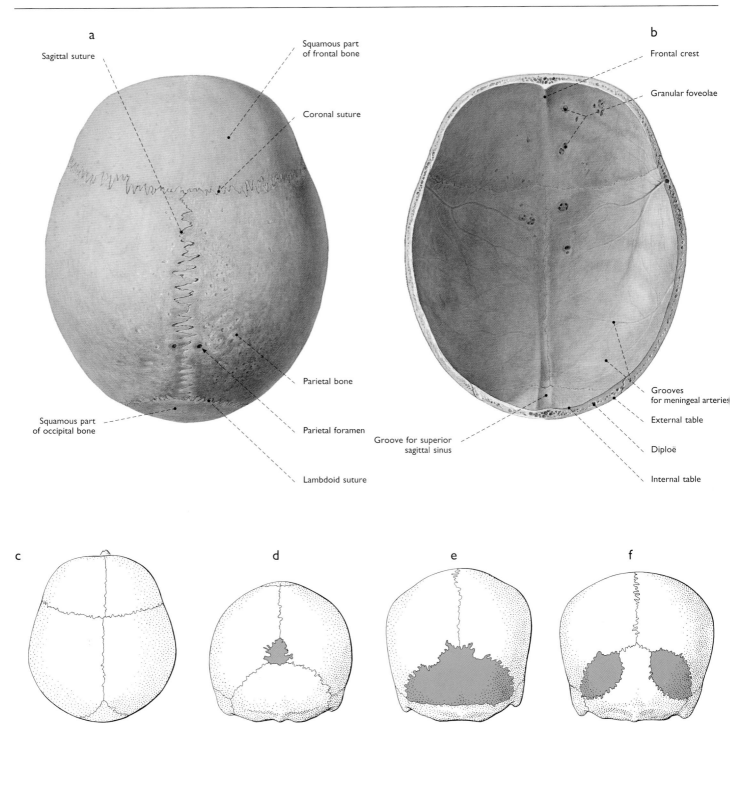

a

Sagittal suture

Squamous part
of frontal bone

Coronal suture

Parietal bone

Parietal foramen

Squamous part
of occipital bone

Lambdoid suture

Parietal foramen

b

Frontal crest

Granular foveolae

Grooves
for meningeal arteries

External table

Diploë

Internal table

Groove for superior
sagittal sinus

c

d

e

f

152 Skull cap (= calvaria) and sutural bones

a   Superior aspect (vertical aspect) (50%)
b   Inferior aspect (internal aspect) (50%)
c   Persisting frontal suture
d   Sutural bone (gray) in the sagittal suture
e   Inca bone (interparietal bone) (gray)
f   Sutural bones (gray) in the lambdoid suture

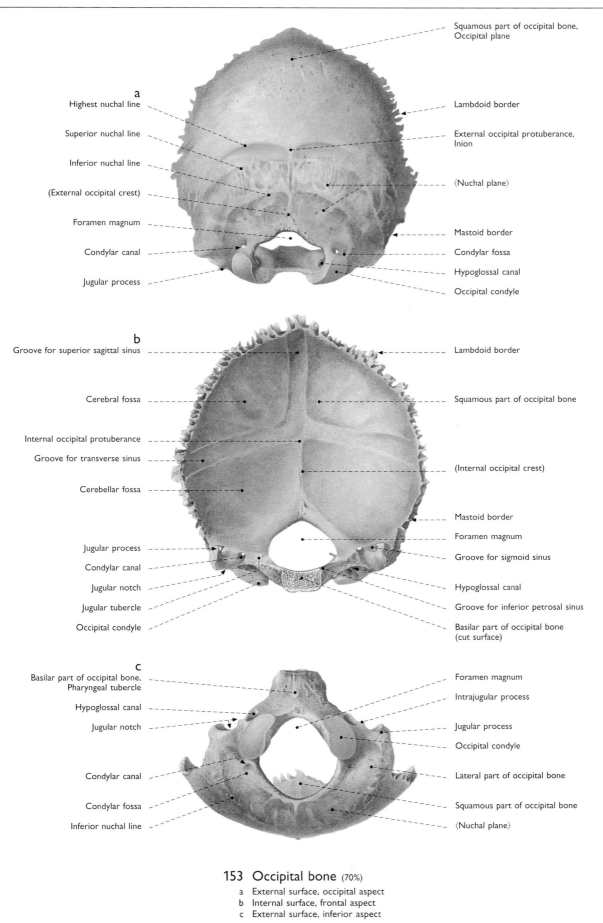

a

Highest nuchal line

Superior nuchal line

Inferior nuchal line

(External occipital crest)

Foramen magnum

Condylar canal

Jugular process

Squamous part of occipital bone,
Occipital plane

Lambdoid border

External occipital protuberance,
Inion

⟨Nuchal plane⟩

Mastoid border

Condylar fossa

Hypoglossal canal

Occipital condyle

b

Groove for superior sagittal sinus

Cerebral fossa

Internal occipital protuberance

Groove for transverse sinus

Cerebellar fossa

Jugular process

Condylar canal

Jugular notch

Jugular tubercle

Occipital condyle

Lambdoid border

Squamous part of occipital bone

(Internal occipital crest)

Mastoid border

Foramen magnum

Groove for sigmoid sinus

Hypoglossal canal

Groove for inferior petrosal sinus

Basilar part of occipital bone
(cut surface)

c

Basilar part of occipital bone,
Pharyngeal tubercle

Hypoglossal canal

Jugular notch

Condylar canal

Condylar fossa

Inferior nuchal line

Foramen magnum

Intrajugular process

Jugular process

Occipital condyle

Lateral part of occipital bone

Squamous part of occipital bone

⟨Nuchal plane⟩

153 Occipital bone (70%)
a External surface, occipital aspect
b Internal surface, frontal aspect
c External surface, inferior aspect

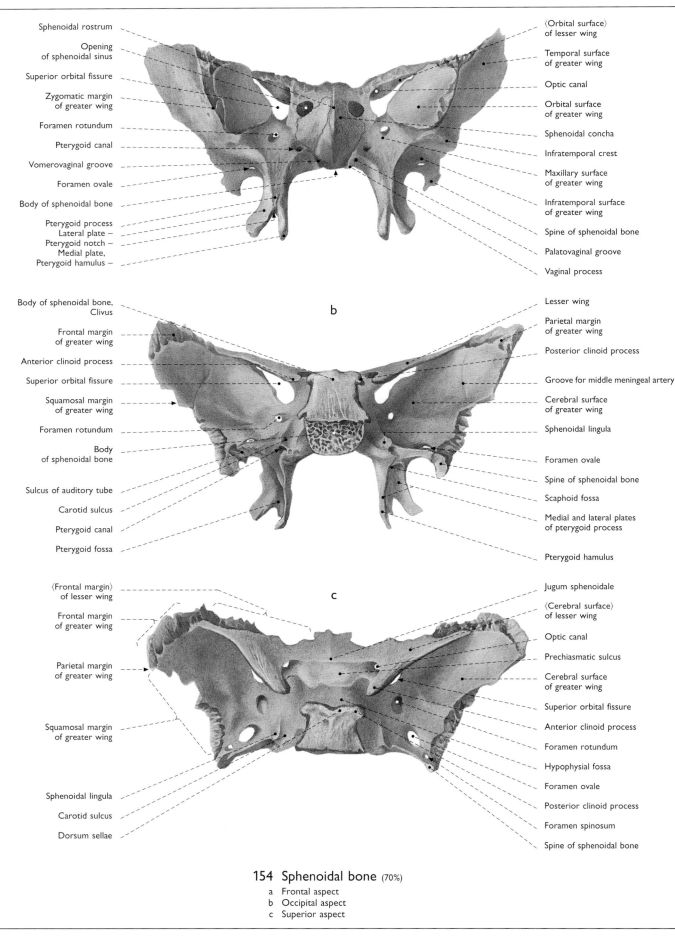

Sphenoidal rostrum

Opening of sphenoidal sinus

Superior orbital fissure

Zygomatic margin of greater wing

Foramen rotundum

Pterygoid canal

Vomerovaginal groove

Foramen ovale

Body of sphenoidal bone

Pterygoid process
Lateral plate –
Pterygoid notch –
Medial plate,
Pterygoid hamulus –

⟨Orbital surface⟩ of lesser wing

Temporal surface of greater wing

Optic canal

Orbital surface of greater wing

Sphenoidal concha

Infratemporal crest

Maxillary surface of greater wing

Infratemporal surface of greater wing

Spine of sphenoidal bone

Palatovaginal groove

Vaginal process

b

Body of sphenoidal bone, Clivus

Frontal margin of greater wing

Anterior clinoid process

Superior orbital fissure

Squamosal margin of greater wing

Foramen rotundum

Body of sphenoidal bone

Sulcus of auditory tube

Carotid sulcus

Pterygoid canal

Pterygoid fossa

Lesser wing

Parietal margin of greater wing

Posterior clinoid process

Groove for middle meningeal artery

Cerebral surface of greater wing

Sphenoidal lingula

Foramen ovale

Spine of sphenoidal bone

Scaphoid fossa

Medial and lateral plates of pterygoid process

Pterygoid hamulus

c

⟨Frontal margin⟩ of lesser wing

Frontal margin of greater wing

Parietal margin of greater wing

Squamosal margin of greater wing

Sphenoidal lingula

Carotid sulcus

Dorsum sellae

Jugum sphenoidale

⟨Cerebral surface⟩ of lesser wing

Optic canal

Prechiasmatic sulcus

Cerebral surface of greater wing

Superior orbital fissure

Anterior clinoid process

Foramen rotundum

Hypophysial fossa

Foramen ovale

Posterior clinoid process

Foramen spinosum

Spine of sphenoidal bone

154 Sphenoidal bone (70%)
a Frontal aspect
b Occipital aspect
c Superior aspect

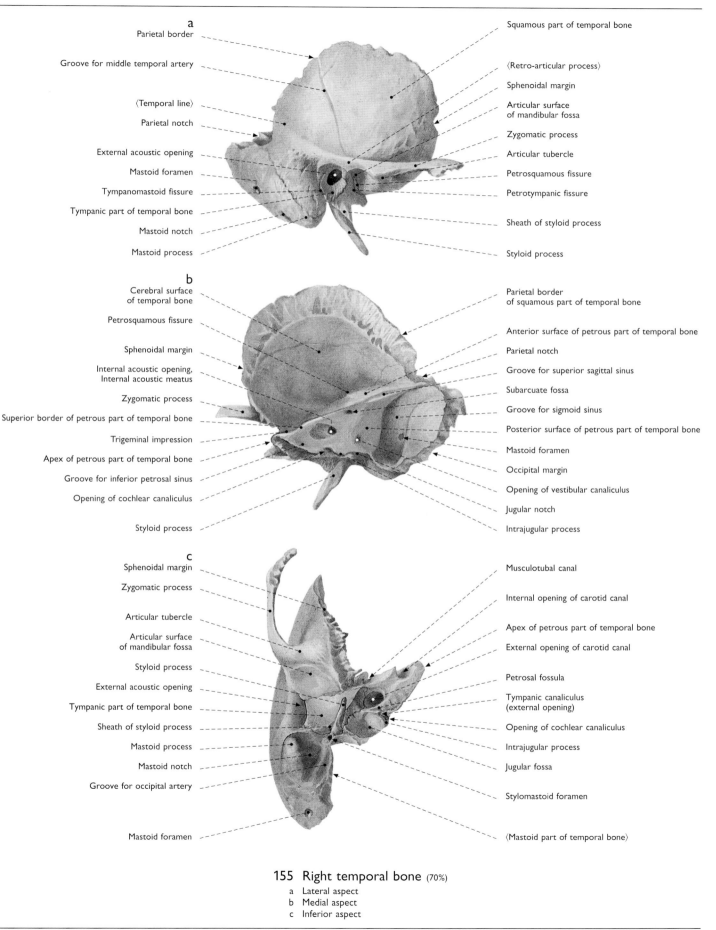

a

Parietal border

Groove for middle temporal artery

〈Temporal line〉

Parietal notch

External acoustic opening

Mastoid foramen

Tympanomastoid fissure

Tympanic part of temporal bone

Mastoid notch

Mastoid process

Squamous part of temporal bone

〈Retro-articular process〉

Sphenoidal margin

Articular surface
of mandibular fossa

Zygomatic process

Articular tubercle

Petrosquamous fissure

Petrotympanic fissure

Sheath of styloid process

Styloid process

b

Cerebral surface
of temporal bone

Petrosquamous fissure

Sphenoidal margin

Internal acoustic opening,
Internal acoustic meatus

Zygomatic process

Superior border of petrous part of temporal bone

Trigeminal impression

Apex of petrous part of temporal bone

Groove for inferior petrosal sinus

Opening of cochlear canaliculus

Styloid process

Parietal border
of squamous part of temporal bone

Anterior surface of petrous part of temporal bone

Parietal notch

Groove for superior sagittal sinus

Subarcuate fossa

Groove for sigmoid sinus

Posterior surface of petrous part of temporal bone

Mastoid foramen

Occipital margin

Opening of vestibular canaliculus

Jugular notch

Intrajugular process

c

Sphenoidal margin

Zygomatic process

Articular tubercle

Articular surface
of mandibular fossa

Styloid process

External acoustic opening

Tympanic part of temporal bone

Sheath of styloid process

Mastoid process

Mastoid notch

Groove for occipital artery

Mastoid foramen

Musculotubal canal

Internal opening of carotid canal

Apex of petrous part of temporal bone

External opening of carotid canal

Petrosal fossula

Tympanic canaliculus
(external opening)

Opening of cochlear canaliculus

Intrajugular process

Jugular fossa

Stylomastoid foramen

〈Mastoid part of temporal bone〉

155 Right temporal bone (70%)
a Lateral aspect
b Medial aspect
c Inferior aspect

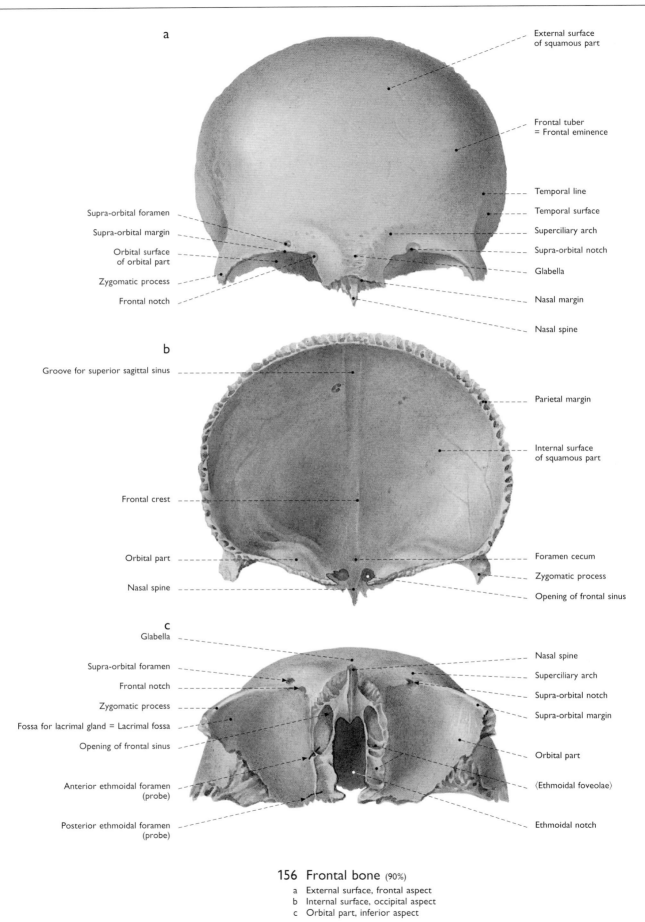

a

External surface
of squamous part

Frontal tuber
= Frontal eminence

Temporal line

Temporal surface

Superciliary arch

Supra-orbital notch

Glabella

Nasal margin

Nasal spine

Supra-orbital foramen

Supra-orbital margin

Orbital surface
of orbital part

Zygomatic process

Frontal notch

b

Groove for superior sagittal sinus

Parietal margin

Internal surface
of squamous part

Frontal crest

Orbital part

Foramen cecum

Nasal spine

Zygomatic process

Opening of frontal sinus

c

Glabella

Nasal spine

Supra-orbital foramen

Superciliary arch

Frontal notch

Supra-orbital notch

Zygomatic process

Supra-orbital margin

Fossa for lacrimal gland = Lacrimal fossa

Opening of frontal sinus

Orbital part

Anterior ethmoidal foramen
(probe)

(Ethmoidal foveolae)

Posterior ethmoidal foramen
(probe)

Ethmoidal notch

156 Frontal bone (90%)
a External surface, frontal aspect
b Internal surface, occipital aspect
c Orbital part, inferior aspect

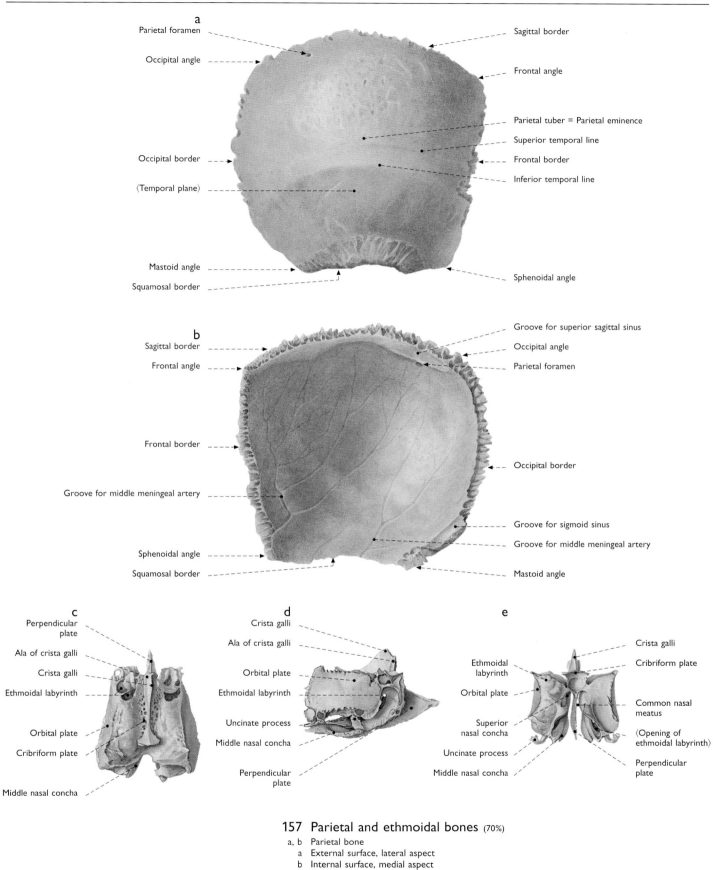

a
Parietal foramen
Occipital angle
Occipital border
⟨Temporal plane⟩
Mastoid angle
Squamosal border
Sagittal border
Frontal angle
Parietal tuber = Parietal eminence
Superior temporal line
Frontal border
Inferior temporal line
Sphenoidal angle

b
Sagittal border
Frontal angle
Frontal border
Groove for middle meningeal artery
Sphenoidal angle
Squamosal border
Groove for superior sagittal sinus
Occipital angle
Parietal foramen
Occipital border
Groove for sigmoid sinus
Groove for middle meningeal artery
Mastoid angle

c
Perpendicular plate
Ala of crista galli
Crista galli
Ethmoidal labyrinth
Orbital plate
Cribriform plate
Middle nasal concha

d
Crista galli
Ala of crista galli
Orbital plate
Ethmoidal labyrinth
Uncinate process
Middle nasal concha
Perpendicular plate

e
Ethmoidal labyrinth
Orbital plate
Superior nasal concha
Uncinate process
Middle nasal concha
Crista galli
Cribriform plate
Common nasal meatus
⟨Opening of ethmoidal labyrinth⟩
Perpendicular plate

157 Parietal and ethmoidal bones (70%)
a, b   Parietal bone
   a   External surface, lateral aspect
   b   Internal surface, medial aspect
c–e   Ethmoidal bone
   c   Superior aspect
   d   Lateral aspect
   e   Occipital aspect

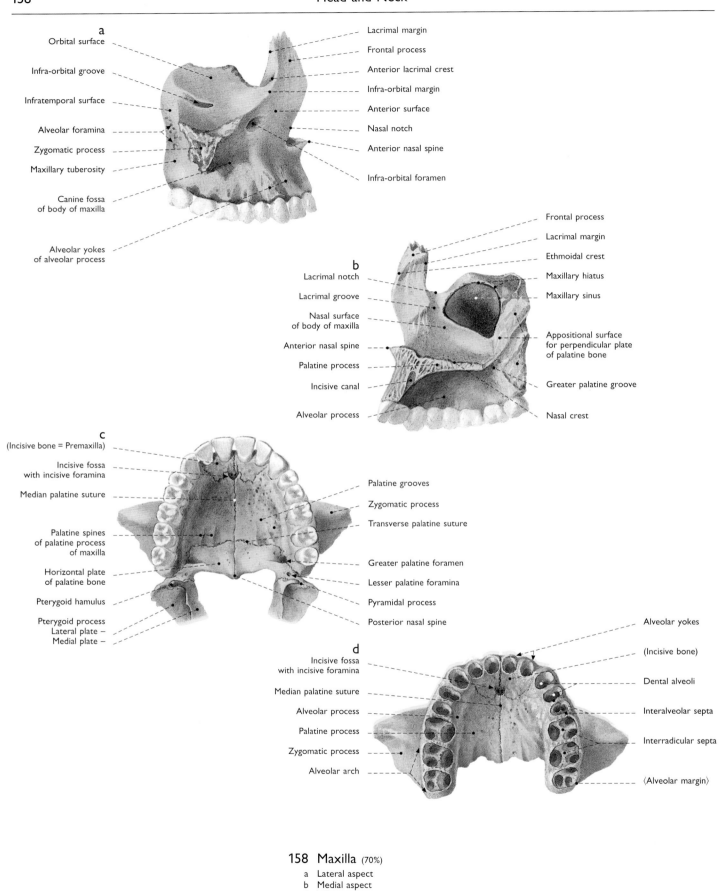

a

Orbital surface

Infra-orbital groove

Infratemporal surface

Alveolar foramina

Zygomatic process

Maxillary tuberosity

Canine fossa
of body of maxilla

Alveolar yokes
of alveolar process

Lacrimal margin

Frontal process

Anterior lacrimal crest

Infra-orbital margin

Anterior surface

Nasal notch

Anterior nasal spine

Infra-orbital foramen

b

Lacrimal notch

Lacrimal groove

Nasal surface
of body of maxilla

Anterior nasal spine

Palatine process

Incisive canal

Alveolar process

Frontal process

Lacrimal margin

Ethmoidal crest

Maxillary hiatus

Maxillary sinus

Appositional surface
for perpendicular plate
of palatine bone

Greater palatine groove

Nasal crest

c

(Incisive bone = Premaxilla)

Incisive fossa
with incisive foramina

Median palatine suture

Palatine spines
of palatine process
of maxilla

Horizontal plate
of palatine bone

Pterygoid hamulus

Pterygoid process
Lateral plate –
Medial plate –

Palatine grooves

Zygomatic process

Transverse palatine suture

Greater palatine foramen

Lesser palatine foramina

Pyramidal process

Posterior nasal spine

d

Incisive fossa
with incisive foramina

Median palatine suture

Alveolar process

Palatine process

Zygomatic process

Alveolar arch

Alveolar yokes

(Incisive bone)

Dental alveoli

Interalveolar septa

Interradicular septa

⟨Alveolar margin⟩

158　Maxilla (70%)

  a Lateral aspect
  b Medial aspect
c, d Inferior aspect
  c Alveolar arch with teeth
  d Alveolar arch after extraction of teeth

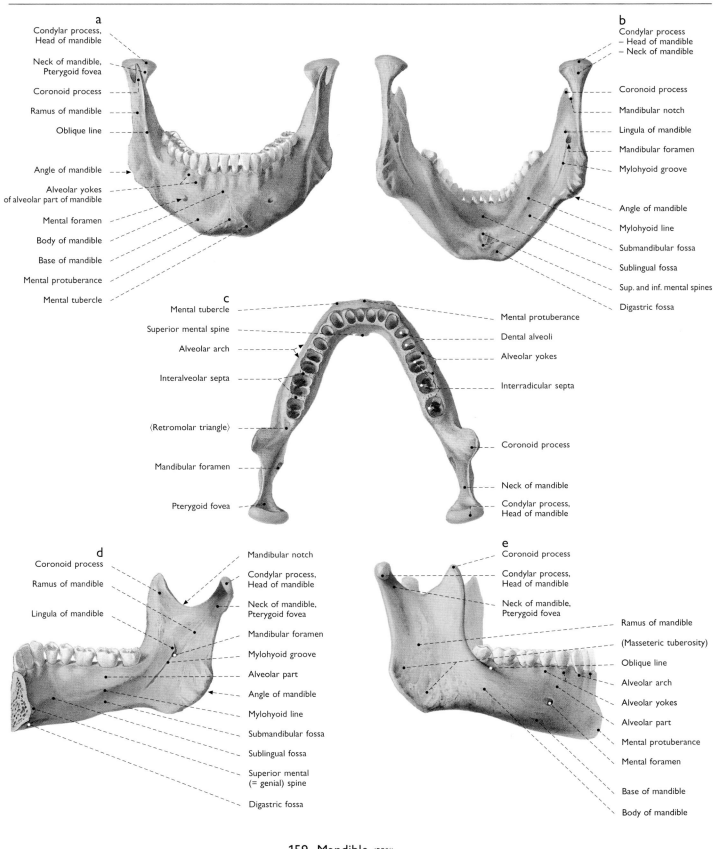

a

Condylar process, Head of mandible
Neck of mandible, Pterygoid fovea
Coronoid process
Ramus of mandible
Oblique line

Angle of mandible
Alveolar yokes of alveolar part of mandible
Mental foramen
Body of mandible
Base of mandible
Mental protuberance
Mental tubercle

b

Condylar process
– Head of mandible
– Neck of mandible
Coronoid process
Mandibular notch
Lingula of mandible
Mandibular foramen
Mylohyoid groove
Angle of mandible
Mylohyoid line
Submandibular fossa
Sublingual fossa
Sup. and inf. mental spines
Digastric fossa

c

Mental tubercle
Superior mental spine
Alveolar arch
Interalveolar septa
⟨Retromolar triangle⟩
Mandibular foramen
Pterygoid fovea

Mental protuberance
Dental alveoli
Alveolar yokes
Interradicular septa
Coronoid process
Neck of mandible
Condylar process, Head of mandible

d

Coronoid process
Ramus of mandible
Lingula of mandible

Mandibular notch
Condylar process, Head of mandible
Neck of mandible, Pterygoid fovea
Mandibular foramen
Mylohyoid groove
Alveolar part
Angle of mandible
Mylohyoid line
Submandibular fossa
Sublingual fossa
Superior mental (= genial) spine
Digastric fossa

e

Coronoid process
Condylar process, Head of mandible
Neck of mandible, Pterygoid fovea
Ramus of mandible
⟨Masseteric tuberosity⟩
Oblique line
Alveolar arch
Alveolar yokes
Alveolar part
Mental protuberance
Mental foramen
Base of mandible
Body of mandible

159 Mandible (55%)

a Frontal aspect
b Occipital aspect
c Superior aspect, alveolar arch after extraction of teeth
d Medial aspect of the right half of the mandible
e Right lateral aspect

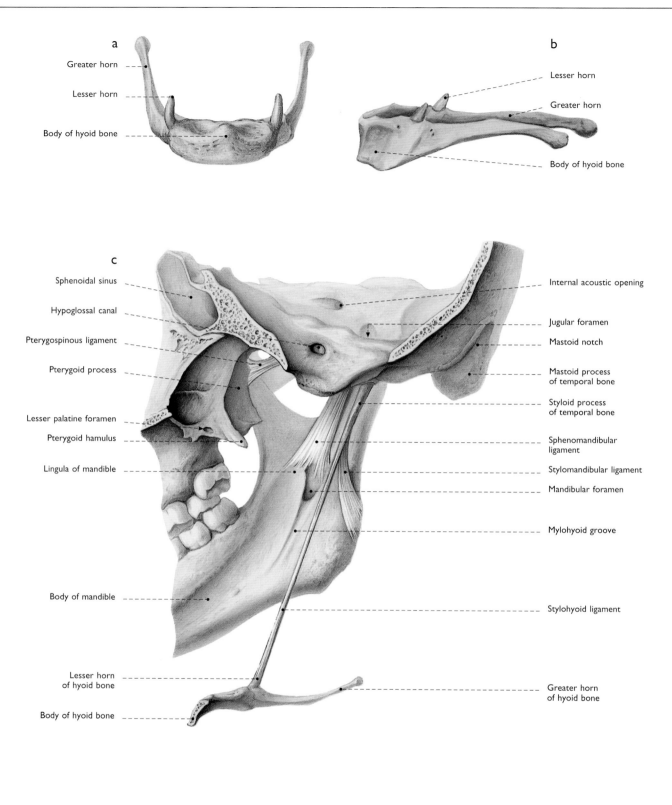

a

Greater horn

Lesser horn

Body of hyoid bone

b

Lesser horn

Greater horn

Body of hyoid bone

c

Sphenoidal sinus

Hypoglossal canal

Pterygospinous ligament

Pterygoid process

Lesser palatine foramen

Pterygoid hamulus

Lingula of mandible

Body of mandible

Lesser horn
of hyoid bone

Body of hyoid bone

Internal acoustic opening

Jugular foramen

Mastoid notch

Mastoid process
of temporal bone

Styloid process
of temporal bone

Sphenomandibular
ligament

Stylomandibular ligament

Mandibular foramen

Mylohyoid groove

Stylohyoid ligament

Greater horn
of hyoid bone

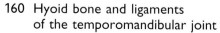

**160　Hyoid bone and ligaments
　　　of the temporomandibular joint**

a, b　Hyoid bone (80%)
　　a　Ventral aspect
　　b　Left lateral aspect
　　c　Ligaments of the right temporomandibular joint (100%),
　　　medial aspect

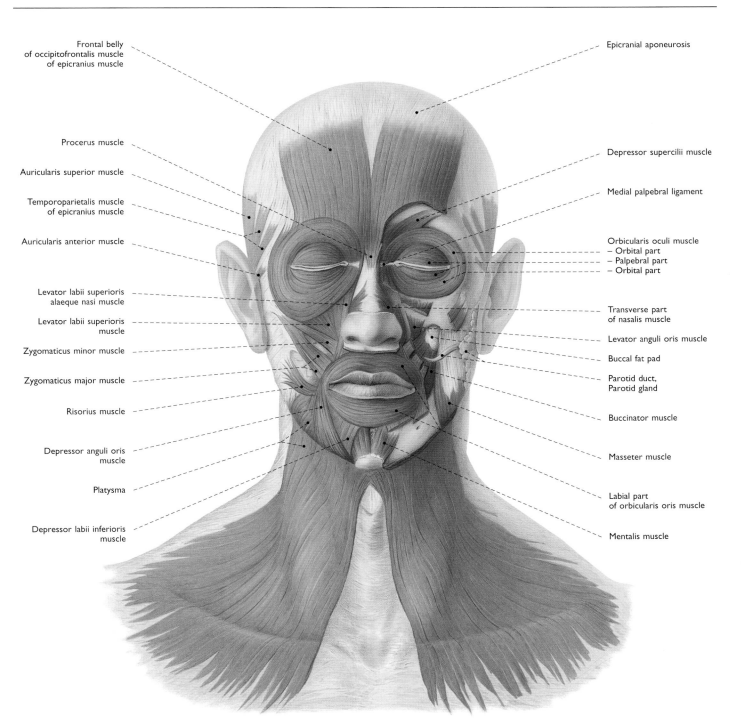

Frontal belly
of occipitofrontalis muscle
of epicranius muscle

Procerus muscle

Auricularis superior muscle

Temporoparietalis muscle
of epicranius muscle

Auricularis anterior muscle

Levator labii superioris
alaeque nasi muscle

Levator labii superioris
muscle

Zygomaticus minor muscle

Zygomaticus major muscle

Risorius muscle

Depressor anguli oris
muscle

Platysma

Depressor labii inferioris
muscle

Epicranial aponeurosis

Depressor supercilii muscle

Medial palpebral ligament

Orbicularis oculi muscle
– Orbital part
– Palpebral part
– Orbital part

Transverse part
of nasalis muscle

Levator anguli oris muscle

Buccal fat pad

Parotid duct,
Parotid gland

Buccinator muscle

Masseter muscle

Labial part
of orbicularis oris muscle

Mentalis muscle

161 Muscles of the scalp and face (50%)
On the right side of the face, the superficial layer of
the facial musculature is demonstrated, the deep layer and
the masseter muscle are shown on the left side. Frontal aspect

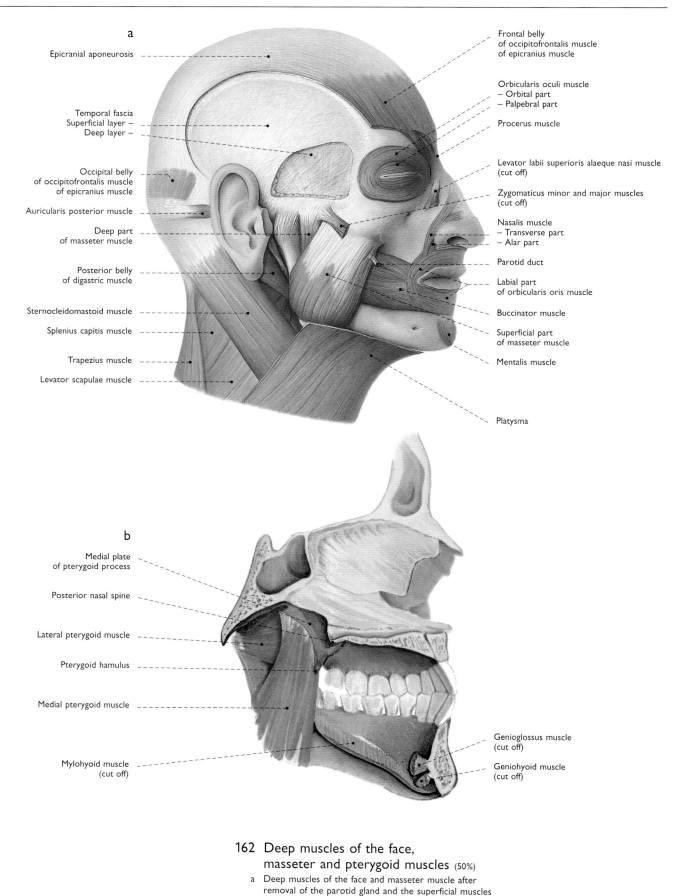

a

Epicranial aponeurosis

Temporal fascia
Superficial layer –
Deep layer –

Occipital belly
of occipitofrontalis muscle
of epicranius muscle

Auricularis posterior muscle

Deep part
of masseter muscle

Posterior belly
of digastric muscle

Sternocleidomastoid muscle

Splenius capitis muscle

Trapezius muscle

Levator scapulae muscle

Frontal belly
of occipitofrontalis muscle
of epicranius muscle

Orbicularis oculi muscle
– Orbital part
– Palpebral part

Procerus muscle

Levator labii superioris alaeque nasi muscle
(cut off)

Zygomaticus minor and major muscles
(cut off)

Nasalis muscle
– Transverse part
– Alar part

Parotid duct

Labial part
of orbicularis oris muscle

Buccinator muscle

Superficial part
of masseter muscle

Mentalis muscle

Platysma

b

Medial plate
of pterygoid process

Posterior nasal spine

Lateral pterygoid muscle

Pterygoid hamulus

Medial pterygoid muscle

Mylohyoid muscle
(cut off)

Genioglossus muscle
(cut off)

Geniohyoid muscle
(cut off)

**162 Deep muscles of the face,
masseter and pterygoid muscles** (50%)

a Deep muscles of the face and masseter muscle after
removal of the parotid gland and the superficial muscles
of the face, right lateral aspect
b Pterygoid muscles, medial aspect

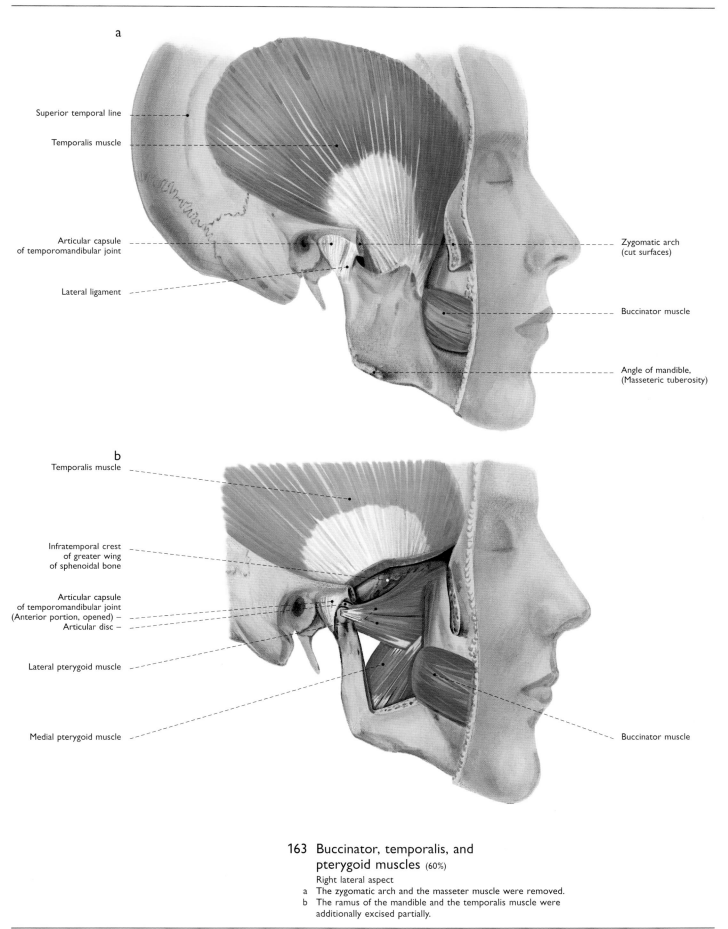

a

Superior temporal line

Temporalis muscle

Articular capsule
of temporomandibular joint

Lateral ligament

Zygomatic arch
(cut surfaces)

Buccinator muscle

Angle of mandible,
(Masseteric tuberosity)

b

Temporalis muscle

Infratemporal crest
of greater wing
of sphenoidal bone

Articular capsule
of temporomandibular joint
(Anterior portion, opened) –
Articular disc –

Lateral pterygoid muscle

Medial pterygoid muscle

Buccinator muscle

**163 Buccinator, temporalis, and
pterygoid muscles** (60%)
Right lateral aspect
a The zygomatic arch and the masseter muscle were removed.
b The ramus of the mandible and the temporalis muscle were
additionally excised partially.

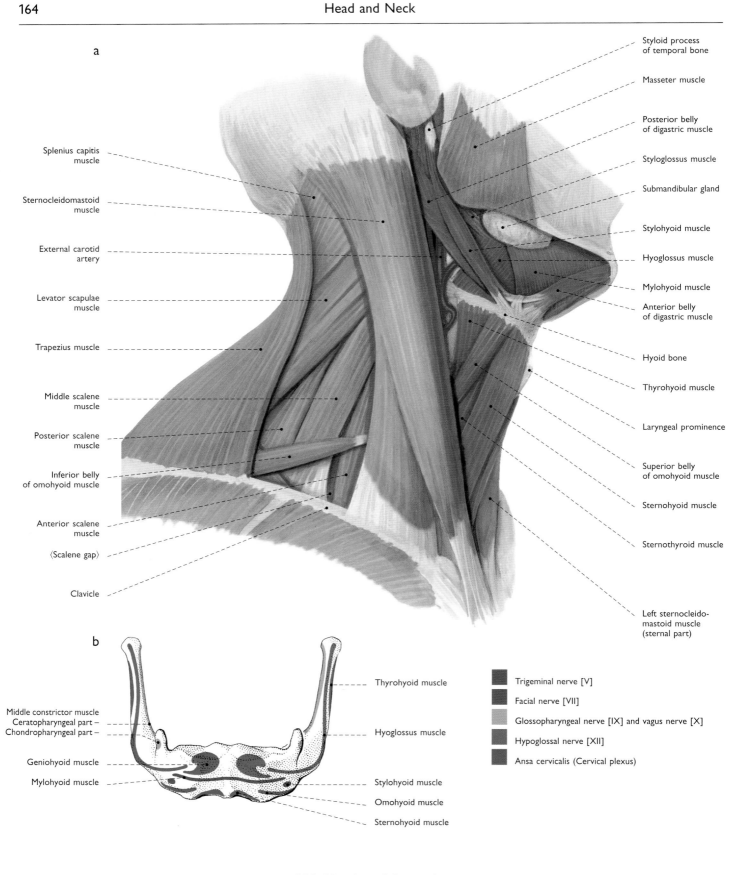

a

Styloid process
of temporal bone

Masseter muscle

Posterior belly
of digastric muscle

Styloglossus muscle

Submandibular gland

Stylohyoid muscle

Hyoglossus muscle

Mylohyoid muscle

Anterior belly
of digastric muscle

Hyoid bone

Thyrohyoid muscle

Laryngeal prominence

Superior belly
of omohyoid muscle

Sternohyoid muscle

Sternothyroid muscle

Left sternocleido-
mastoid muscle
(sternal part)

Splenius capitis
muscle

Sternocleidomastoid
muscle

External carotid
artery

Levator scapulae
muscle

Trapezius muscle

Middle scalene
muscle

Posterior scalene
muscle

Inferior belly
of omohyoid muscle

Anterior scalene
muscle

⟨Scalene gap⟩

Clavicle

b

Middle constrictor muscle
Ceratopharyngeal part –
Chondropharyngeal part –

Geniohyoid muscle

Mylohyoid muscle

Thyrohyoid muscle

Hyoglossus muscle

Stylohyoid muscle

Omohyoid muscle

Sternohyoid muscle

Trigeminal nerve [V]

Facial nerve [VII]

Glossopharyngeal nerve [IX] and vagus nerve [X]

Hypoglossal nerve [XII]

Ansa cervicalis (Cervical plexus)

164 Muscles of the neck

a The platysma and the cervical fascia were removed (60%).
  Right lateral aspect
b Muscle attachments on the ventral and superior surfaces
  of hyoid bone

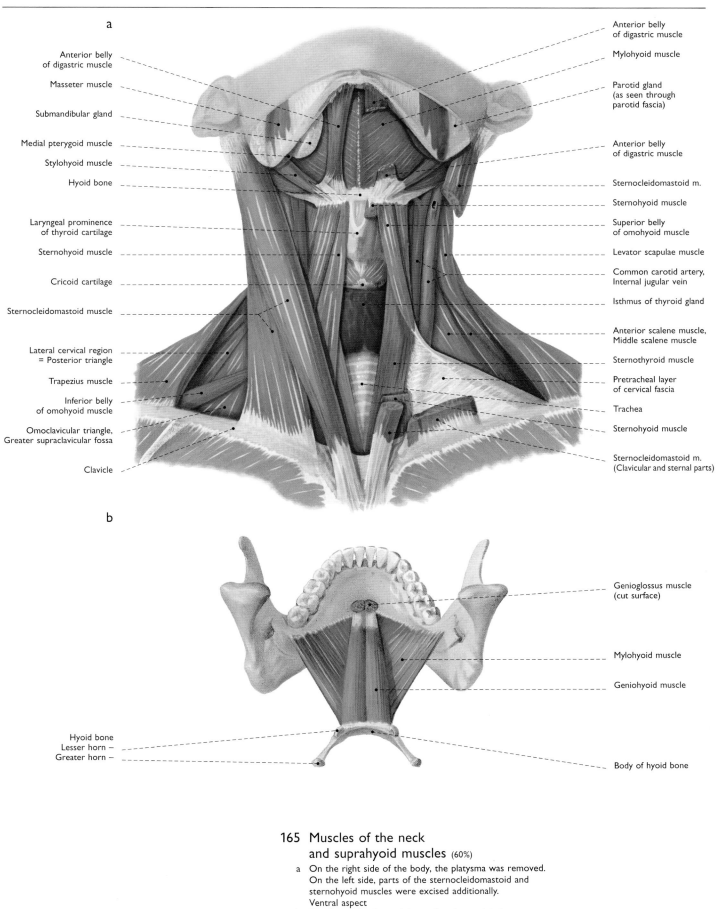

a

Anterior belly
of digastric muscle

Masseter muscle

Submandibular gland

Medial pterygoid muscle

Stylohyoid muscle

Hyoid bone

Laryngeal prominence
of thyroid cartilage

Sternohyoid muscle

Cricoid cartilage

Sternocleidomastoid muscle

Lateral cervical region
= Posterior triangle

Trapezius muscle

Inferior belly
of omohyoid muscle

Omoclavicular triangle,
Greater supraclavicular fossa

Clavicle

Anterior belly
of digastric muscle

Mylohyoid muscle

Parotid gland
(as seen through
parotid fascia)

Anterior belly
of digastric muscle

Sternocleidomastoid m.

Sternohyoid muscle

Superior belly
of omohyoid muscle

Levator scapulae muscle

Common carotid artery,
Internal jugular vein

Isthmus of thyroid gland

Anterior scalene muscle,
Middle scalene muscle

Sternothyroid muscle

Pretracheal layer
of cervical fascia

Trachea

Sternohyoid muscle

Sternocleidomastoid m.
(Clavicular and sternal parts)

b

Genioglossus muscle
(cut surface)

Mylohyoid muscle

Geniohyoid muscle

Hyoid bone
Lesser horn —
Greater horn —

Body of hyoid bone

**165 Muscles of the neck
and suprahyoid muscles** (60%)

a On the right side of the body, the platysma was removed.
On the left side, parts of the sternocleidomastoid and
sternohyoid muscles were excised additionally.
Ventral aspect
b Muscles of the floor of the oral cavity, occipital aspect

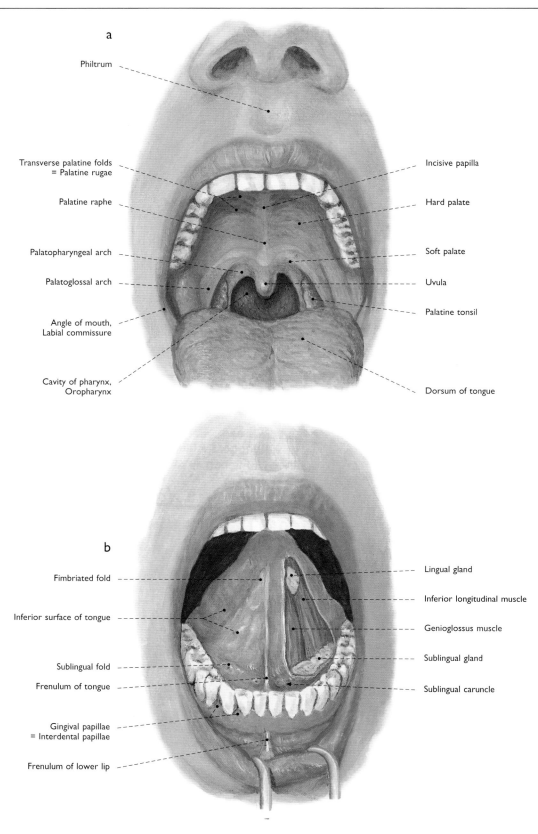

a

Philtrum

Transverse palatine folds
= Palatine rugae

Palatine raphe

Palatopharyngeal arch

Palatoglossal arch

Angle of mouth,
Labial commissure

Cavity of pharynx,
Oropharynx

Incisive papilla

Hard palate

Soft palate

Uvula

Palatine tonsil

Dorsum of tongue

b

Fimbriated fold

Inferior surface of tongue

Sublingual fold

Frenulum of tongue

Gingival papillae
= Interdental papillae

Frenulum of lower lip

Lingual gland

Inferior longitudinal muscle

Genioglossus muscle

Sublingual gland

Sublingual caruncle

166 Oral cavity (100%)

Frontal aspect
a   The mouth is wide open and the tongue extended.
b   The apex of tongue is turned upwards. Fenestration of the mucosa
    on the left side exposes the tongue muscles and the glands.

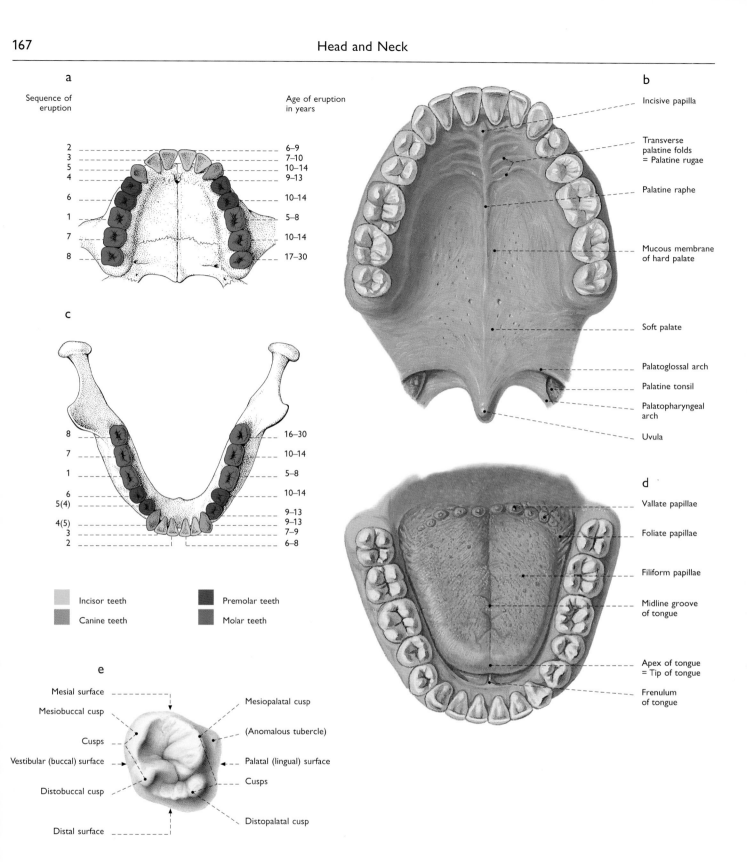

a

Sequence of
eruption

Age of eruption
in years

2	6–9
3	7–10
5	10–14
4	9–13
6	10–14
1	5–8
7	10–14
8	17–30

c

8	16–30
7	10–14
1	5–8
6	10–14
5(4)	9–13
4(5)	9–13
3	7–9
2	6–8

☐ Incisor teeth ☐ Premolar teeth

☐ Canine teeth ☐ Molar teeth

b

Incisive papilla

Transverse
palatine folds
= Palatine rugae

Palatine raphe

Mucous membrane
of hard palate

Soft palate

Palatoglossal arch

Palatine tonsil

Palatopharyngeal
arch

Uvula

d

Vallate papillae

Foliate papillae

Filiform papillae

Midline groove
of tongue

Apex of tongue
= Tip of tongue

Frenulum
of tongue

e

Mesial surface

Mesiobuccal cusp

Cusps

Vestibular (buccal) surface

Distobuccal cusp

Distal surface

Mesiopalatal cusp

(Anomalous tubercle)

Palatal (lingual) surface

Cusps

Distopalatal cusp

167 Permanent dentition and oral cavity
- a, c Permanent teeth
- a Upper jaw, inferior aspect
- c Lower jaw, superior aspect
- b, d Oral cavity proper (100%)
- b Roof of the oral cavity proper, inferior aspect
- d Floor of the oral cavity proper, superior aspect
- e Occlusal surface of the first upper molar (400%)

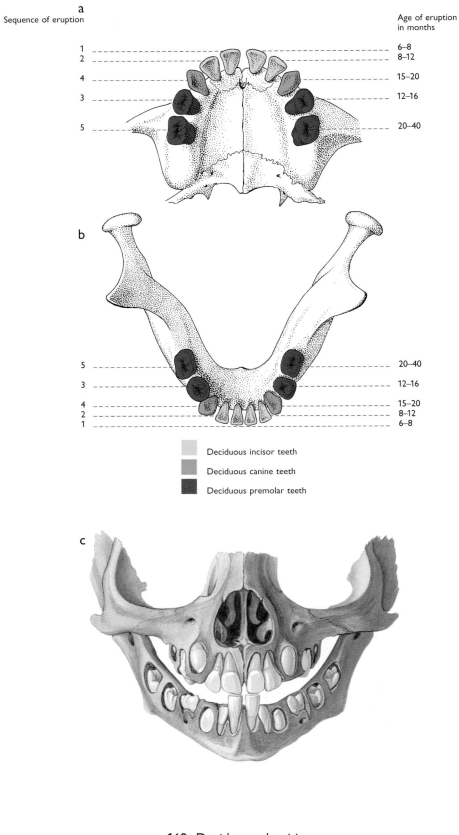

a Sequence of eruption

Age of eruption in months

1	6–8
2	8–12
4	15–20
3	12–16
5	20–40

b

5	20–40
3	12–16
4	15–20
2	8–12
1	6–8

Deciduous incisor teeth

Deciduous canine teeth

Deciduous premolar teeth

c

168 Deciduous dentition

a　Upper jaw, inferior aspect
b　Lower jaw, superior aspect
c　Partially erupted deciduous dentition of a 1-year-old child (100%), frontal aspect

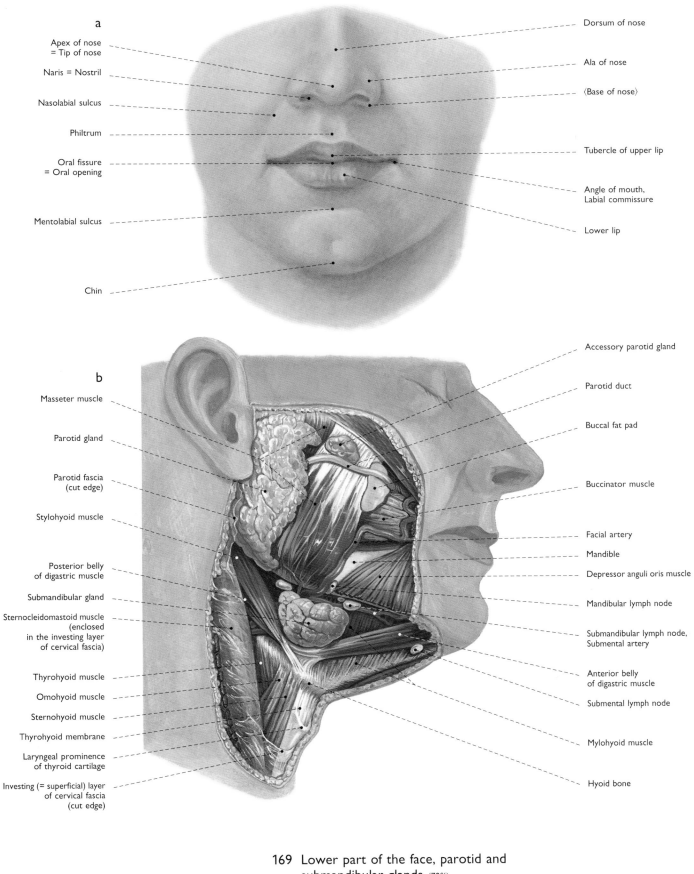

a

Apex of nose = Tip of nose

Naris = Nostril

Nasolabial sulcus

Philtrum

Oral fissure = Oral opening

Mentolabial sulcus

Chin

Dorsum of nose

Ala of nose

⟨Base of nose⟩

Tubercle of upper lip

Angle of mouth, Labial commissure

Lower lip

b

Masseter muscle

Parotid gland

Parotid fascia (cut edge)

Stylohyoid muscle

Posterior belly of digastric muscle

Submandibular gland

Sternocleidomastoid muscle (enclosed in the investing layer of cervical fascia)

Thyrohyoid muscle

Omohyoid muscle

Sternohyoid muscle

Thyrohyoid membrane

Laryngeal prominence of thyroid cartilage

Investing (= superficial) layer of cervical fascia (cut edge)

Accessory parotid gland

Parotid duct

Buccal fat pad

Buccinator muscle

Facial artery

Mandible

Depressor anguli oris muscle

Mandibular lymph node

Submandibular lymph node, Submental artery

Anterior belly of digastric muscle

Submental lymph node

Mylohyoid muscle

Hyoid bone

169 Lower part of the face, parotid and submandibular glands (70%)

a Lower part of the face, frontal aspect
b Parotid and submandibular glands, right lateral aspect

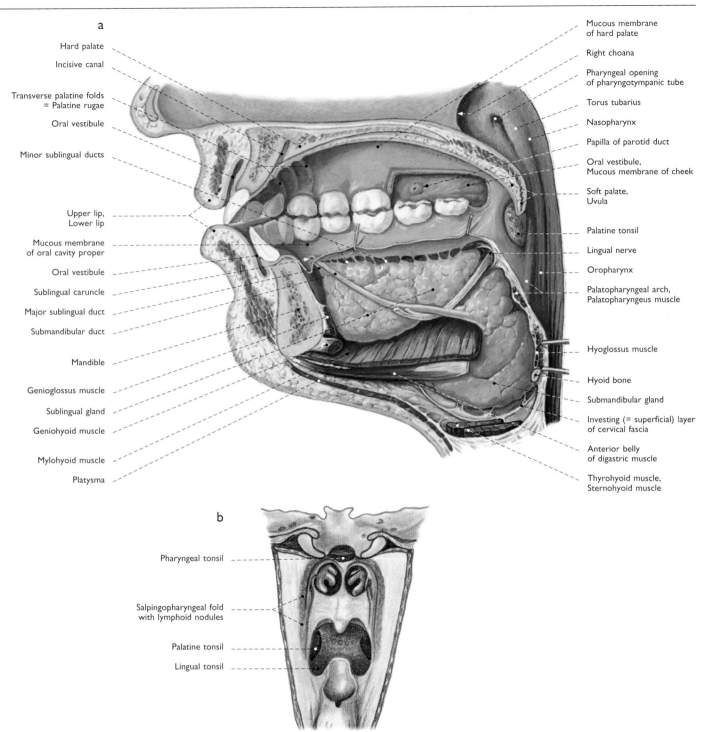

a

Hard palate

Incisive canal

Transverse palatine folds
= Palatine rugae

Oral vestibule

Minor sublingual ducts

Upper lip,
Lower lip

Mucous membrane
of oral cavity proper

Oral vestibule

Sublingual caruncle

Major sublingual duct

Submandibular duct

Mandible

Genioglossus muscle

Sublingual gland

Geniohyoid muscle

Mylohyoid muscle

Platysma

Mucous membrane
of hard palate

Right choana

Pharyngeal opening
of pharyngotympanic tube

Torus tubarius

Nasopharynx

Papilla of parotid duct

Oral vestibule,
Mucous membrane of cheek

Soft palate,
Uvula

Palatine tonsil

Lingual nerve

Oropharynx

Palatopharyngeal arch,
Palatopharyngeus muscle

Hyoglossus muscle

Hyoid bone

Submandibular gland

Investing (= superficial) layer
of cervical fascia

Anterior belly
of digastric muscle

Thyrohyoid muscle,
Sternohyoid muscle

b

Pharyngeal tonsil

Salpingopharyngeal fold
with lymphoid nodules

Palatine tonsil

Lingual tonsil

170 Parotid, submandibular, and sublingual glands

a Medial aspect of the right half of a head, bisected in the median
   plane (100%). A window was cut into the posterior part
   of the lateral wall of the palate in order to demonstrate
   the papilla of the parotid duct. The ventrolateral wall
   of the pharynx together with the posterior part of the hyoid bone
   are retracted dorsomedially with hooks in order to expose
   the submandibular and sublingual glands.
b Pharyngeal lymphoid ring (Waldeyer's ring), consisting of the lingual,
   palatine, and pharyngeal tonsils and minor lymphoid masses,
   dorsal aspect

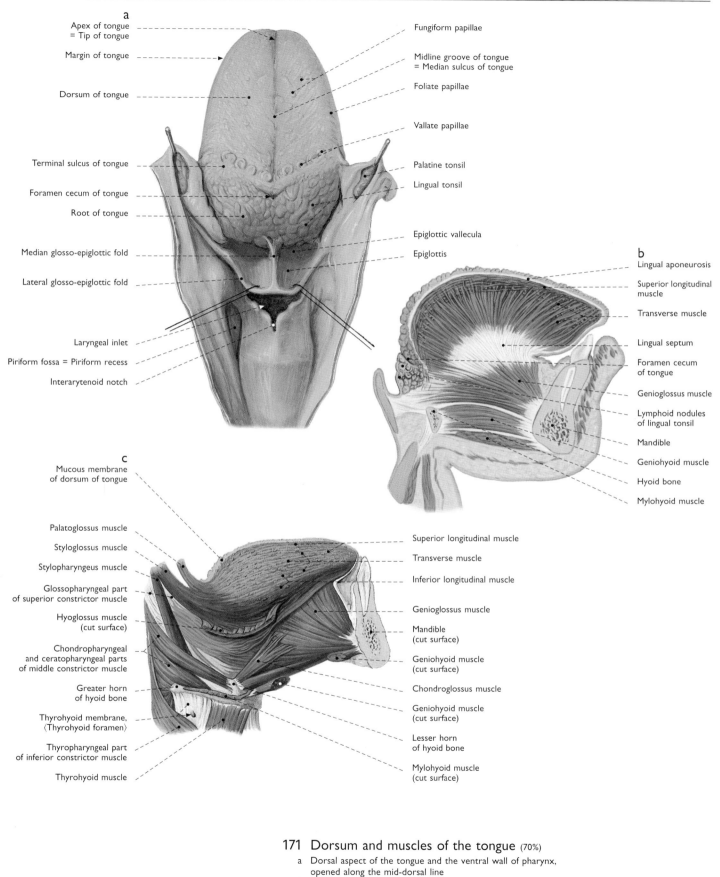

a

Apex of tongue
= Tip of tongue

Margin of tongue

Dorsum of tongue

Terminal sulcus of tongue

Foramen cecum of tongue

Root of tongue

Median glosso-epiglottic fold

Lateral glosso-epiglottic fold

Laryngeal inlet

Piriform fossa = Piriform recess

Interarytenoid notch

Fungiform papillae

Midline groove of tongue
= Median sulcus of tongue

Foliate papillae

Vallate papillae

Palatine tonsil

Lingual tonsil

Epiglottic vallecula

Epiglottis

b

Lingual aponeurosis

Superior longitudinal muscle

Transverse muscle

Lingual septum

Foramen cecum of tongue

Genioglossus muscle

Lymphoid nodules of lingual tonsil

Mandible

Geniohyoid muscle

Hyoid bone

Mylohyoid muscle

c

Mucous membrane of dorsum of tongue

Palatoglossus muscle

Styloglossus muscle

Stylopharyngeus muscle

Glossopharyngeal part of superior constrictor muscle

Hyoglossus muscle (cut surface)

Chondropharyngeal and ceratopharyngeal parts of middle constrictor muscle

Greater horn of hyoid bone

Thyrohyoid membrane, (Thyrohyoid foramen)

Thyropharyngeal part of inferior constrictor muscle

Thyrohyoid muscle

Superior longitudinal muscle

Transverse muscle

Inferior longitudinal muscle

Genioglossus muscle

Mandible (cut surface)

Geniohyoid muscle (cut surface)

Chondroglossus muscle

Geniohyoid muscle (cut surface)

Lesser horn of hyoid bone

Mylohyoid muscle (cut surface)

171 Dorsum and muscles of the tongue (70%)

a Dorsal aspect of the tongue and the ventral wall of pharynx, opened along the mid-dorsal line
b Medial aspect of the tongue sectioned in the sagittal plane
c Lateral aspect of the tongue after removal of the mucous membrane and the lingual aponeurosis

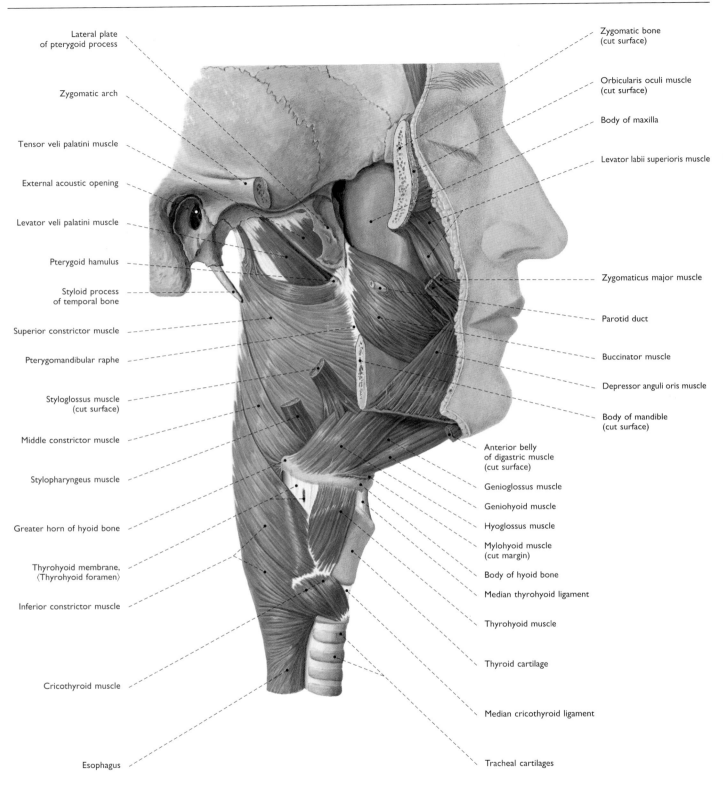

Lateral plate
of pterygoid process

Zygomatic arch

Tensor veli palatini muscle

External acoustic opening

Levator veli palatini muscle

Pterygoid hamulus

Styloid process
of temporal bone

Superior constrictor muscle

Pterygomandibular raphe

Styloglossus muscle
(cut surface)

Middle constrictor muscle

Stylopharyngeus muscle

Greater horn of hyoid bone

Thyrohyoid membrane,
(Thyrohyoid foramen)

Inferior constrictor muscle

Cricothyroid muscle

Esophagus

Zygomatic bone
(cut surface)

Orbicularis oculi muscle
(cut surface)

Body of maxilla

Levator labii superioris muscle

Zygomaticus major muscle

Parotid duct

Buccinator muscle

Depressor anguli oris muscle

Body of mandible
(cut surface)

Anterior belly
of digastric muscle
(cut surface)

Genioglossus muscle

Geniohyoid muscle

Hyoglossus muscle

Mylohyoid muscle
(cut margin)

Body of hyoid bone

Median thyrohyoid ligament

Thyrohyoid muscle

Thyroid cartilage

Median cricothyroid ligament

Tracheal cartilages

172 Muscles of the pharynx (75%)
Right lateral aspect

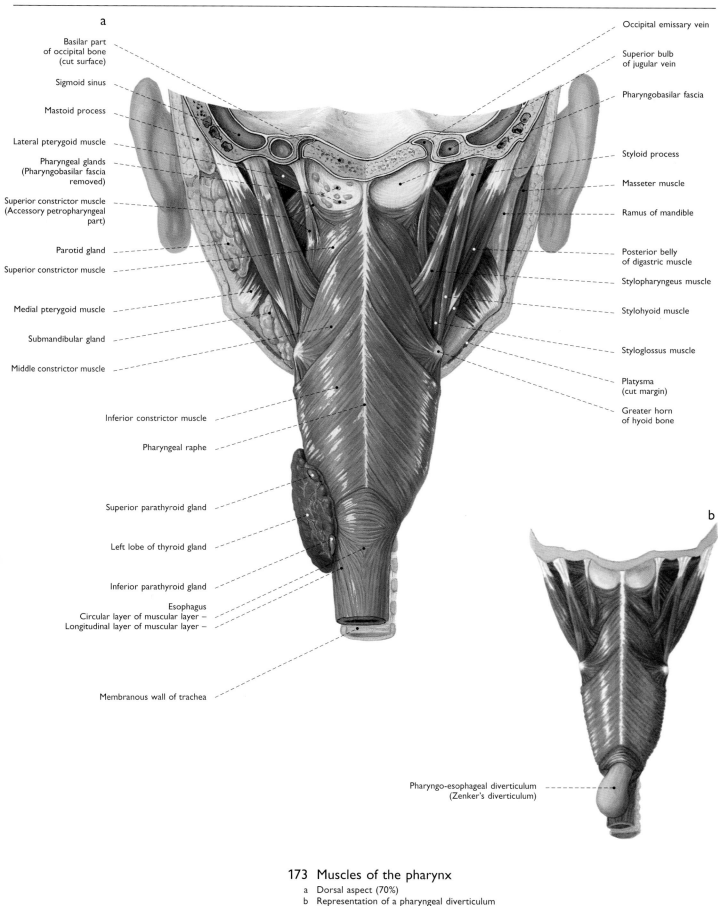

a

Basilar part
of occipital bone
(cut surface)

Sigmoid sinus

Mastoid process

Lateral pterygoid muscle

Pharyngeal glands
(Pharyngobasilar fascia
removed)

Superior constrictor muscle
(Accessory petropharyngeal
part)

Parotid gland

Superior constrictor muscle

Medial pterygoid muscle

Submandibular gland

Middle constrictor muscle

Inferior constrictor muscle

Pharyngeal raphe

Superior parathyroid gland

Left lobe of thyroid gland

Inferior parathyroid gland

Esophagus
Circular layer of muscular layer –
Longitudinal layer of muscular layer –

Membranous wall of trachea

Occipital emissary vein

Superior bulb
of jugular vein

Pharyngobasilar fascia

Styloid process

Masseter muscle

Ramus of mandible

Posterior belly
of digastric muscle

Stylopharyngeus muscle

Stylohyoid muscle

Styloglossus muscle

Platysma
(cut margin)

Greater horn
of hyoid bone

b

Pharyngo-esophageal diverticulum
(Zenker's diverticulum)

173 Muscles of the pharynx

a   Dorsal aspect (70%)
b   Representation of a pharyngeal diverticulum
in the Laimer's triangle, dorsal aspect

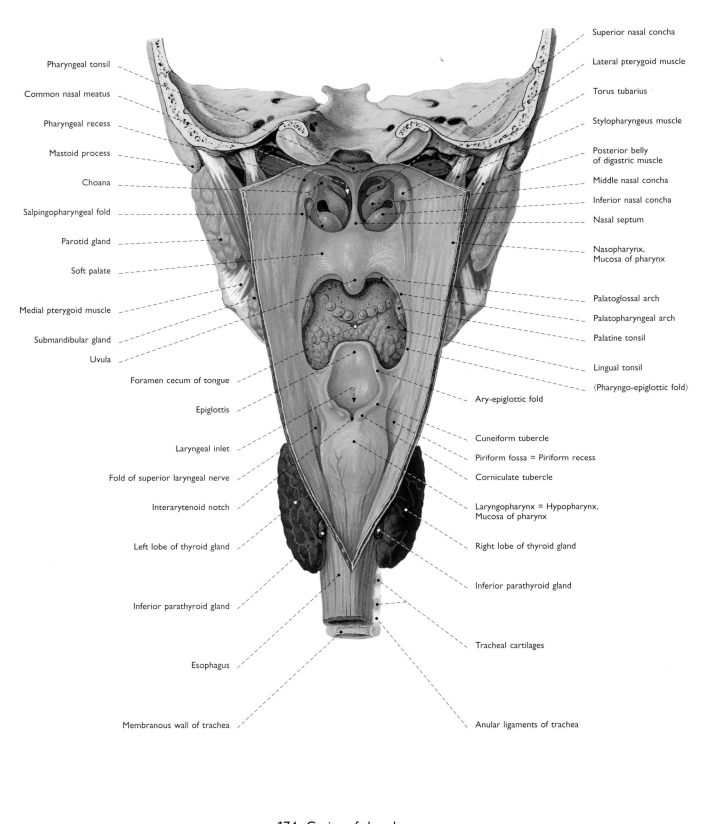

Pharyngeal tonsil

Common nasal meatus

Pharyngeal recess

Mastoid process

Choana

Salpingopharyngeal fold

Parotid gland

Soft palate

Medial pterygoid muscle

Submandibular gland

Uvula

Foramen cecum of tongue

Epiglottis

Laryngeal inlet

Fold of superior laryngeal nerve

Interarytenoid notch

Left lobe of thyroid gland

Inferior parathyroid gland

Esophagus

Membranous wall of trachea

Superior nasal concha

Lateral pterygoid muscle

Torus tubarius

Stylopharyngeus muscle

Posterior belly
of digastric muscle

Middle nasal concha

Inferior nasal concha

Nasal septum

Nasopharynx,
Mucosa of pharynx

Palatoglossal arch

Palatopharyngeal arch

Palatine tonsil

Lingual tonsil

(Pharyngo-epiglottic fold)

Ary-epiglottic fold

Cuneiform tubercle

Piriform fossa = Piriform recess

Corniculate tubercle

Laryngopharynx = Hypopharynx,
Mucosa of pharynx

Right lobe of thyroid gland

Inferior parathyroid gland

Tracheal cartilages

Anular ligaments of trachea

174 Cavity of the pharynx (70%)
The dorsal wall of the pharynx was cut along
the mid-dorsal line and opened. Dorsal aspect

Highest nasal concha

Frontal sinus

Superior nasal concha

Middle nasal concha

Middle nasal meatus

Inferior nasal concha

Limen nasi

Nasal vestibule

Inferior nasal meatus

Hard palate

Upper lip

Oral cavity proper

Lower lip,
Oral vestibule

Tongue,
Palatoglossal arch

Palatine tonsil,
Palatopharyngeal arch

Body of hyoid bone

Thyroid cartilage

Vestibular fold,
Laryngeal ventricle,
Vocal fold

Arch of cricoid cartilage

Hypophysial fossa

Sphenoidal sinus

Straight sinus

Pharyngeal opening
of pharyngotympanic tube

Torus tubarius

Nasopharynx

Soft palate

Anterior arch of atlas

Dens axis

Uvula

Oropharynx

Epiglottis

Laryngopharynx
= Hypopharynx

Ary-epiglottic fold

Cuneiform tubercle

Corniculate tubercle

Oblique and transverse
arytenoid muscles

Lamina
of cricoid cartilage

Esophagus

Trachea

Isthmus of thyroid gland

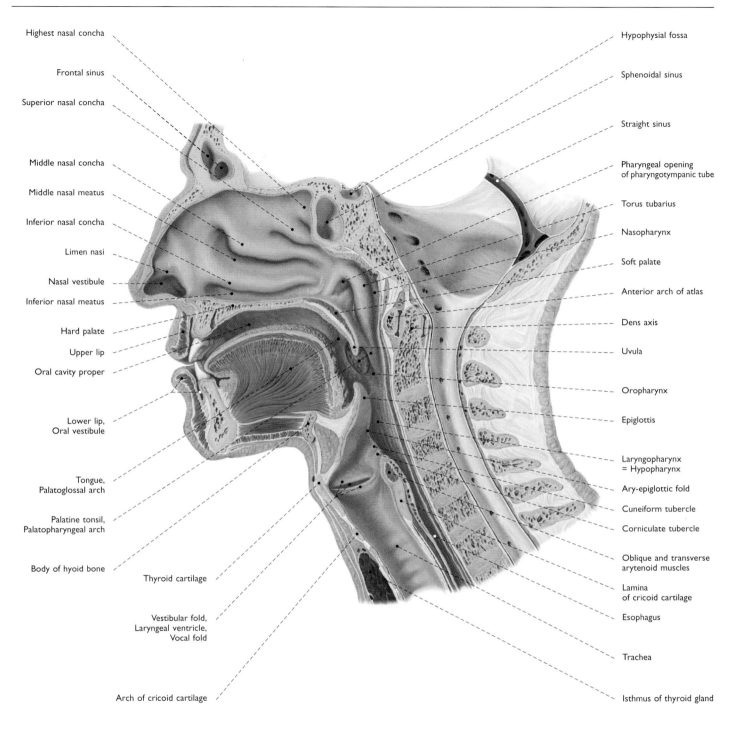

**175 Alimentary and respiratory systems
in the head and neck** (70%)
Median sagittal section through the head and neck,
medial aspect of the right half

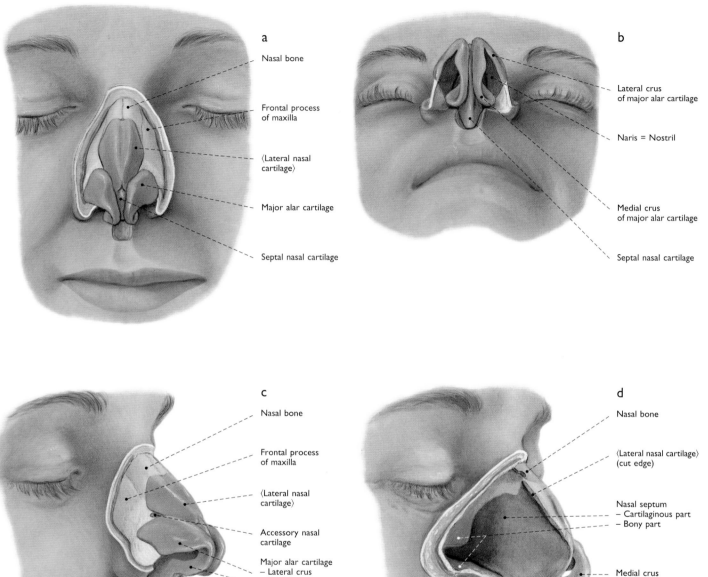

a

Nasal bone

Frontal process
of maxilla

〈Lateral nasal
cartilage〉

Major alar cartilage

Septal nasal cartilage

b

Lateral crus
of major alar cartilage

Naris = Nostril

Medial crus
of major alar cartilage

Septal nasal cartilage

c

Nasal bone

Frontal process
of maxilla

〈Lateral nasal
cartilage〉

Accessory nasal
cartilage

Major alar cartilage
– Lateral crus
– Medial crus

d

Nasal bone

〈Lateral nasal cartilage〉
(cut edge)

Nasal septum
– Cartilaginous part
– Bony part

Medial crus
of major alar cartilage

176 Skeleton of the external nose (80%)
a Frontal aspect
b Inferior aspect
c Lateral aspect
d The nasal septum. The right lateral wall
of the external nose was removed. Lateral aspect

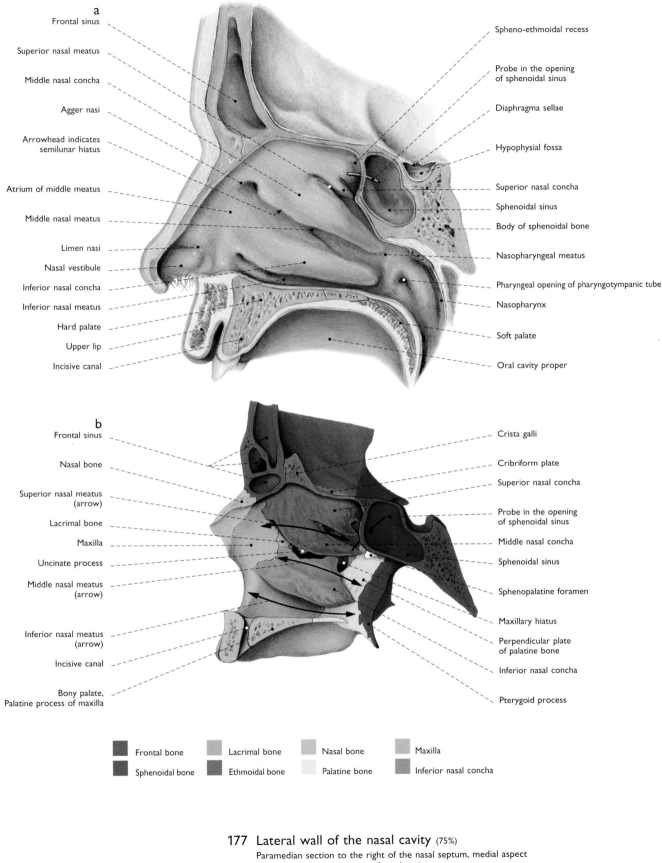

a
Frontal sinus
Superior nasal meatus
Middle nasal concha
Agger nasi
Arrowhead indicates
semilunar hiatus
Atrium of middle meatus
Middle nasal meatus
Limen nasi
Nasal vestibule
Inferior nasal concha
Inferior nasal meatus
Hard palate
Upper lip
Incisive canal

Spheno-ethmoidal recess
Probe in the opening
of sphenoidal sinus
Diaphragma sellae
Hypophysial fossa
Superior nasal concha
Sphenoidal sinus
Body of sphenoidal bone
Nasopharyngeal meatus
Pharyngeal opening of pharyngotympanic tube
Nasopharynx
Soft palate
Oral cavity proper

b
Frontal sinus
Nasal bone
Superior nasal meatus
(arrow)
Lacrimal bone
Maxilla
Uncinate process
Middle nasal meatus
(arrow)
Inferior nasal meatus
(arrow)
Incisive canal
Bony palate,
Palatine process of maxilla

Crista galli
Cribriform plate
Superior nasal concha
Probe in the opening
of sphenoidal sinus
Middle nasal concha
Sphenoidal sinus
Sphenopalatine foramen
Maxillary hiatus
Perpendicular plate
of palatine bone
Inferior nasal concha
Pterygoid process

Frontal bone   Lacrimal bone   Nasal bone   Maxilla

Sphenoidal bone   Ethmoidal bone   Palatine bone   Inferior nasal concha

177 Lateral wall of the nasal cavity (75%)
Paramedian section to the right of the nasal septum, medial aspect
a Nasal vestibule and mucosa of nasal cavity
b Bony lateral wall of the nasal cavity.
The individual bones are indicated by different colors.

a

Superior oblique muscle
Levator palpebrae superioris muscle
Superior rectus muscle
Medial rectus muscle
Periorbita
Temporalis muscle
Lateral rectus muscle
Inferior rectus muscle
Temporal fascia
Ethmoidal cells
Zygomatic arch
Maxillary sinus
Middle nasal concha
Bony nasal septum
Inferior nasal concha
Masseter muscle, Buccinator muscle
Mucous membrane of palate
Tongue

Longitudinal cerebral fissure
Roof of orbit
Frontal nerve
Falx cerebri
Retrobulbar fat
Lacrimal nerve
Lateral wall of orbit
Ophthalmic artery, Superior ophthalmic vein
Optic nerve [II] with central retinal artery
Inferior ophthalmic vein
Medial wall of orbit
Infra-orbital nerve and artery
Floor of orbit
Ramus of mandible
Middle nasal meatus
Parotid gland
Inferior nasal meatus
Oral cavity proper

b

Ethmoidal labyrinth
Orbital surface of frontal bone
Temporal surface of greater wing of sphenoidal bone
Orbital surface of lesser wing of sphenoidal bone
Superior orbital fissure
Orbital surface of greater wing of sphenoidal bone
Lateral surface
Inferior orbital fissure
Infra-orbital groove
Superior nasal concha
Middle nasal concha
Inferior nasal concha
Alveolar process of maxilla

Crista galli
Frontal sinus
Zygomatic process
Optic canal
Perpendicular plate
Orbital surface
Orbital surface of maxilla
⟨Maxillary process⟩
Maxillary sinus
Left choana, Common nasal meatus
Vomer
Nasal surface of horizontal plate of palatine bone
Molar tooth

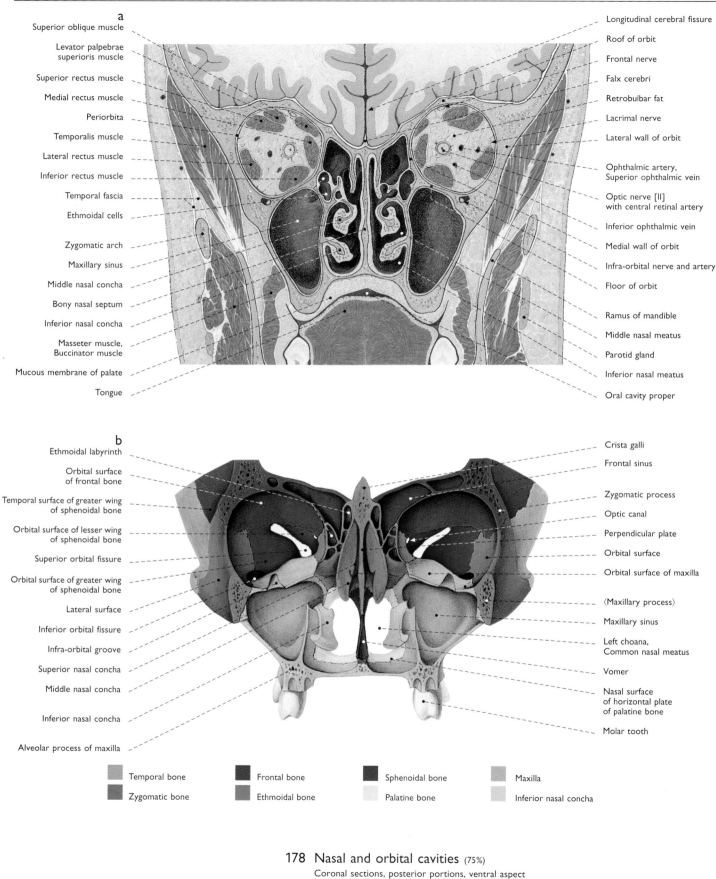

Temporal bone   Frontal bone   Sphenoidal bone   Maxilla
Zygomatic bone   Ethmoidal bone   Palatine bone   Inferior nasal concha

178 Nasal and orbital cavities (75%)

Coronal sections, posterior portions, ventral aspect
a Section of the head of an adult male behind the crista galli
b Section of the skull (= cranium) passing the crista galli.
   The individual bones are indicated by different colors.

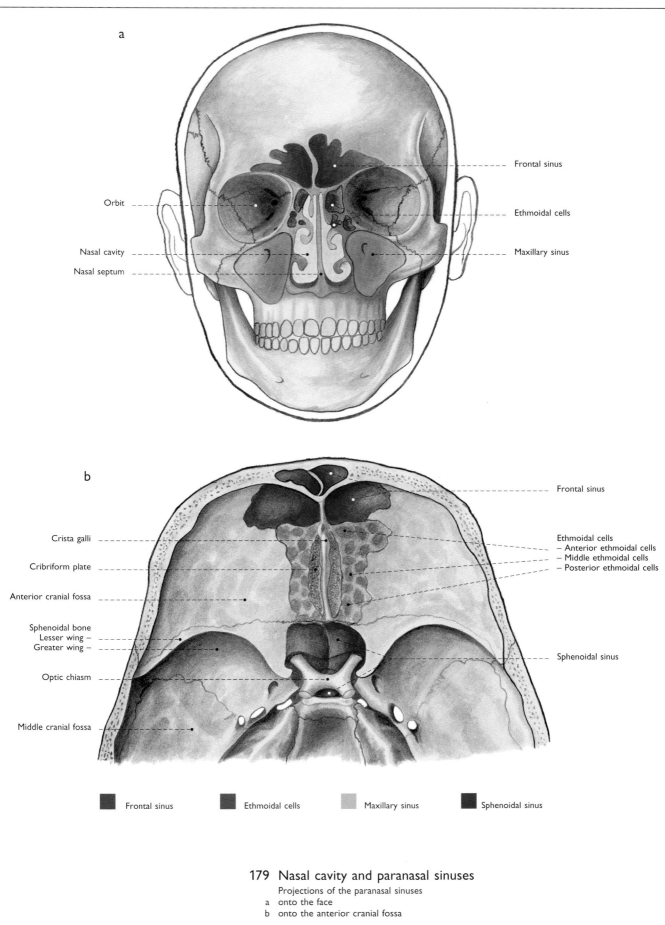

a

Orbit

Nasal cavity

Nasal septum

Frontal sinus

Ethmoidal cells

Maxillary sinus

b

Crista galli

Cribriform plate

Anterior cranial fossa

Sphenoidal bone
Lesser wing –
Greater wing –

Optic chiasm

Middle cranial fossa

Frontal sinus

Ethmoidal cells
– Anterior ethmoidal cells
– Middle ethmoidal cells
– Posterior ethmoidal cells

Sphenoidal sinus

Frontal sinus   Ethmoidal cells   Maxillary sinus   Sphenoidal sinus

179 Nasal cavity and paranasal sinuses

Projections of the paranasal sinuses
a onto the face
b onto the anterior cranial fossa

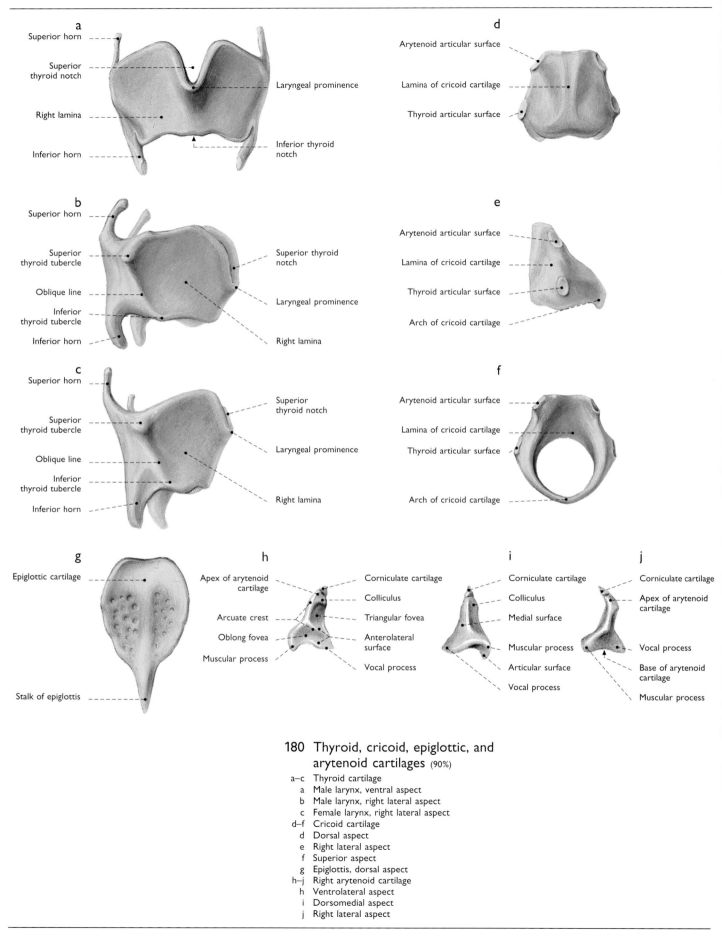

a
Superior horn
Superior thyroid notch
Right lamina
Inferior horn
Laryngeal prominence
Inferior thyroid notch

d
Arytenoid articular surface
Lamina of cricoid cartilage
Thyroid articular surface

b
Superior horn
Superior thyroid tubercle
Oblique line
Inferior thyroid tubercle
Inferior horn
Superior thyroid notch
Laryngeal prominence
Right lamina

e
Arytenoid articular surface
Lamina of cricoid cartilage
Thyroid articular surface
Arch of cricoid cartilage

c
Superior horn
Superior thyroid tubercle
Oblique line
Inferior thyroid tubercle
Inferior horn
Superior thyroid notch
Laryngeal prominence
Right lamina

f
Arytenoid articular surface
Lamina of cricoid cartilage
Thyroid articular surface
Arch of cricoid cartilage

g
Epiglottic cartilage
Stalk of epiglottis

h
Apex of arytenoid cartilage
Arcuate crest
Oblong fovea
Muscular process
Corniculate cartilage
Colliculus
Triangular fovea
Anterolateral surface
Vocal process

i
Corniculate cartilage
Colliculus
Medial surface
Muscular process
Articular surface
Vocal process

j
Corniculate cartilage
Apex of arytenoid cartilage
Vocal process
Base of arytenoid cartilage
Muscular process

180 Thyroid, cricoid, epiglottic, and arytenoid cartilages (90%)

a–c Thyroid cartilage
  a Male larynx, ventral aspect
  b Male larynx, right lateral aspect
  c Female larynx, right lateral aspect
d–f Cricoid cartilage
  d Dorsal aspect
  e Right lateral aspect
  f Superior aspect
  g Epiglottis, dorsal aspect
h–j Right arytenoid cartilage
  h Ventrolateral aspect
  i Dorsomedial aspect
  j Right lateral aspect

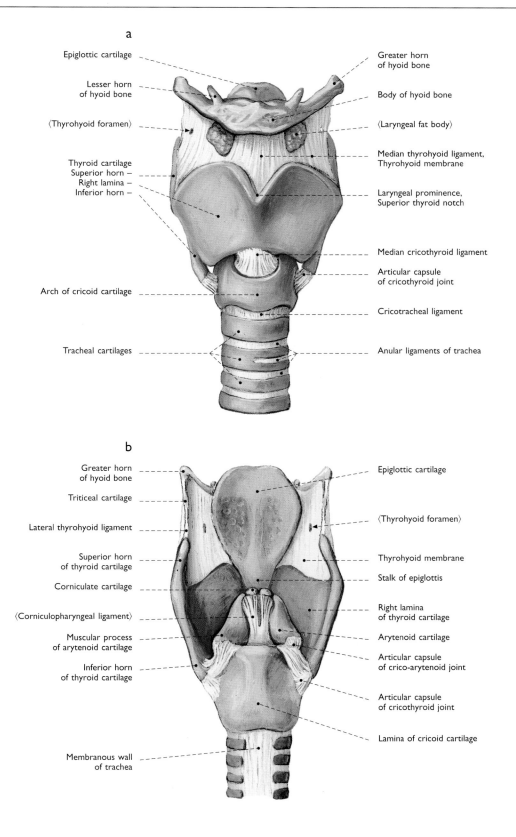

a

Epiglottic cartilage

Lesser horn
of hyoid bone

〈Thyrohyoid foramen〉

Thyroid cartilage
Superior horn –
Right lamina –
Inferior horn –

Arch of cricoid cartilage

Tracheal cartilages

Greater horn
of hyoid bone

Body of hyoid bone

〈Laryngeal fat body〉

Median thyrohyoid ligament,
Thyrohyoid membrane

Laryngeal prominence,
Superior thyroid notch

Median cricothyroid ligament

Articular capsule
of cricothyroid joint

Cricotracheal ligament

Anular ligaments of trachea

b

Greater horn
of hyoid bone

Triticeal cartilage

Lateral thyrohyoid ligament

Superior horn
of thyroid cartilage

Corniculate cartilage

〈Corniculopharyngeal ligament〉

Muscular process
of arytenoid cartilage

Inferior horn
of thyroid cartilage

Membranous wall
of trachea

Epiglottic cartilage

〈Thyrohyoid foramen〉

Thyrohyoid membrane

Stalk of epiglottis

Right lamina
of thyroid cartilage

Arytenoid cartilage

Articular capsule
of crico-arytenoid joint

Articular capsule
of cricothyroid joint

Lamina of cricoid cartilage

181 Laryngeal cartilages and joints (100%)
  a   Ventral aspect
  b   Dorsal aspect

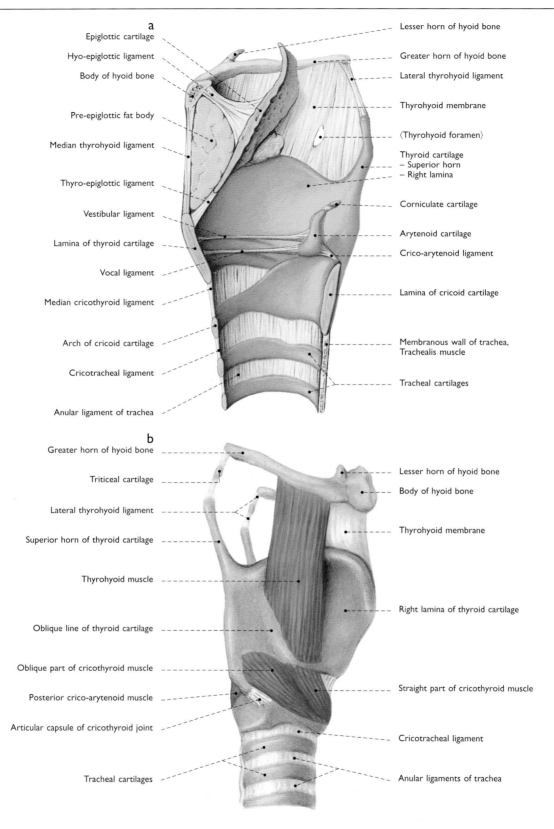

a

Epiglottic cartilage

Hyo-epiglottic ligament

Body of hyoid bone

Pre-epiglottic fat body

Median thyrohyoid ligament

Thyro-epiglottic ligament

Vestibular ligament

Lamina of thyroid cartilage

Vocal ligament

Median cricothyroid ligament

Arch of cricoid cartilage

Cricotracheal ligament

Anular ligament of trachea

Lesser horn of hyoid bone

Greater horn of hyoid bone

Lateral thyrohyoid ligament

Thyrohyoid membrane

⟨Thyrohyoid foramen⟩

Thyroid cartilage
– Superior horn
– Right lamina

Corniculate cartilage

Arytenoid cartilage

Crico-arytenoid ligament

Lamina of cricoid cartilage

Membranous wall of trachea,
Trachealis muscle

Tracheal cartilages

b

Greater horn of hyoid bone

Triticeal cartilage

Lateral thyrohyoid ligament

Superior horn of thyroid cartilage

Thyrohyoid muscle

Oblique line of thyroid cartilage

Oblique part of cricothyroid muscle

Posterior crico-arytenoid muscle

Articular capsule of cricothyroid joint

Tracheal cartilages

Lesser horn of hyoid bone

Body of hyoid bone

Thyrohyoid membrane

Right lamina of thyroid cartilage

Straight part of cricothyroid muscle

Cricotracheal ligament

Anular ligaments of trachea

182 Ligaments and muscles of the larynx (100%)
  a Median section through the hyoid bone and the skeleton
     of the larynx, medial aspect
  b Thyrohyoid and cricothyroid muscles, right lateral aspect

a

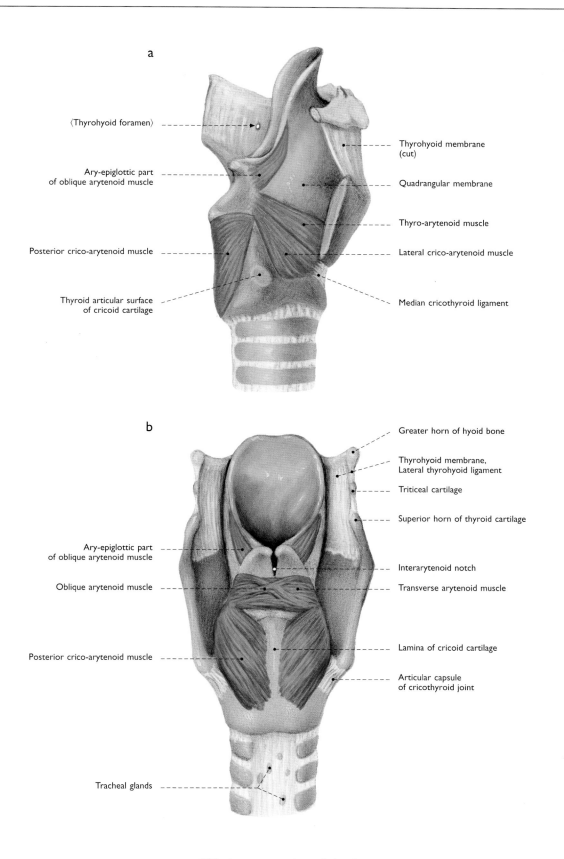

(Thyrohyoid foramen)

Ary-epiglottic part
of oblique arytenoid muscle

Posterior crico-arytenoid muscle

Thyroid articular surface
of cricoid cartilage

Thyrohyoid membrane
(cut)

Quadrangular membrane

Thyro-arytenoid muscle

Lateral crico-arytenoid muscle

Median cricothyroid ligament

b

Ary-epiglottic part
of oblique arytenoid muscle

Oblique arytenoid muscle

Posterior crico-arytenoid muscle

Tracheal glands

Greater horn of hyoid bone

Thyrohyoid membrane,
Lateral thyrohyoid ligament

Triticeal cartilage

Superior horn of thyroid cartilage

Interarytenoid notch

Transverse arytenoid muscle

Lamina of cricoid cartilage

Articular capsule
of cricothyroid joint

183 Inner muscles of the larynx (100%)
a The right lamina of the thyroid cartilage was partially removed,
the oblique and transverse arytenoid muscles are omitted.
Right lateral aspect
b The pharyngeal mucosa was completely removed. Dorsal aspect

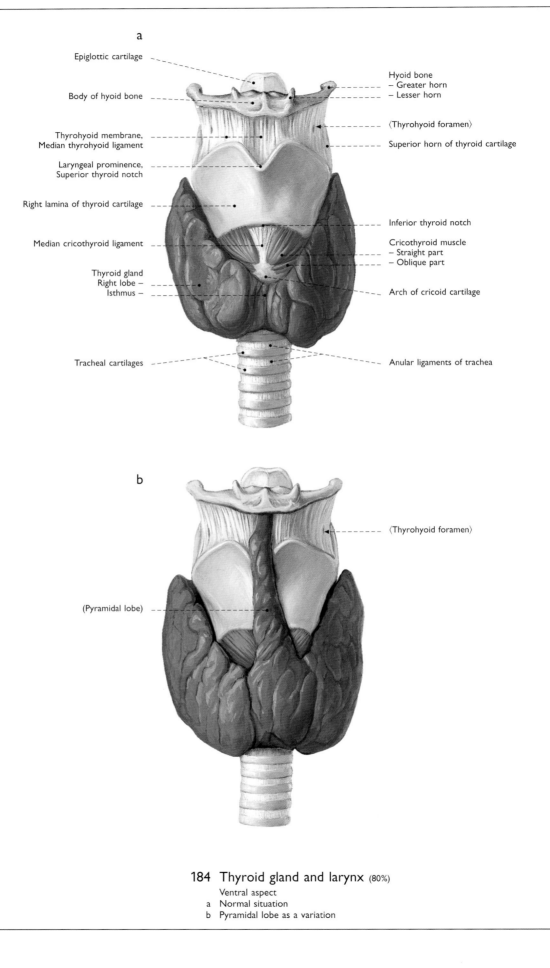

a

Epiglottic cartilage

Body of hyoid bone

Hyoid bone
– Greater horn
– Lesser horn

Thyrohyoid membrane,
Median thyrohyoid ligament

⟨Thyrohyoid foramen⟩

Laryngeal prominence,
Superior thyroid notch

Superior horn of thyroid cartilage

Right lamina of thyroid cartilage

Median cricothyroid ligament

Inferior thyroid notch

Thyroid gland
Right lobe –
Isthmus –

Cricothyroid muscle
– Straight part
– Oblique part

Tracheal cartilages

Arch of cricoid cartilage

Anular ligaments of trachea

b

⟨Thyrohyoid foramen⟩

⟨Pyramidal lobe⟩

184 Thyroid gland and larynx (80%)

Ventral aspect
a Normal situation
b Pyramidal lobe as a variation

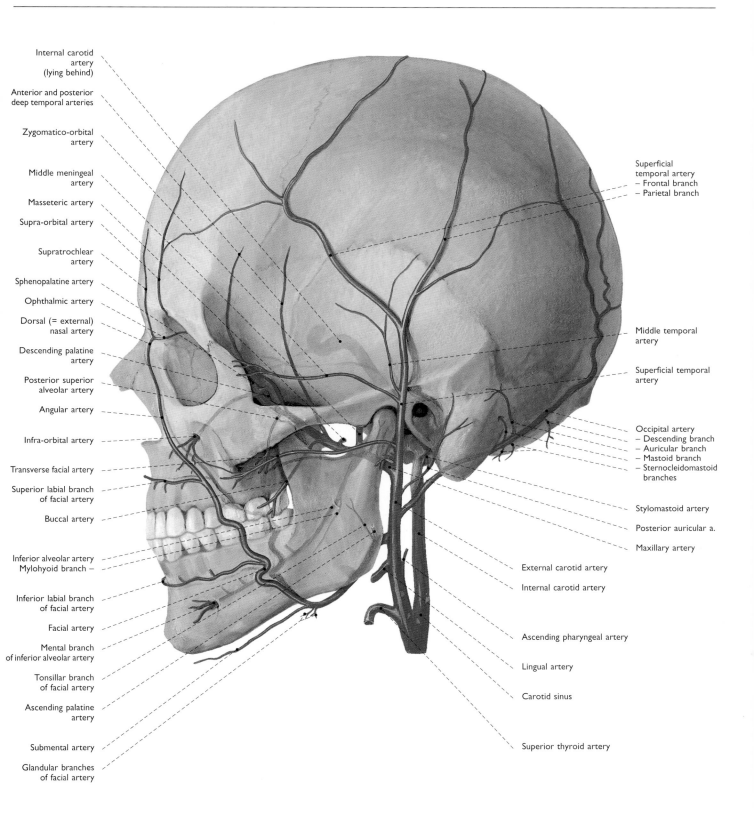

Internal carotid artery (lying behind)

Anterior and posterior deep temporal arteries

Zygomatico-orbital artery

Middle meningeal artery

Masseteric artery

Supra-orbital artery

Supratrochlear artery

Sphenopalatine artery

Ophthalmic artery

Dorsal (= external) nasal artery

Descending palatine artery

Posterior superior alveolar artery

Angular artery

Infra-orbital artery

Transverse facial artery

Superior labial branch of facial artery

Buccal artery

Inferior alveolar artery
Mylohyoid branch –

Inferior labial branch of facial artery

Facial artery

Mental branch of inferior alveolar artery

Tonsillar branch of facial artery

Ascending palatine artery

Submental artery

Glandular branches of facial artery

Superficial temporal artery
– Frontal branch
– Parietal branch

Middle temporal artery

Superficial temporal artery

Occipital artery
– Descending branch
– Auricular branch
– Mastoid branch
– Sternocleidomastoid branches

Stylomastoid artery

Posterior auricular a.

Maxillary artery

External carotid artery

Internal carotid artery

Ascending pharyngeal artery

Lingual artery

Carotid sinus

Superior thyroid artery

185 Arteries of the skull (75%)
Schematic representation, left lateral aspect

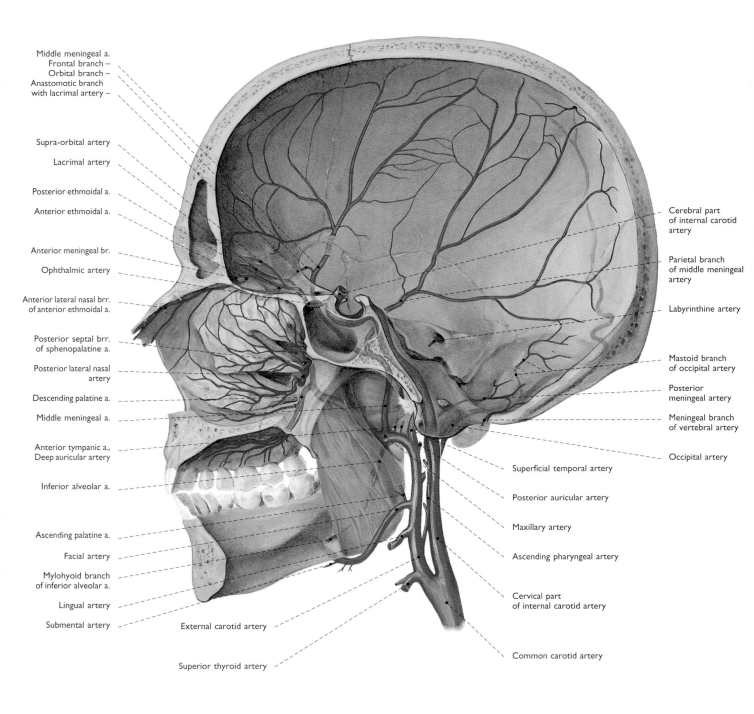

Middle meningeal a.
Frontal branch –
Orbital branch –
Anastomotic branch
with lacrimal artery –

Supra-orbital artery

Lacrimal artery

Posterior ethmoidal a.

Anterior ethmoidal a.

Anterior meningeal br.

Ophthalmic artery

Anterior lateral nasal brr.
of anterior ethmoidal a.

Posterior septal brr.
of sphenopalatine a.

Posterior lateral nasal
artery

Descending palatine a.

Middle meningeal a.

Anterior tympanic a.,
Deep auricular artery

Inferior alveolar a.

Ascending palatine a.

Facial artery

Mylohyoid branch
of inferior alveolar a.

Lingual artery

Submental artery

External carotid artery

Superior thyroid artery

Cerebral part
of internal carotid
artery

Parietal branch
of middle meningeal
artery

Labyrinthine artery

Mastoid branch
of occipital artery

Posterior
meningeal artery

Meningeal branch
of vertebral artery

Occipital artery

Superficial temporal artery

Posterior auricular artery

Maxillary artery

Ascending pharyngeal artery

Cervical part
of internal carotid artery

Common carotid artery

186 Arteries of the skull (75%)
Schematic representation, median section,
medial aspect of the right half of the skull

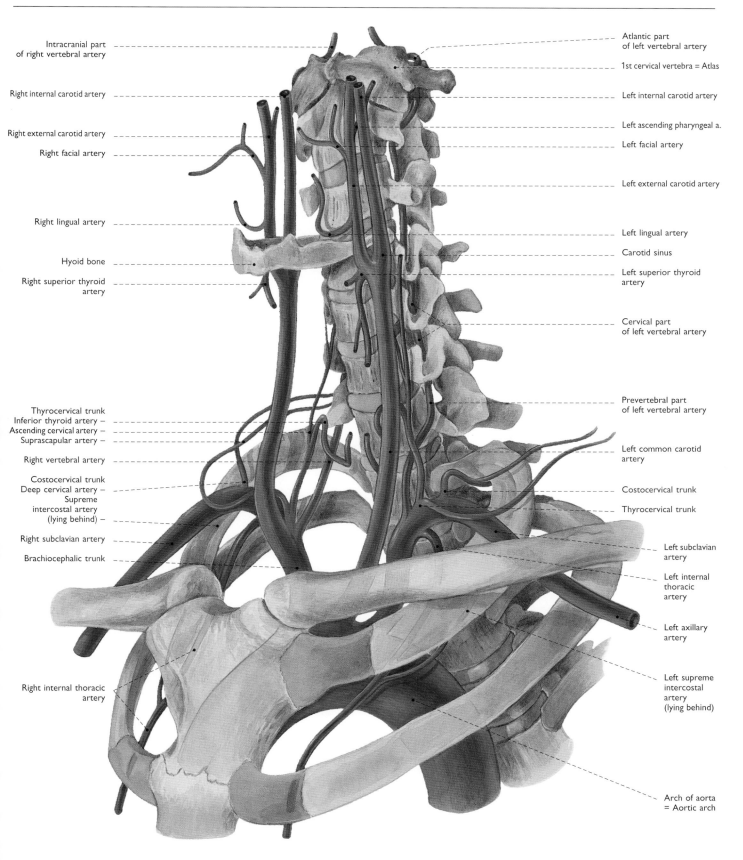

Intracranial part
of right vertebral artery

Right internal carotid artery

Right external carotid artery

Right facial artery

Right lingual artery

Hyoid bone

Right superior thyroid
artery

Thyrocervical trunk
Inferior thyroid artery –
Ascending cervical artery –
Suprascapular artery –

Right vertebral artery

Costocervical trunk
Deep cervical artery –
Supreme
intercostal artery
(lying behind) –

Right subclavian artery

Brachiocephalic trunk

Right internal thoracic
artery

Atlantic part
of left vertebral artery

1st cervical vertebra = Atlas

Left internal carotid artery

Left ascending pharyngeal a.

Left facial artery

Left external carotid artery

Left lingual artery

Carotid sinus

Left superior thyroid
artery

Cervical part
of left vertebral artery

Prevertebral part
of left vertebral artery

Left common carotid
artery

Costocervical trunk

Thyrocervical trunk

Left subclavian
artery

Left internal
thoracic
artery

Left axillary
artery

Left supreme
intercostal
artery
(lying behind)

Arch of aorta
= Aortic arch

187 Arteries of the neck and head (90%)
Left anterior oblique view (LAO projection)

a

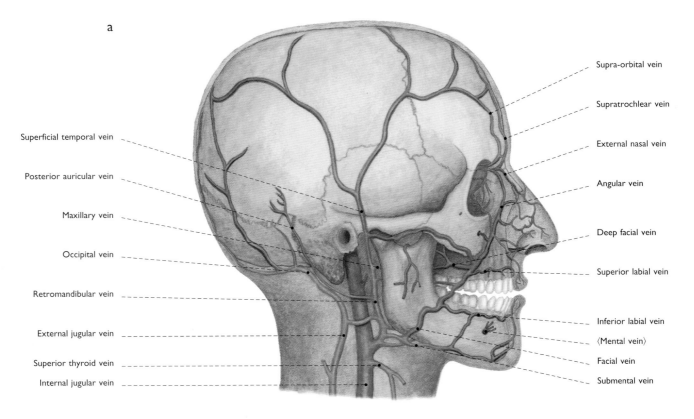

Superficial temporal vein

Posterior auricular vein

Maxillary vein

Occipital vein

Retromandibular vein

External jugular vein

Superior thyroid vein

Internal jugular vein

Supra-orbital vein

Supratrochlear vein

External nasal vein

Angular vein

Deep facial vein

Superior labial vein

Inferior labial vein

⟨Mental vein⟩

Facial vein

Submental vein

b

Superior ophthalmic vein

Inferior ophthalmic vein

Cavernous sinus

Superior petrosal sinus

Middle meningeal veins

Lateral pterygoid muscle
(cut surface)

Superficial temporal vein

Maxillary vein

Medial pterygoid muscle

⟨Inferior alveolar vein⟩

⟨Buccal vein⟩

Internal jugular vein

Retromandibular vein

Supratrochlear vein

Nasofrontal vein

Lacrimal gland

Angular vein

External nasal veins

⟨Infra-orbital vein⟩

Maxillary sinus

Pterygoid plexus

Deep facial vein

Superior labial vein

Buccinator muscle

Facial vein

Inferior labial veins

⟨Mental vein⟩

Submental vein

188 Veins of the head
Right lateral aspect
a Superficial veins (50%)
b Deep veins (65%)

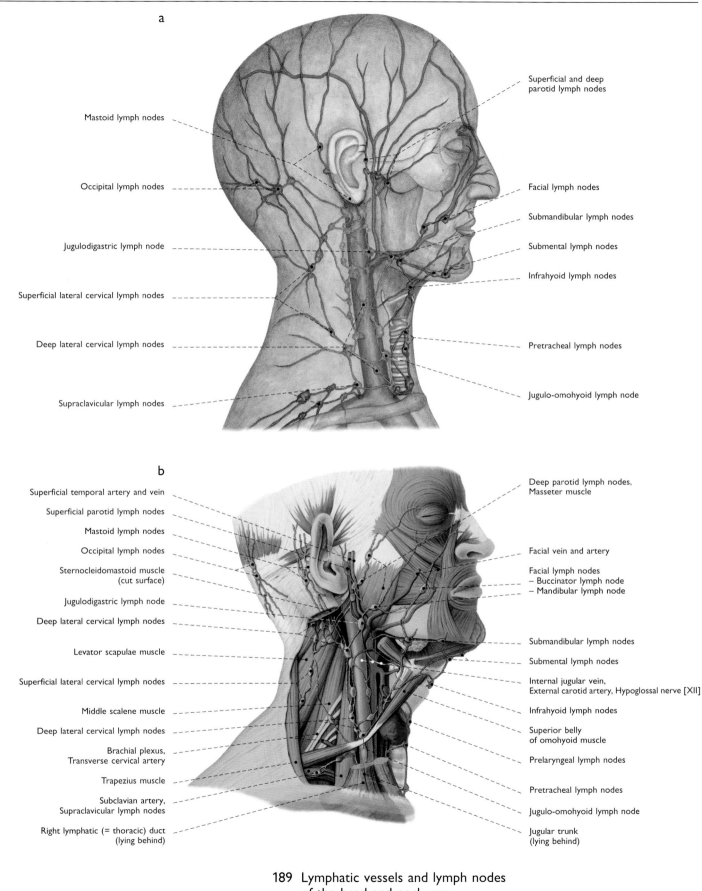

a

Mastoid lymph nodes

Occipital lymph nodes

Jugulodigastric lymph node

Superficial lateral cervical lymph nodes

Deep lateral cervical lymph nodes

Supraclavicular lymph nodes

Superficial and deep
parotid lymph nodes

Facial lymph nodes

Submandibular lymph nodes

Submental lymph nodes

Infrahyoid lymph nodes

Pretracheal lymph nodes

Jugulo-omohyoid lymph node

b

Superficial temporal artery and vein

Superficial parotid lymph nodes

Mastoid lymph nodes

Occipital lymph nodes

Sternocleidomastoid muscle
(cut surface)

Jugulodigastric lymph node

Deep lateral cervical lymph nodes

Levator scapulae muscle

Superficial lateral cervical lymph nodes

Middle scalene muscle

Deep lateral cervical lymph nodes

Brachial plexus,
Transverse cervical artery

Trapezius muscle

Subclavian artery,
Supraclavicular lymph nodes

Right lymphatic (= thoracic) duct
(lying behind)

Deep parotid lymph nodes,
Masseter muscle

Facial vein and artery

Facial lymph nodes
– Buccinator lymph node
– Mandibular lymph node

Submandibular lymph nodes

Submental lymph nodes

Internal jugular vein,
External carotid artery, Hypoglossal nerve [XII]

Infrahyoid lymph nodes

Superior belly
of omohyoid muscle

Prelaryngeal lymph nodes

Pretracheal lymph nodes

Jugulo-omohyoid lymph node

Jugular trunk
(lying behind)

**189 Lymphatic vessels and lymph nodes
of the head and neck** (40%)
Schematic representations, right lateral aspect

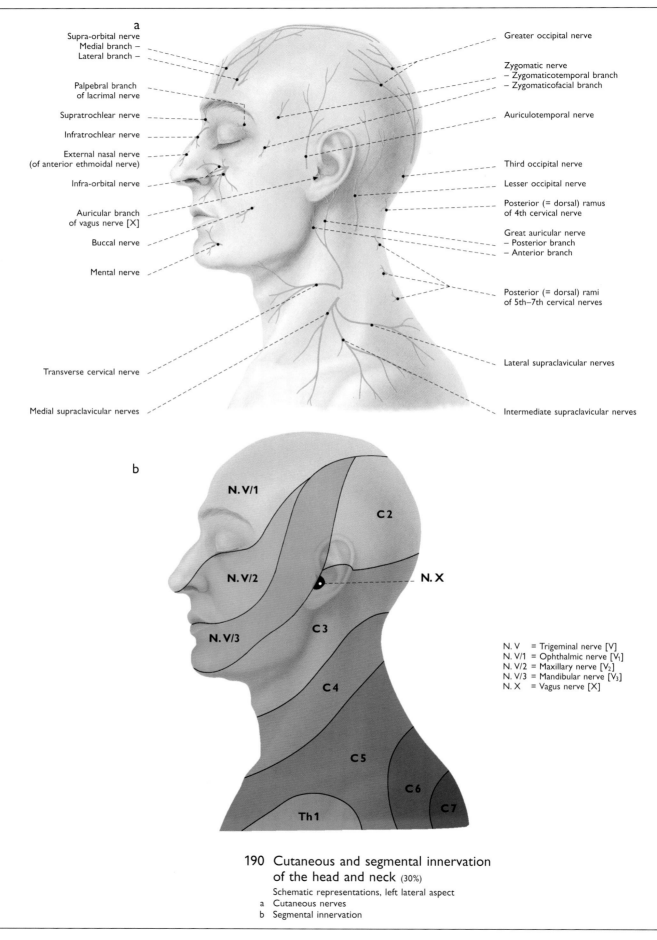

a
Supra-orbital nerve
Medial branch –
Lateral branch –

Palpebral branch
of lacrimal nerve

Supratrochlear nerve

Infratrochlear nerve

External nasal nerve
(of anterior ethmoidal nerve)

Infra-orbital nerve

Auricular branch
of vagus nerve [X]

Buccal nerve

Mental nerve

Transverse cervical nerve

Medial supraclavicular nerves

Greater occipital nerve

Zygomatic nerve
– Zygomaticotemporal branch
– Zygomaticofacial branch

Auriculotemporal nerve

Third occipital nerve

Lesser occipital nerve

Posterior (= dorsal) ramus
of 4th cervical nerve

Great auricular nerve
– Posterior branch
– Anterior branch

Posterior (= dorsal) rami
of 5th–7th cervical nerves

Lateral supraclavicular nerves

Intermediate supraclavicular nerves

b

N. V/1

C 2

N. V/2

N. X

N. V/3

C 3

C 4

C 5

C 6

C 7

Th 1

N. V   = Trigeminal nerve [V]
N. V/1 = Ophthalmic nerve [V₁]
N. V/2 = Maxillary nerve [V₂]
N. V/3 = Mandibular nerve [V₃]
N. X   = Vagus nerve [X]

**190 Cutaneous and segmental innervation
of the head and neck** (30%)

Schematic representations, left lateral aspect
a Cutaneous nerves
b Segmental innervation

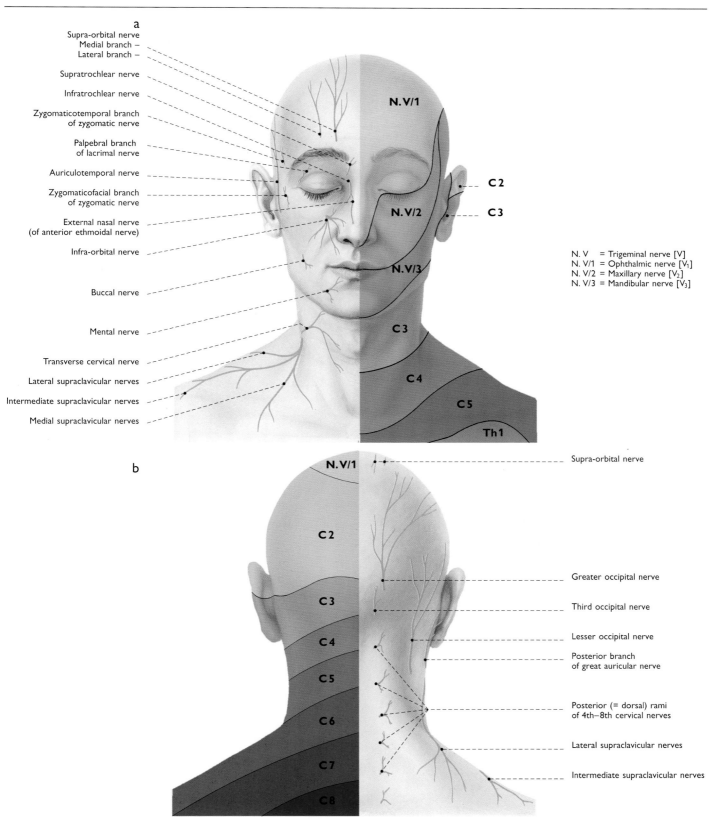

a
Supra-orbital nerve
Medial branch –
Lateral branch –

Supratrochlear nerve

Infratrochlear nerve

Zygomaticotemporal branch
of zygomatic nerve

Palpebral branch
of lacrimal nerve

Auriculotemporal nerve

Zygomaticofacial branch
of zygomatic nerve

External nasal nerve
(of anterior ethmoidal nerve)

Infra-orbital nerve

Buccal nerve

Mental nerve

Transverse cervical nerve

Lateral supraclavicular nerves

Intermediate supraclavicular nerves

Medial supraclavicular nerves

N. V/1

C 2

C 3

N. V/2

N. V/3

C 3

C 4

C 5

Th 1

N. V   = Trigeminal nerve [V]
N. V/1 = Ophthalmic nerve [V₁]
N. V/2 = Maxillary nerve [V₂]
N. V/3 = Mandibular nerve [V₃]

b

N. V/1

C 2

C 3

C 4

C 5

C 6

C 7

C 8

Supra-orbital nerve

Greater occipital nerve

Third occipital nerve

Lesser occipital nerve

Posterior branch
of great auricular nerve

Posterior (= dorsal) rami
of 4th–8th cervical nerves

Lateral supraclavicular nerves

Intermediate supraclavicular nerves

191 Cutaneous and segmental innervation
of the head and neck (30%)
Schematic representations
a Anterior aspect
b Posterior aspect

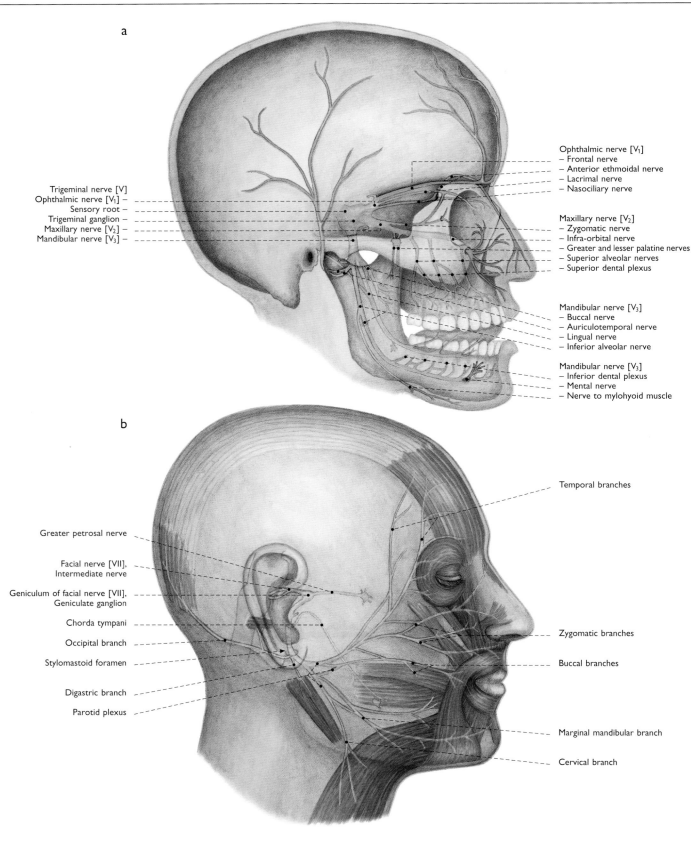

a

Ophthalmic nerve [V₁]
– Frontal nerve
– Anterior ethmoidal nerve
– Lacrimal nerve
– Nasociliary nerve

Trigeminal nerve [V]
Ophthalmic nerve [V₁] –
Sensory root –
Trigeminal ganglion –
Maxillary nerve [V₂] –
Mandibular nerve [V₃] –

Maxillary nerve [V₂]
– Zygomatic nerve
– Infra-orbital nerve
– Greater and lesser palatine nerves
– Superior alveolar nerves
– Superior dental plexus

Mandibular nerve [V₃]
– Buccal nerve
– Auriculotemporal nerve
– Lingual nerve
– Inferior alveolar nerve

Mandibular nerve [V₃]
– Inferior dental plexus
– Mental nerve
– Nerve to mylohyoid muscle

b

Greater petrosal nerve

Facial nerve [VII],
Intermediate nerve

Geniculum of facial nerve [VII],
Geniculate ganglion

Chorda tympani

Occipital branch

Stylomastoid foramen

Digastric branch

Parotid plexus

Temporal branches

Zygomatic branches

Buccal branches

Marginal mandibular branch

Cervical branch

192 Nerves of the head (50%)
Schematic representations, right lateral aspect
a Ramification of the trigeminal nerve (N. V) in the deep facial region
b Ramification of the facial nerve (N. VII) in the superficial facial region

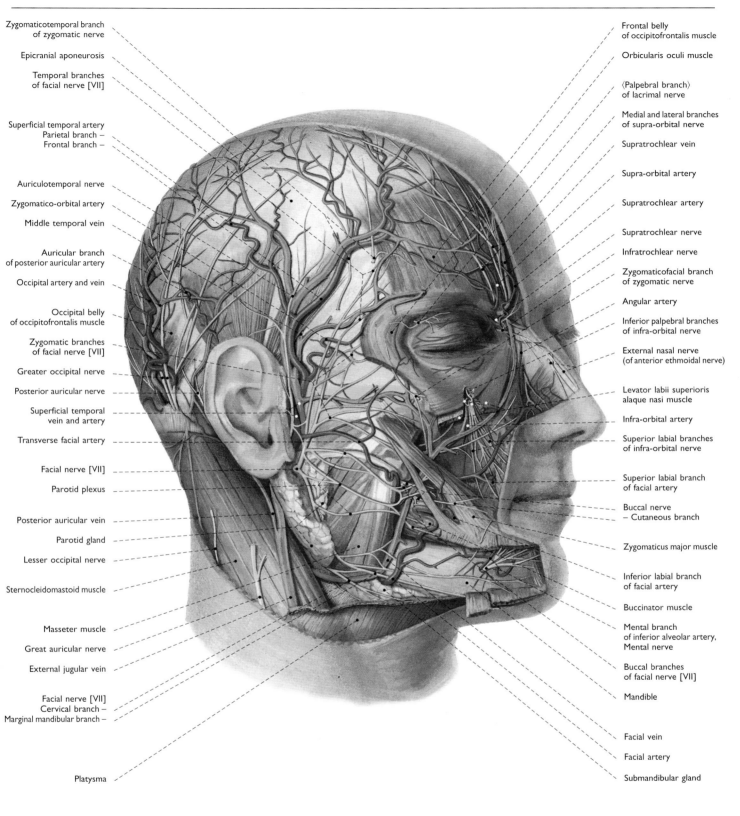

Zygomaticotemporal branch
of zygomatic nerve

Epicranial aponeurosis

Temporal branches
of facial nerve [VII]

Superficial temporal artery
Parietal branch –
Frontal branch –

Auriculotemporal nerve

Zygomatico-orbital artery

Middle temporal vein

Auricular branch
of posterior auricular artery

Occipital artery and vein

Occipital belly
of occipitofrontalis muscle

Zygomatic branches
of facial nerve [VII]

Greater occipital nerve

Posterior auricular nerve

Superficial temporal
vein and artery

Transverse facial artery

Facial nerve [VII]

Parotid plexus

Posterior auricular vein

Parotid gland

Lesser occipital nerve

Sternocleidomastoid muscle

Masseter muscle

Great auricular nerve

External jugular vein

Facial nerve [VII]
Cervical branch –
Marginal mandibular branch –

Platysma

Frontal belly
of occipitofrontalis muscle

Orbicularis oculi muscle

⟨Palpebral branch⟩
of lacrimal nerve

Medial and lateral branches
of supra-orbital nerve

Supratrochlear vein

Supra-orbital artery

Supratrochlear artery

Supratrochlear nerve

Infratrochlear nerve

Zygomaticofacial branch
of zygomatic nerve

Angular artery

Inferior palpebral branches
of infra-orbital nerve

External nasal nerve
(of anterior ethmoidal nerve)

Levator labii superioris
alaque nasi muscle

Infra-orbital artery

Superior labial branches
of infra-orbital nerve

Superior labial branch
of facial artery

Buccal nerve
– Cutaneous branch

Zygomaticus major muscle

Inferior labial branch
of facial artery

Buccinator muscle

Mental branch
of inferior alveolar artery,
Mental nerve

Buccal branches
of facial nerve [VII]

Mandible

Facial vein

Facial artery

Submandibular gland

193 Superficial blood vessels and nerves
of the head (60%)
Right lateral aspect

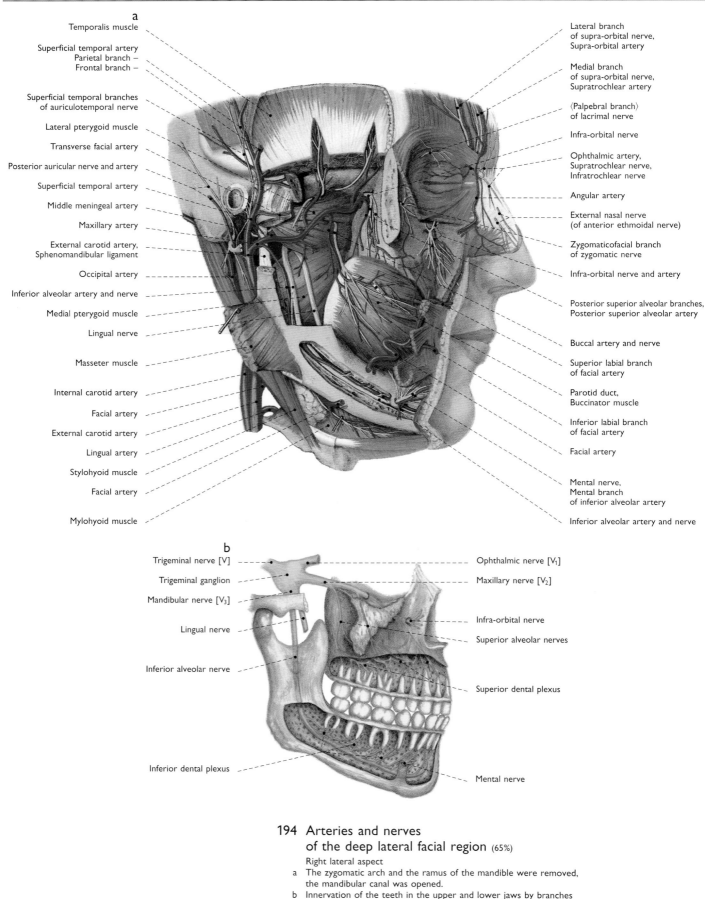

a

Temporalis muscle

Superficial temporal artery
Parietal branch –
Frontal branch –

Superficial temporal branches
of auriculotemporal nerve

Lateral pterygoid muscle

Transverse facial artery

Posterior auricular nerve and artery

Superficial temporal artery

Middle meningeal artery

Maxillary artery

External carotid artery,
Sphenomandibular ligament

Occipital artery

Inferior alveolar artery and nerve

Medial pterygoid muscle

Lingual nerve

Masseter muscle

Internal carotid artery

Facial artery

External carotid artery

Lingual artery

Stylohyoid muscle

Facial artery

Mylohyoid muscle

Lateral branch
of supra-orbital nerve,
Supra-orbital artery

Medial branch
of supra-orbital nerve,
Supratrochlear artery

⟨Palpebral branch⟩
of lacrimal nerve

Infra-orbital nerve

Ophthalmic artery,
Supratrochlear nerve,
Infratrochlear nerve

Angular artery

External nasal nerve
(of anterior ethmoidal nerve)

Zygomaticofacial branch
of zygomatic nerve

Infra-orbital nerve and artery

Posterior superior alveolar branches,
Posterior superior alveolar artery

Buccal artery and nerve

Superior labial branch
of facial artery

Parotid duct,
Buccinator muscle

Inferior labial branch
of facial artery

Facial artery

Mental nerve,
Mental branch
of inferior alveolar artery

Inferior alveolar artery and nerve

b

Trigeminal nerve [V]

Trigeminal ganglion

Mandibular nerve [V₃]

Lingual nerve

Inferior alveolar nerve

Inferior dental plexus

Ophthalmic nerve [V₁]

Maxillary nerve [V₂]

Infra-orbital nerve

Superior alveolar nerves

Superior dental plexus

Mental nerve

**194 Arteries and nerves
of the deep lateral facial region** (65%)

Right lateral aspect
a The zygomatic arch and the ramus of the mandible were removed,
the mandibular canal was opened.
b Innervation of the teeth in the upper and lower jaws by branches
of the trigeminal nerve (N. V), schematic representation

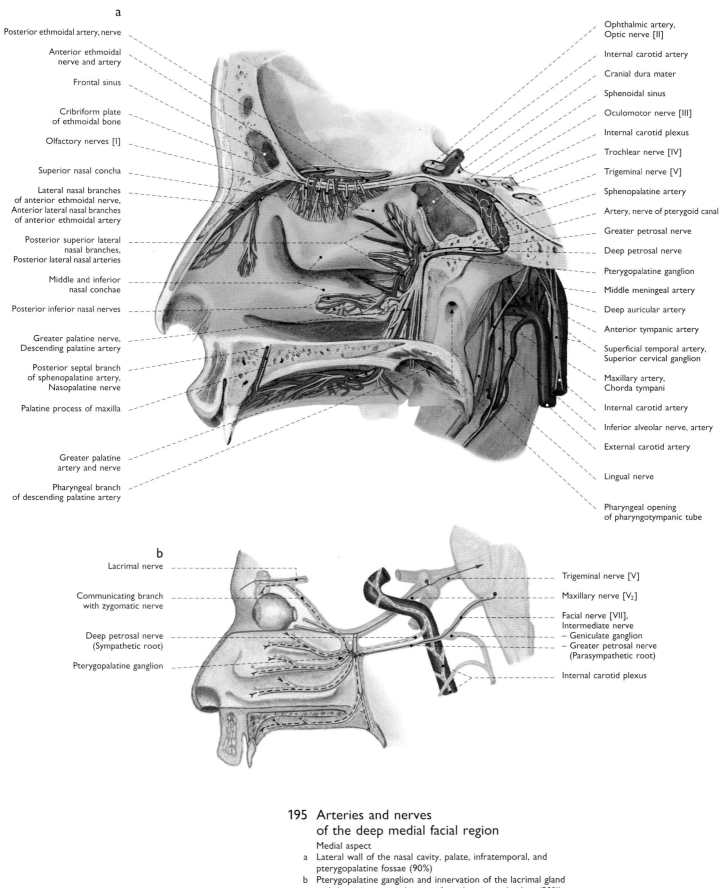

a

Posterior ethmoidal artery, nerve

Anterior ethmoidal nerve and artery

Frontal sinus

Cribriform plate of ethmoidal bone

Olfactory nerves [I]

Superior nasal concha

Lateral nasal branches of anterior ethmoidal nerve, Anterior lateral nasal branches of anterior ethmoidal artery

Posterior superior lateral nasal branches, Posterior lateral nasal arteries

Middle and inferior nasal conchae

Posterior inferior nasal nerves

Greater palatine nerve, Descending palatine artery

Posterior septal branch of sphenopalatine artery, Nasopalatine nerve

Palatine process of maxilla

Greater palatine artery and nerve

Pharyngeal branch of descending palatine artery

Ophthalmic artery, Optic nerve [II]

Internal carotid artery

Cranial dura mater

Sphenoidal sinus

Oculomotor nerve [III]

Internal carotid plexus

Trochlear nerve [IV]

Trigeminal nerve [V]

Sphenopalatine artery

Artery, nerve of pterygoid canal

Greater petrosal nerve

Deep petrosal nerve

Pterygopalatine ganglion

Middle meningeal artery

Deep auricular artery

Anterior tympanic artery

Superficial temporal artery, Superior cervical ganglion

Maxillary artery, Chorda tympani

Internal carotid artery

Inferior alveolar nerve, artery

External carotid artery

Lingual nerve

Pharyngeal opening of pharyngotympanic tube

b

Lacrimal nerve

Communicating branch with zygomatic nerve

Deep petrosal nerve (Sympathetic root)

Pterygopalatine ganglion

Trigeminal nerve [V]

Maxillary nerve [V₂]

Facial nerve [VII], Intermediate nerve
– Geniculate ganglion
– Greater petrosal nerve (Parasympathetic root)

Internal carotid plexus

195 Arteries and nerves
of the deep medial facial region
Medial aspect
a Lateral wall of the nasal cavity, palate, infratemporal, and
   pterygopalatine fossae (90%)
b Pterygopalatine ganglion and innervation of the lacrimal gland
   and the mucous membranes of nasal cavity and palate (50%),
   schematic representation

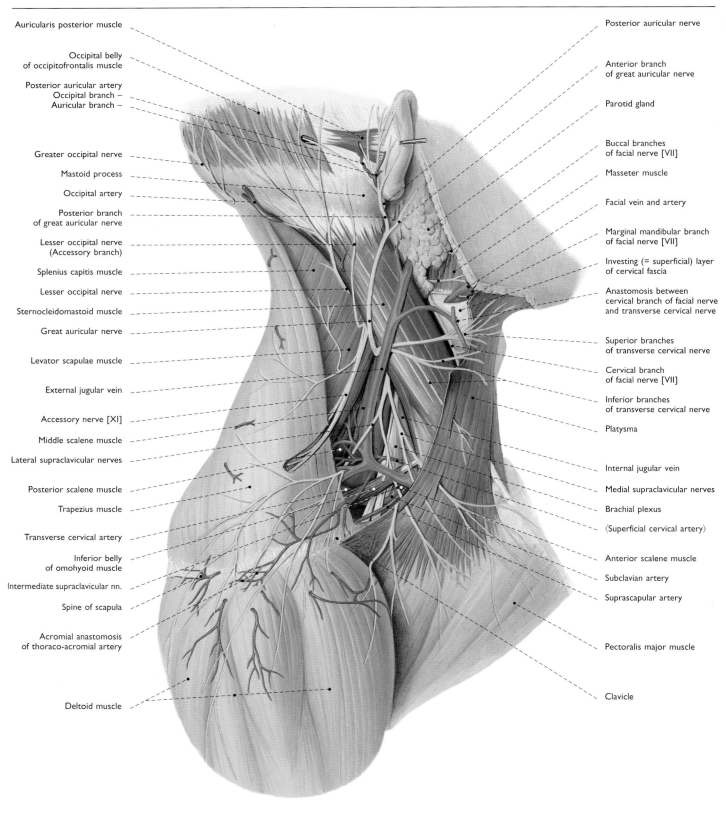

Auricularis posterior muscle

Occipital belly
of occipitofrontalis muscle

Posterior auricular artery
Occipital branch —
Auricular branch —

Greater occipital nerve

Mastoid process

Occipital artery

Posterior branch
of great auricular nerve

Lesser occipital nerve
(Accessory branch)

Splenius capitis muscle

Lesser occipital nerve

Sternocleidomastoid muscle

Great auricular nerve

Levator scapulae muscle

External jugular vein

Accessory nerve [XI]

Middle scalene muscle

Lateral supraclavicular nerves

Posterior scalene muscle

Trapezius muscle

Transverse cervical artery

Inferior belly
of omohyoid muscle

Intermediate supraclavicular nn.

Spine of scapula

Acromial anastomosis
of thoraco-acromial artery

Deltoid muscle

Posterior auricular nerve

Anterior branch
of great auricular nerve

Parotid gland

Buccal branches
of facial nerve [VII]

Masseter muscle

Facial vein and artery

Marginal mandibular branch
of facial nerve [VII]

Investing (= superficial) layer
of cervical fascia

Anastomosis between
cervical branch of facial nerve
and transverse cervical nerve

Superior branches
of transverse cervical nerve

Cervical branch
of facial nerve [VII]

Inferior branches
of transverse cervical nerve

Platysma

Internal jugular vein

Medial supraclavicular nerves

Brachial plexus

⟨Superficial cervical artery⟩

Anterior scalene muscle

Subclavian artery

Suprascapular artery

Pectoralis major muscle

Clavicle

196 Superficial blood vessels and nerves
of the neck region (50%)
Right lateral aspect

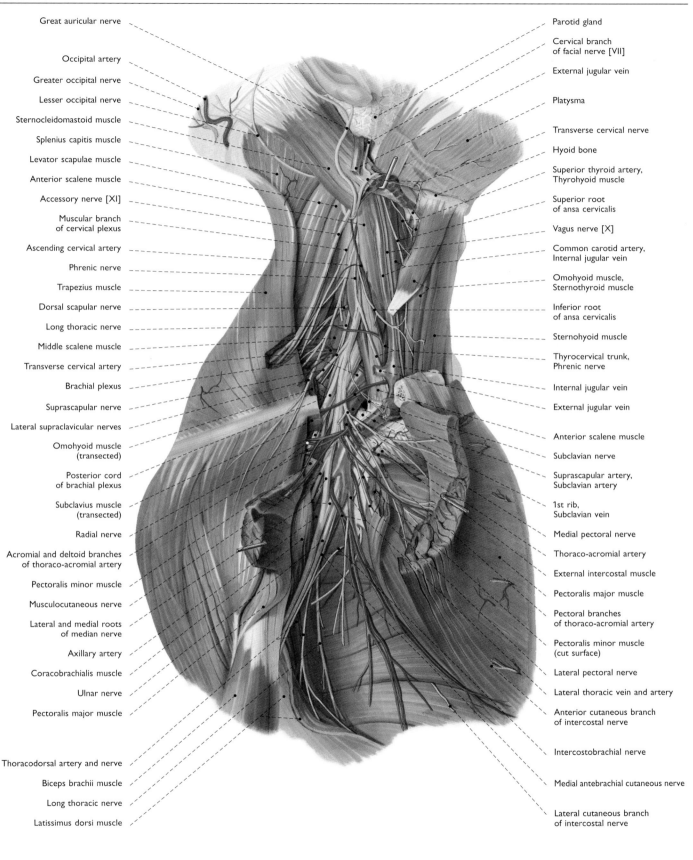

Great auricular nerve

Occipital artery

Greater occipital nerve

Lesser occipital nerve

Sternocleidomastoid muscle

Splenius capitis muscle

Levator scapulae muscle

Anterior scalene muscle

Accessory nerve [XI]

Muscular branch
of cervical plexus

Ascending cervical artery

Phrenic nerve

Trapezius muscle

Dorsal scapular nerve

Long thoracic nerve

Middle scalene muscle

Transverse cervical artery

Brachial plexus

Suprascapular nerve

Lateral supraclavicular nerves

Omohyoid muscle
(transected)

Posterior cord
of brachial plexus

Subclavius muscle
(transected)

Radial nerve

Acromial and deltoid branches
of thoraco-acromial artery

Pectoralis minor muscle

Musculocutaneous nerve

Lateral and medial roots
of median nerve

Axillary artery

Coracobrachialis muscle

Ulnar nerve

Pectoralis major muscle

Thoracodorsal artery and nerve

Biceps brachii muscle

Long thoracic nerve

Latissimus dorsi muscle

Parotid gland

Cervical branch
of facial nerve [VII]

External jugular vein

Platysma

Transverse cervical nerve

Hyoid bone

Superior thyroid artery,
Thyrohyoid muscle

Superior root
of ansa cervicalis

Vagus nerve [X]

Common carotid artery,
Internal jugular vein

Omohyoid muscle,
Sternothyroid muscle

Inferior root
of ansa cervicalis

Sternohyoid muscle

Thyrocervical trunk,
Phrenic nerve

Internal jugular vein

External jugular vein

Anterior scalene muscle

Subclavian nerve

Suprascapular artery,
Subclavian artery

1st rib,
Subclavian vein

Medial pectoral nerve

Thoraco-acromial artery

External intercostal muscle

Pectoralis major muscle

Pectoral branches
of thoraco-acromial artery

Pectoralis minor muscle
(cut surface)

Lateral pectoral nerve

Lateral thoracic vein and artery

Anterior cutaneous branch
of intercostal nerve

Intercostobrachial nerve

Medial antebrachial cutaneous nerve

Lateral cutaneous branch
of intercostal nerve

**197 Blood vessels and nerves
of the neck, axilla, and thorax** (50%)
The platysma and the sternocleidomastoid muscle
were removed. Right lateral aspect

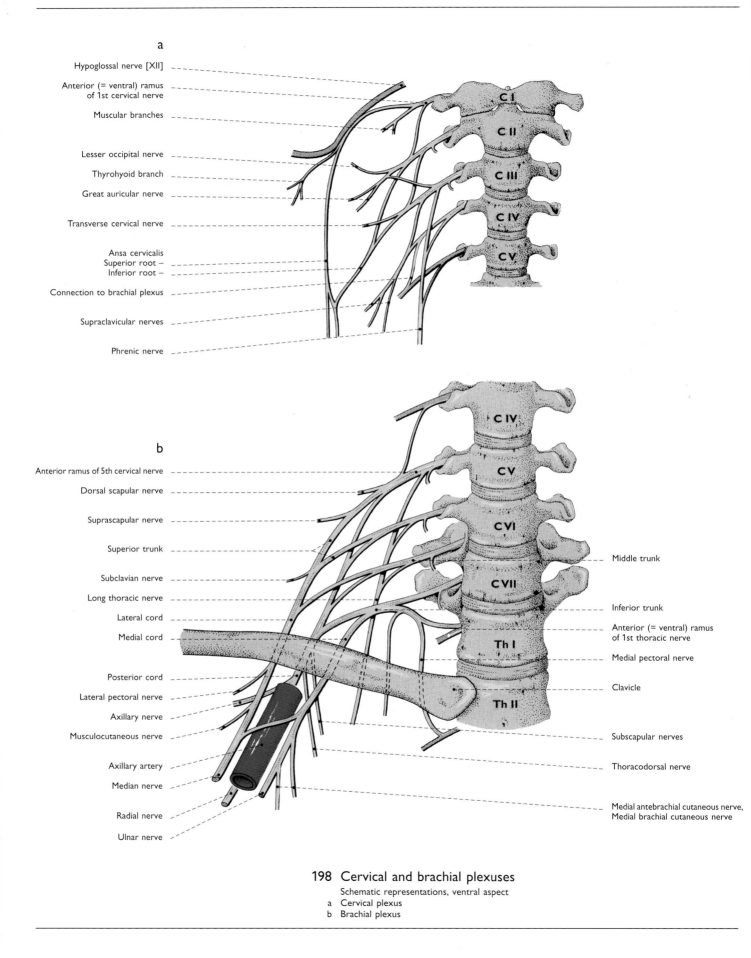

a

Hypoglossal nerve [XII]

Anterior (= ventral) ramus of 1st cervical nerve

Muscular branches

Lesser occipital nerve

Thyrohyoid branch

Great auricular nerve

Transverse cervical nerve

Ansa cervicalis
Superior root –
Inferior root –

Connection to brachial plexus

Supraclavicular nerves

Phrenic nerve

C I
C II
C III
C IV
C V

b

Anterior ramus of 5th cervical nerve

Dorsal scapular nerve

Suprascapular nerve

Superior trunk

Subclavian nerve

Long thoracic nerve

Lateral cord

Medial cord

Posterior cord

Lateral pectoral nerve

Axillary nerve

Musculocutaneous nerve

Axillary artery

Median nerve

Radial nerve

Ulnar nerve

C IV
C V
C VI
C VII
Th I
Th II

Middle trunk

Inferior trunk

Anterior (= ventral) ramus of 1st thoracic nerve

Medial pectoral nerve

Clavicle

Subscapular nerves

Thoracodorsal nerve

Medial antebrachial cutaneous nerve, Medial brachial cutaneous nerve

198 Cervical and brachial plexuses
Schematic representations, ventral aspect
a Cervical plexus
b Brachial plexus

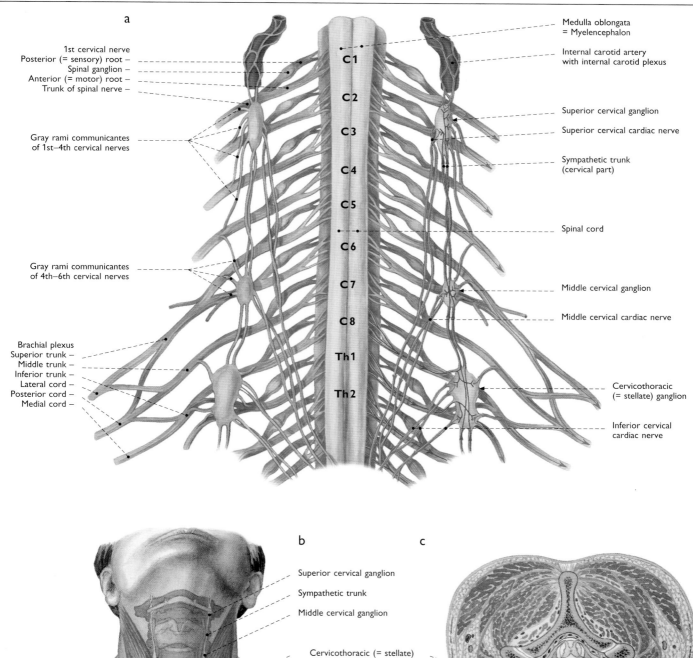

a

1st cervical nerve
Posterior (= sensory) root –
Spinal ganglion –
Anterior (= motor) root –
Trunk of spinal nerve –

Gray rami communicantes
of 1st–4th cervical nerves

Gray rami communicantes
of 4th–6th cervical nerves

Brachial plexus
Superior trunk –
Middle trunk –
Inferior trunk –
Lateral cord –
Posterior cord –
Medial cord –

C1
C2
C3
C4
C5
C6
C7
C8
Th1
Th2

Medulla oblongata
= Myelencephalon

Internal carotid artery
with internal carotid plexus

Superior cervical ganglion

Superior cervical cardiac nerve

Sympathetic trunk
(cervical part)

Spinal cord

Middle cervical ganglion

Middle cervical cardiac nerve

Cervicothoracic
(= stellate) ganglion

Inferior cervical
cardiac nerve

b

Superior cervical ganglion

Sympathetic trunk

Middle cervical ganglion

Cervicothoracic (= stellate)
ganglion

Sternocleidomastoid muscle

c

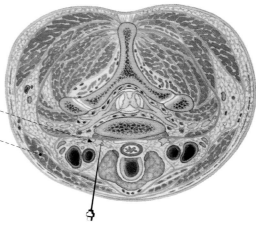

199 Sympathetic part of the autonomic division of peripheral nervous system in the neck

a Essential circuitry and origins of the sympathetic nervous system
in the neck and upper thorax. On the right side of the picture,
the preganglionic neurons of the sympathetic efferent nervous system
are indicated by blue lines, the postganglionic neurons by green ones (60%).
Schematic representation, ventral aspect

b, c Puncture of the cervicothoracic (= stellate) ganglion
b Ventral aspect (25%)
c Transverse section at the level of the first thoracic vertebra (Th1) (50%),
superior aspect

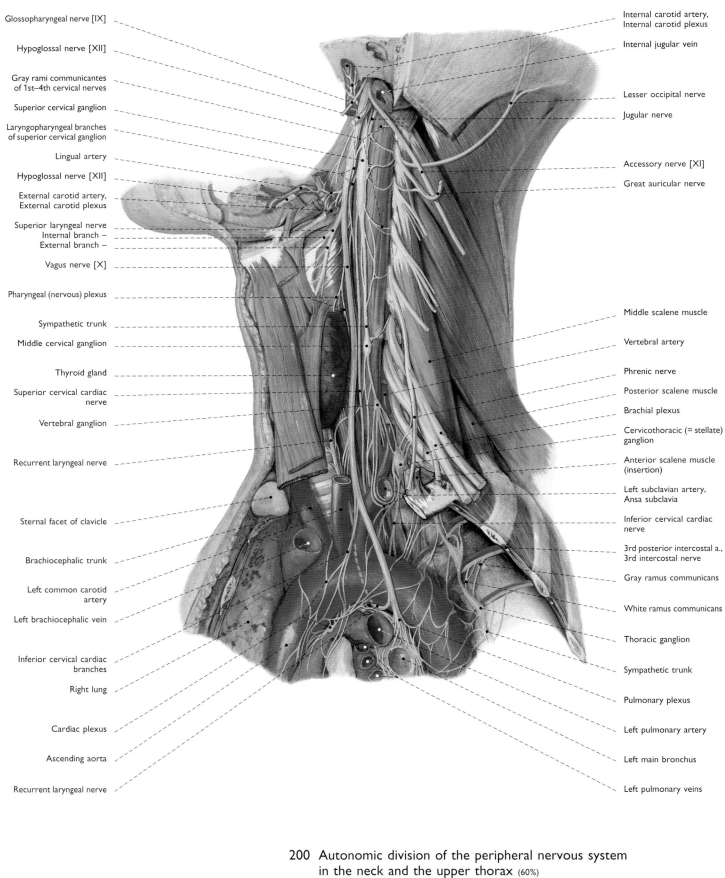

Glossopharyngeal nerve [IX]

Hypoglossal nerve [XII]

Gray rami communicantes of 1st–4th cervical nerves

Superior cervical ganglion

Laryngopharyngeal branches of superior cervical ganglion

Lingual artery

Hypoglossal nerve [XII]

External carotid artery, External carotid plexus

Superior laryngeal nerve
Internal branch –
External branch –

Vagus nerve [X]

Pharyngeal (nervous) plexus

Sympathetic trunk

Middle cervical ganglion

Thyroid gland

Superior cervical cardiac nerve

Vertebral ganglion

Recurrent laryngeal nerve

Sternal facet of clavicle

Brachiocephalic trunk

Left common carotid artery

Left brachiocephalic vein

Inferior cervical cardiac branches

Right lung

Cardiac plexus

Ascending aorta

Recurrent laryngeal nerve

Internal carotid artery, Internal carotid plexus

Internal jugular vein

Lesser occipital nerve

Jugular nerve

Accessory nerve [XI]

Great auricular nerve

Middle scalene muscle

Vertebral artery

Phrenic nerve

Posterior scalene muscle

Brachial plexus

Cervicothoracic (= stellate) ganglion

Anterior scalene muscle (insertion)

Left subclavian artery, Ansa subclavia

Inferior cervical cardiac nerve

3rd posterior intercostal a., 3rd intercostal nerve

Gray ramus communicans

White ramus communicans

Thoracic ganglion

Sympathetic trunk

Pulmonary plexus

Left pulmonary artery

Left main bronchus

Left pulmonary veins

200 Autonomic division of the peripheral nervous system in the neck and the upper thorax (60%)
Left lateral aspect

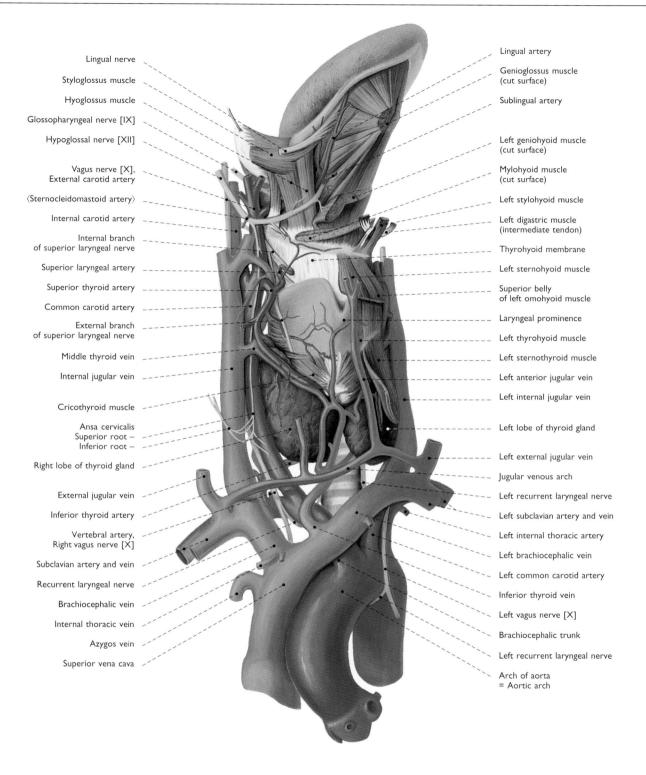

Lingual nerve

Styloglossus muscle

Hyoglossus muscle

Glossopharyngeal nerve [IX]

Hypoglossal nerve [XII]

Vagus nerve [X],
External carotid artery

⟨Sternocleidomastoid artery⟩

Internal carotid artery

Internal branch
of superior laryngeal nerve

Superior laryngeal artery

Superior thyroid artery

Common carotid artery

External branch
of superior laryngeal nerve

Middle thyroid vein

Internal jugular vein

Cricothyroid muscle

Ansa cervicalis
Superior root –
Inferior root –

Right lobe of thyroid gland

External jugular vein

Inferior thyroid artery

Vertebral artery,
Right vagus nerve [X]

Subclavian artery and vein

Recurrent laryngeal nerve

Brachiocephalic vein

Internal thoracic vein

Azygos vein

Superior vena cava

Lingual artery

Genioglossus muscle
(cut surface)

Sublingual artery

Left geniohyoid muscle
(cut surface)

Mylohyoid muscle
(cut surface)

Left stylohyoid muscle

Left digastric muscle
(intermediate tendon)

Thyrohyoid membrane

Left sternohyoid muscle

Superior belly
of left omohyoid muscle

Laryngeal prominence

Left thyrohyoid muscle

Left sternothyroid muscle

Left anterior jugular vein

Left internal jugular vein

Left lobe of thyroid gland

Left external jugular vein

Jugular venous arch

Left recurrent laryngeal nerve

Left subclavian artery and vein

Left internal thoracic artery

Left brachiocephalic vein

Left common carotid artery

Inferior thyroid vein

Left vagus nerve [X]

Brachiocephalic trunk

Left recurrent laryngeal nerve

Arch of aorta
= Aortic arch

**201 Blood vessels and nerves
of the cervical viscera and the tongue** (70%)
Right ventrolateral aspect

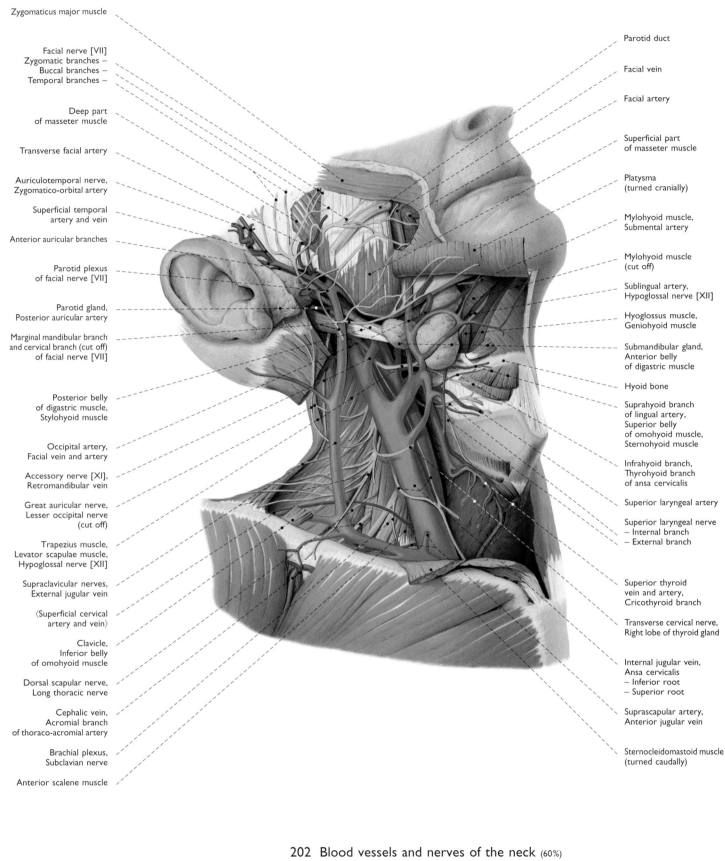

Zygomaticus major muscle

Facial nerve [VII]
Zygomatic branches –
Buccal branches –
Temporal branches –

Deep part
of masseter muscle

Transverse facial artery

Auriculotemporal nerve,
Zygomatico-orbital artery

Superficial temporal
artery and vein

Anterior auricular branches

Parotid plexus
of facial nerve [VII]

Parotid gland,
Posterior auricular artery

Marginal mandibular branch
and cervical branch (cut off)
of facial nerve [VII]

Posterior belly
of digastric muscle,
Stylohyoid muscle

Occipital artery,
Facial vein and artery

Accessory nerve [XI],
Retromandibular vein

Great auricular nerve,
Lesser occipital nerve
(cut off)

Trapezius muscle,
Levator scapulae muscle,
Hypoglossal nerve [XII]

Supraclavicular nerves,
External jugular vein

⟨Superficial cervical
artery and vein⟩

Clavicle,
Inferior belly
of omohyoid muscle

Dorsal scapular nerve,
Long thoracic nerve

Cephalic vein,
Acromial branch
of thoraco-acromial artery

Brachial plexus,
Subclavian nerve

Anterior scalene muscle

Parotid duct

Facial vein

Facial artery

Superficial part
of masseter muscle

Platysma
(turned cranially)

Mylohyoid muscle,
Submental artery

Mylohyoid muscle
(cut off)

Sublingual artery,
Hypoglossal nerve [XII]

Hyoglossus muscle,
Geniohyoid muscle

Submandibular gland,
Anterior belly
of digastric muscle

Hyoid bone

Suprahyoid branch
of lingual artery,
Superior belly
of omohyoid muscle,
Sternohyoid muscle

Infrahyoid branch,
Thyrohyoid branch
of ansa cervicalis

Superior laryngeal artery

Superior laryngeal nerve
– Internal branch
– External branch

Superior thyroid
vein and artery,
Cricothyroid branch

Transverse cervical nerve,
Right lobe of thyroid gland

Internal jugular vein,
Ansa cervicalis
– Inferior root
– Superior root

Suprascapular artery,
Anterior jugular vein

Sternocleidomastoid muscle
(turned caudally)

202 Blood vessels and nerves of the neck (60%)
The parotid gland, the sternocleidomastoid muscle,
the supra- and infrahyoid muscles were partially removed.
Right ventrolateral aspect

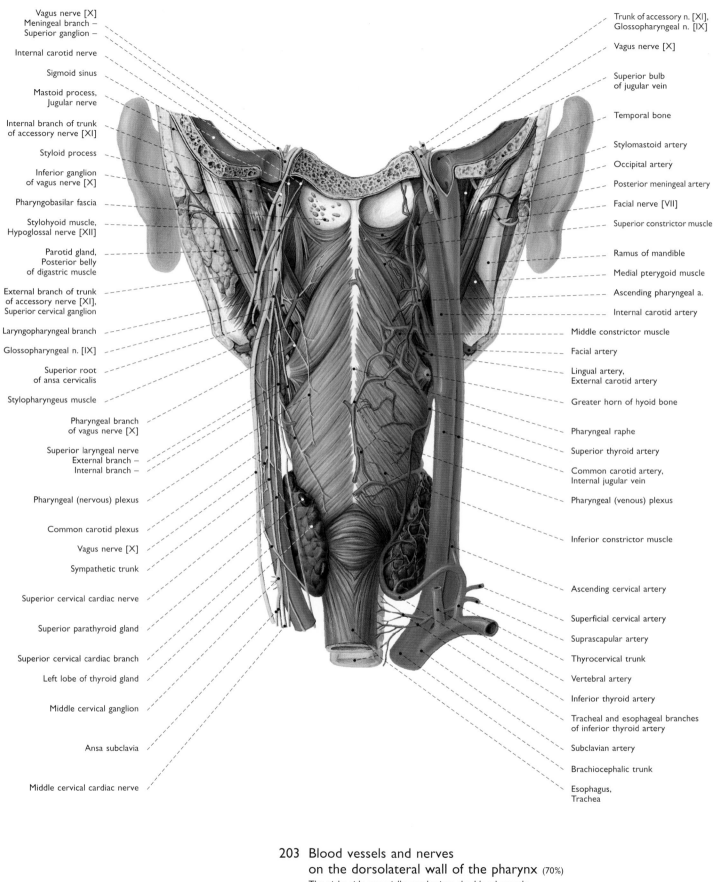

Vagus nerve [X]
Meningeal branch –
Superior ganglion –

Internal carotid nerve

Sigmoid sinus

Mastoid process,
Jugular nerve

Internal branch of trunk
of accessory nerve [XI]

Styloid process

Inferior ganglion
of vagus nerve [X]

Pharyngobasilar fascia

Stylohyoid muscle,
Hypoglossal nerve [XII]

Parotid gland,
Posterior belly
of digastric muscle

External branch of trunk
of accessory nerve [XI],
Superior cervical ganglion

Laryngopharyngeal branch

Glossopharyngeal n. [IX]

Superior root
of ansa cervicalis

Stylopharyngeus muscle

Pharyngeal branch
of vagus nerve [X]

Superior laryngeal nerve
External branch –
Internal branch –

Pharyngeal (nervous) plexus

Common carotid plexus

Vagus nerve [X]

Sympathetic trunk

Superior cervical cardiac nerve

Superior parathyroid gland

Superior cervical cardiac branch

Left lobe of thyroid gland

Middle cervical ganglion

Ansa subclavia

Middle cervical cardiac nerve

Trunk of accessory n. [XI],
Glossopharyngeal n. [IX]

Vagus nerve [X]

Superior bulb
of jugular vein

Temporal bone

Stylomastoid artery

Occipital artery

Posterior meningeal artery

Facial nerve [VII]

Superior constrictor muscle

Ramus of mandible

Medial pterygoid muscle

Ascending pharyngeal a.

Internal carotid artery

Middle constrictor muscle

Facial artery

Lingual artery,
External carotid artery

Greater horn of hyoid bone

Pharyngeal raphe

Superior thyroid artery

Common carotid artery,
Internal jugular vein

Pharyngeal (venous) plexus

Inferior constrictor muscle

Ascending cervical artery

Superficial cervical artery

Suprascapular artery

Thyrocervical trunk

Vertebral artery

Inferior thyroid artery

Tracheal and esophageal branches
of inferior thyroid artery

Subclavian artery

Brachiocephalic trunk

Esophagus,
Trachea

**203 Blood vessels and nerves
on the dorsolateral wall of the pharynx** (70%)
The right side especially emphasizes the blood vessels,
the left side the nerves. Dorsal aspect

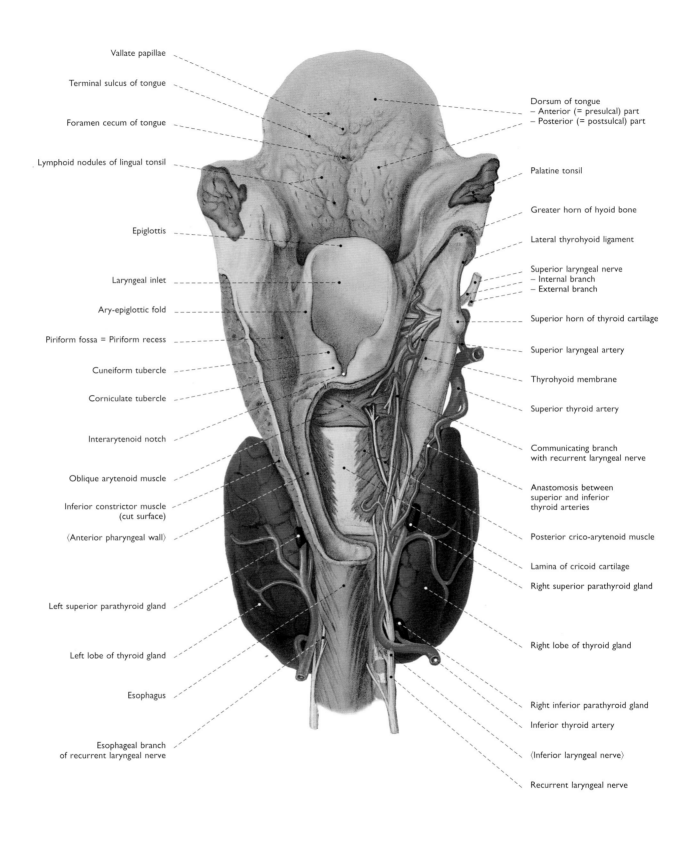

Vallate papillae

Terminal sulcus of tongue

Foramen cecum of tongue

Lymphoid nodules of lingual tonsil

Epiglottis

Laryngeal inlet

Ary-epiglottic fold

Piriform fossa = Piriform recess

Cuneiform tubercle

Corniculate tubercle

Interarytenoid notch

Oblique arytenoid muscle

Inferior constrictor muscle (cut surface)

⟨Anterior pharyngeal wall⟩

Left superior parathyroid gland

Left lobe of thyroid gland

Esophagus

Esophageal branch of recurrent laryngeal nerve

Dorsum of tongue
– Anterior (= presulcal) part
– Posterior (= postsulcal) part

Palatine tonsil

Greater horn of hyoid bone

Lateral thyrohyoid ligament

Superior laryngeal nerve
– Internal branch
– External branch

Superior horn of thyroid cartilage

Superior laryngeal artery

Thyrohyoid membrane

Superior thyroid artery

Communicating branch with recurrent laryngeal nerve

Anastomosis between superior and inferior thyroid arteries

Posterior crico-arytenoid muscle

Lamina of cricoid cartilage

Right superior parathyroid gland

Right lobe of thyroid gland

Right inferior parathyroid gland

Inferior thyroid artery

⟨Inferior laryngeal nerve⟩

Recurrent laryngeal nerve

**204 Arteries and nerves
of the larynx and thyroid gland** (100%)
The dorsal wall of the pharynx was removed completely,
the ventral one partially. Dorsal aspect

Thoracic Viscera

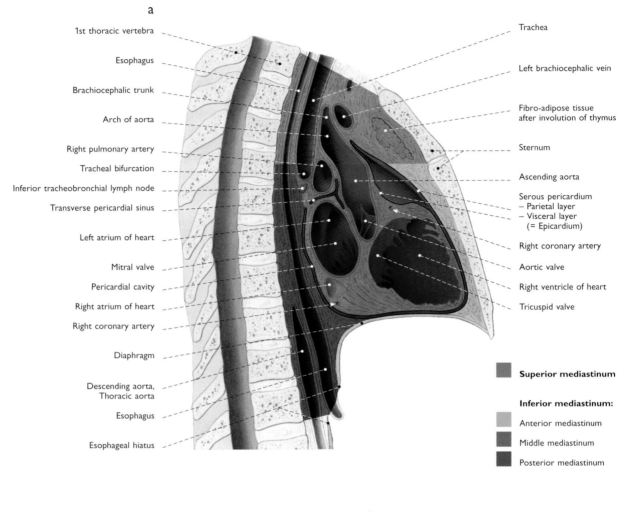

a

1st thoracic vertebra

Esophagus

Brachiocephalic trunk

Arch of aorta

Right pulmonary artery

Tracheal bifurcation

Inferior tracheobronchial lymph node

Transverse pericardial sinus

Left atrium of heart

Mitral valve

Pericardial cavity

Right atrium of heart

Right coronary artery

Diaphragm

Descending aorta, Thoracic aorta

Esophagus

Esophageal hiatus

Trachea

Left brachiocephalic vein

Fibro-adipose tissue after involution of thymus

Sternum

Ascending aorta

Serous pericardium
– Parietal layer
– Visceral layer
  (= Epicardium)

Right coronary artery

Aortic valve

Right ventricle of heart

Tricuspid valve

Superior mediastinum

Inferior mediastinum:

Anterior mediastinum

Middle mediastinum

Posterior mediastinum

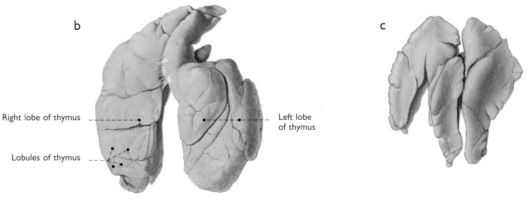

b

Right lobe of thymus

Lobules of thymus

Left lobe of thymus

c

206 Mediastinum and thymus

a Subdivision of the mediastinum, median section (40%), medial aspect of the left half

b Thymus of a 3-year-old child (75%), ventral aspect

c Thymus of a newborn child (75%), ventral aspect

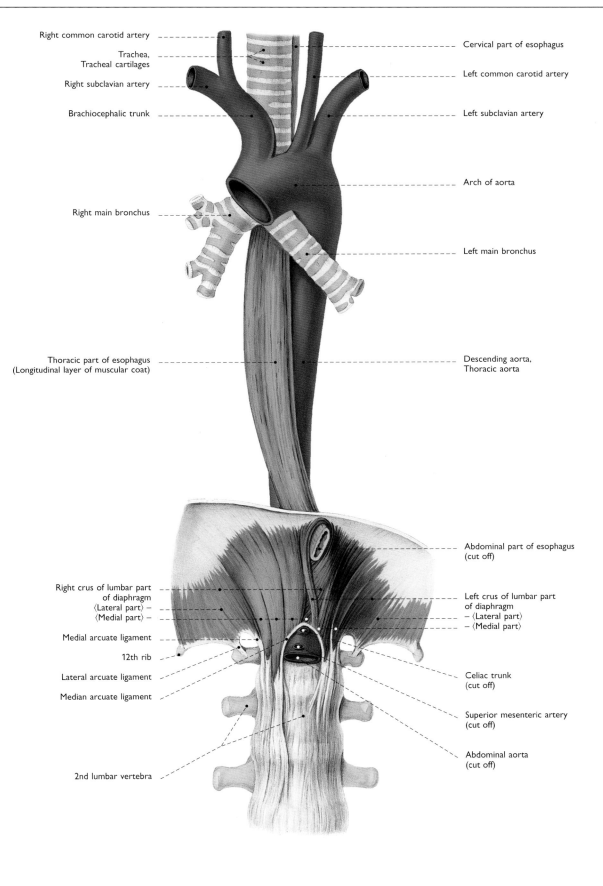

Right common carotid artery

Trachea,
Tracheal cartilages

Right subclavian artery

Brachiocephalic trunk

Right main bronchus

Thoracic part of esophagus
(Longitudinal layer of muscular coat)

Right crus of lumbar part
of diaphragm
⟨Lateral part⟩ —
⟨Medial part⟩ —

Medial arcuate ligament

12th rib

Lateral arcuate ligament

Median arcuate ligament

2nd lumbar vertebra

Cervical part of esophagus

Left common carotid artery

Left subclavian artery

Arch of aorta

Left main bronchus

Descending aorta,
Thoracic aorta

Abdominal part of esophagus
(cut off)

Left crus of lumbar part
of diaphragm
– ⟨Lateral part⟩
– ⟨Medial part⟩

Celiac trunk
(cut off)

Superior mesenteric artery
(cut off)

Abdominal aorta
(cut off)

207 Esophagus and adjacent organs (50%)
Ventral aspect

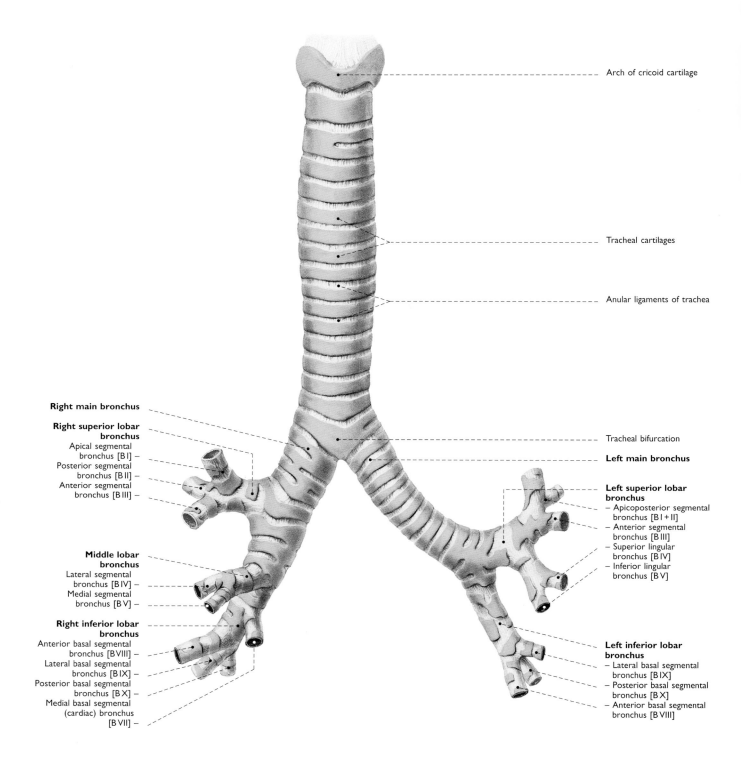

Arch of cricoid cartilage

Tracheal cartilages

Anular ligaments of trachea

Right main bronchus

Right superior lobar bronchus
Apical segmental bronchus [B I] –
Posterior segmental bronchus [B II] –
Anterior segmental bronchus [B III] –

Middle lobar bronchus
Lateral segmental bronchus [B IV] –
Medial segmental bronchus [B V] –

Right inferior lobar bronchus
Anterior basal segmental bronchus [B VIII] –
Lateral basal segmental bronchus [B IX] –
Posterior basal segmental bronchus [B X] –
Medial basal segmental (cardiac) bronchus [B VII] –

Tracheal bifurcation

Left main bronchus

Left superior lobar bronchus
– Apicoposterior segmental bronchus [B I + II]
– Anterior segmental bronchus [B III]
– Superior lingular bronchus [B IV]
– Inferior lingular bronchus [B V]

Left inferior lobar bronchus
– Lateral basal segmental bronchus [B IX]
– Posterior basal segmental bronchus [B X]
– Anterior basal segmental bronchus [B VIII]

208 Trachea and bronchi (90%)
Ventral aspect

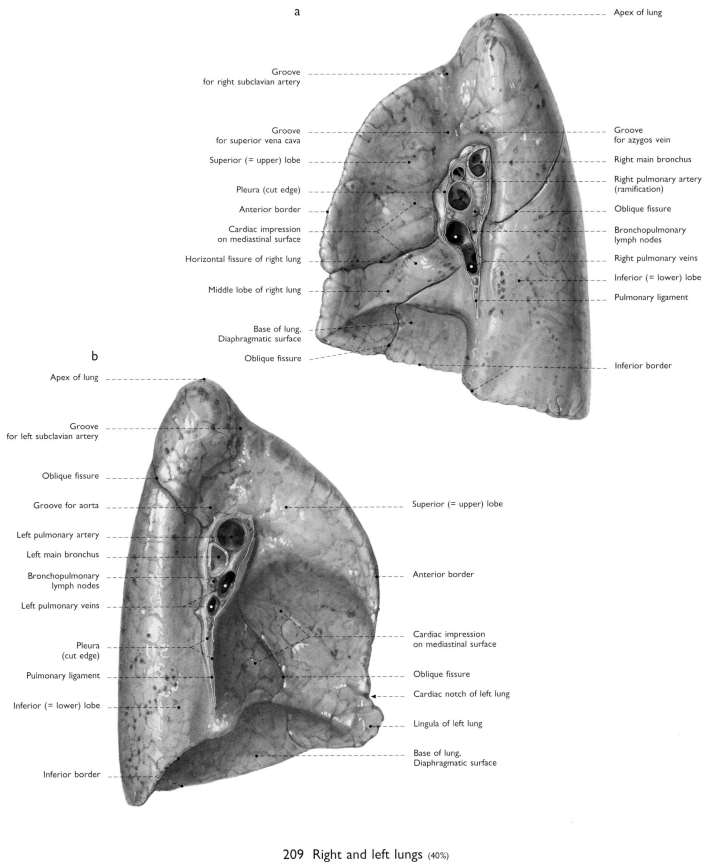

a

Apex of lung

Groove
for right subclavian artery

Groove
for superior vena cava

Superior (= upper) lobe

Pleura (cut edge)

Anterior border

Cardiac impression
on mediastinal surface

Horizontal fissure of right lung

Middle lobe of right lung

Base of lung,
Diaphragmatic surface

Oblique fissure

Groove
for azygos vein

Right main bronchus

Right pulmonary artery
(ramification)

Oblique fissure

Bronchopulmonary
lymph nodes

Right pulmonary veins

Inferior (= lower) lobe

Pulmonary ligament

Inferior border

b

Apex of lung

Groove
for left subclavian artery

Oblique fissure

Groove for aorta

Left pulmonary artery

Left main bronchus

Bronchopulmonary
lymph nodes

Left pulmonary veins

Pleura
(cut edge)

Pulmonary ligament

Inferior (= lower) lobe

Inferior border

Superior (= upper) lobe

Anterior border

Cardiac impression
on mediastinal surface

Oblique fissure

Cardiac notch of left lung

Lingula of left lung

Base of lung,
Diaphragmatic surface

209 Right and left lungs (40%)
    a   Right lung, mediastinal surface
    b   Left lung, mediastinal surface

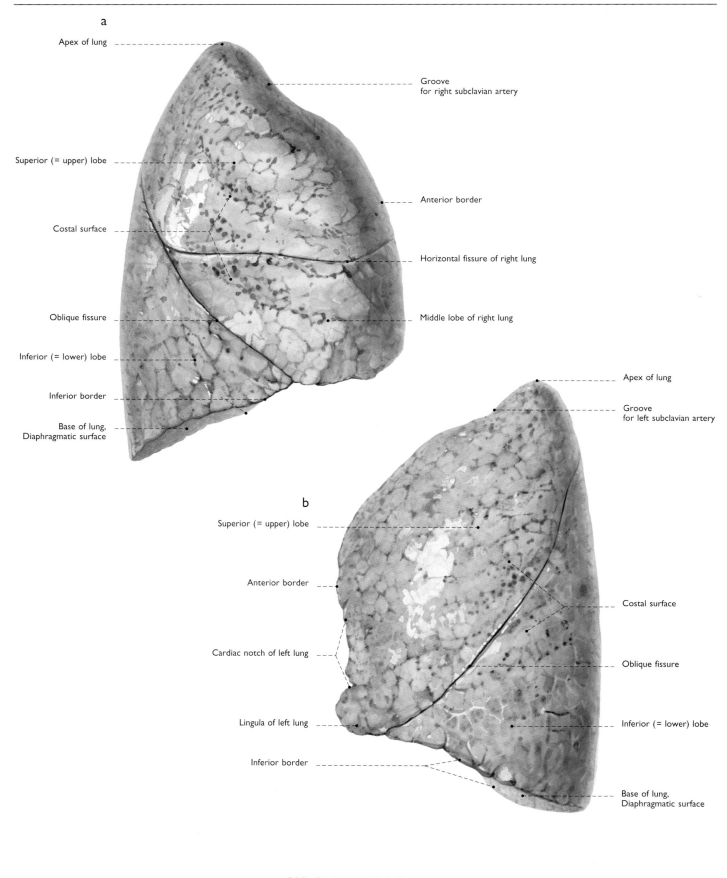

a

Apex of lung

Groove
for right subclavian artery

Superior (= upper) lobe

Anterior border

Costal surface

Horizontal fissure of right lung

Oblique fissure

Middle lobe of right lung

Inferior (= lower) lobe

Inferior border

Base of lung,
Diaphragmatic surface

Apex of lung

Groove
for left subclavian artery

b

Superior (= upper) lobe

Anterior border

Costal surface

Cardiac notch of left lung

Oblique fissure

Lingula of left lung

Inferior (= lower) lobe

Inferior border

Base of lung,
Diaphragmatic surface

210 Right and left lungs (40%)
a Right lung, costal surface
b Left lung, costal surface

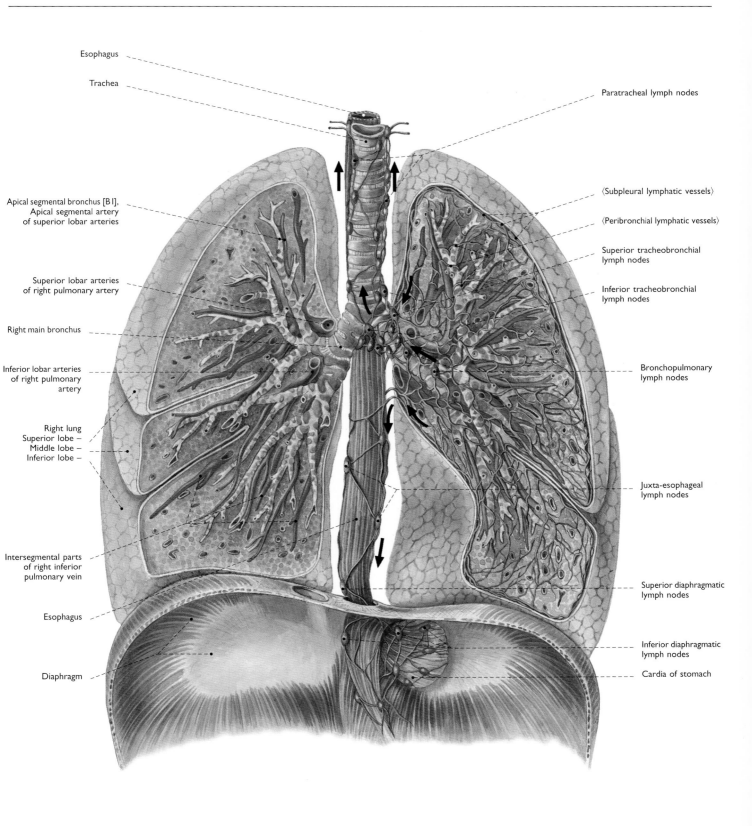

Esophagus

Trachea

Paratracheal lymph nodes

Apical segmental bronchus [B I],
Apical segmental artery
of superior lobar arteries

Superior lobar arteries
of right pulmonary artery

Right main bronchus

Inferior lobar arteries
of right pulmonary
artery

Right lung
Superior lobe –
Middle lobe –
Inferior lobe –

Intersegmental parts
of right inferior
pulmonary vein

Esophagus

Diaphragm

⟨Subpleural lymphatic vessels⟩

⟨Peribronchial lymphatic vessels⟩

Superior tracheobronchial
lymph nodes

Inferior tracheobronchial
lymph nodes

Bronchopulmonary
lymph nodes

Juxta-esophageal
lymph nodes

Superior diaphragmatic
lymph nodes

Inferior diaphragmatic
lymph nodes

Cardia of stomach

211   Right and left lungs (80%)
Lymphatic vessels, lymph nodes, and lymphatic drainage
from the left lung, ventral aspect

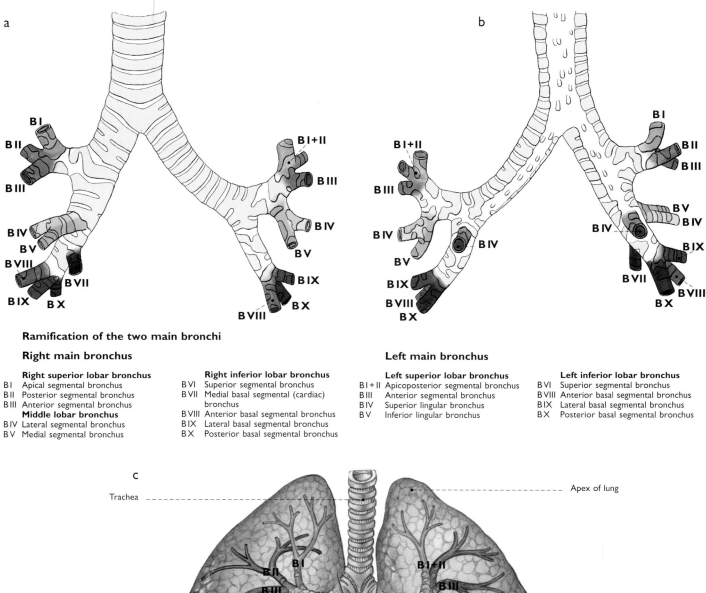

a

b

Ramification of the two main bronchi

Right main bronchus

Right superior lobar bronchus
B I   Apical segmental bronchus
B II   Posterior segmental bronchus
B III Anterior segmental bronchus
Middle lobar bronchus
B IV   Lateral segmental bronchus
B V   Medial segmental bronchus

Right inferior lobar bronchus
B VI   Superior segmental bronchus
B VII Medial basal segmental (cardiac) bronchus
B VIII Anterior basal segmental bronchus
B IX   Lateral basal segmental bronchus
B X   Posterior basal segmental bronchus

Left main bronchus

Left superior lobar bronchus
B I + II Apicoposterior segmental bronchus
B III   Anterior segmental bronchus
B IV   Superior lingular bronchus
B V   Inferior lingular bronchus

Left inferior lobar bronchus
B VI   Superior segmental bronchus
B VIII Anterior basal segmental bronchus
B IX   Lateral basal segmental bronchus
B X   Posterior basal segmental bronchus

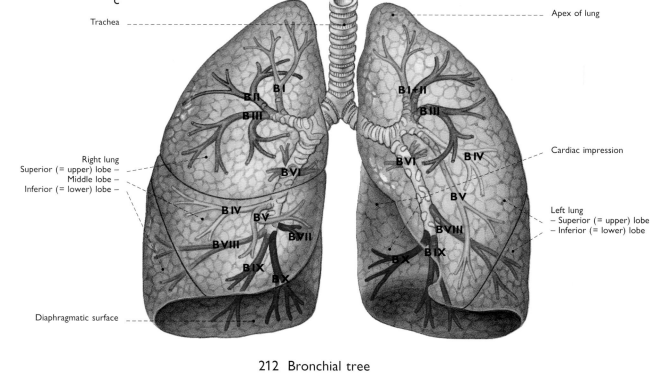

212 Bronchial tree

Schematic representations
a Bronchial tree (70%), ventral aspect
b Bronchial tree (70%), dorsal aspect
c Subdivision of the bronchial tree
   in the right and left lungs (50%), ventral aspect

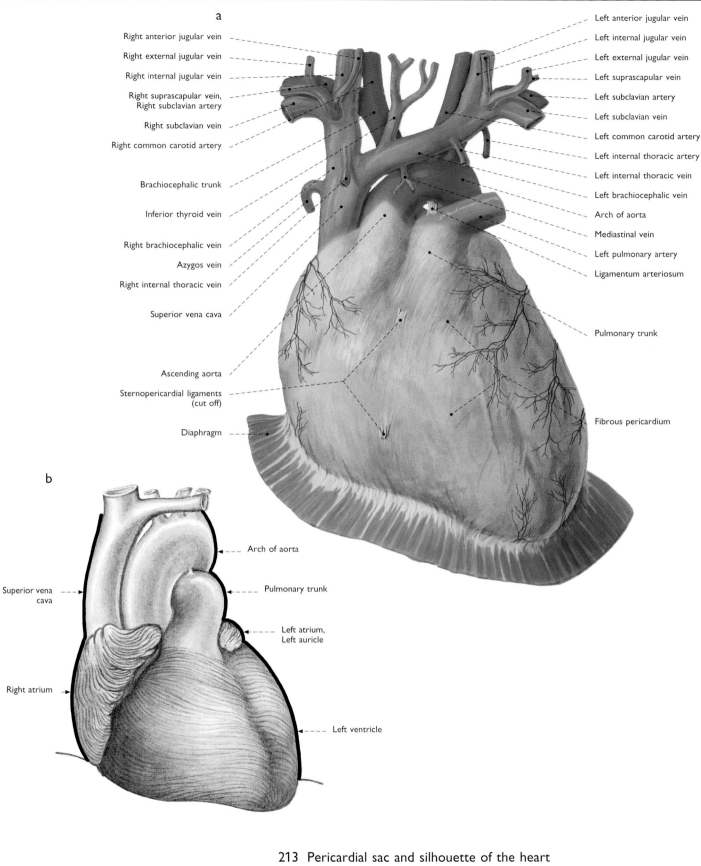

a

Right anterior jugular vein
Right external jugular vein
Right internal jugular vein
Right suprascapular vein,
Right subclavian artery
Right subclavian vein
Right common carotid artery

Brachiocephalic trunk

Inferior thyroid vein

Right brachiocephalic vein
Azygos vein
Right internal thoracic vein

Superior vena cava

Ascending aorta
Sternopericardial ligaments
(cut off)

Diaphragm

Left anterior jugular vein
Left internal jugular vein
Left external jugular vein
Left suprascapular vein
Left subclavian artery
Left subclavian vein
Left common carotid artery
Left internal thoracic artery
Left internal thoracic vein
Left brachiocephalic vein
Arch of aorta
Mediastinal vein
Left pulmonary artery
Ligamentum arteriosum

Pulmonary trunk

Fibrous pericardium

b

Arch of aorta

Superior vena cava

Pulmonary trunk

Left atrium,
Left auricle

Right atrium

Left ventricle

213 Pericardial sac and silhouette of the heart
Ventral aspect
a Pericardial sac and great blood vessels close to the heart (75%)
b Silhouette of the heart marked by the black marginal line (50%),
schematic representation

a

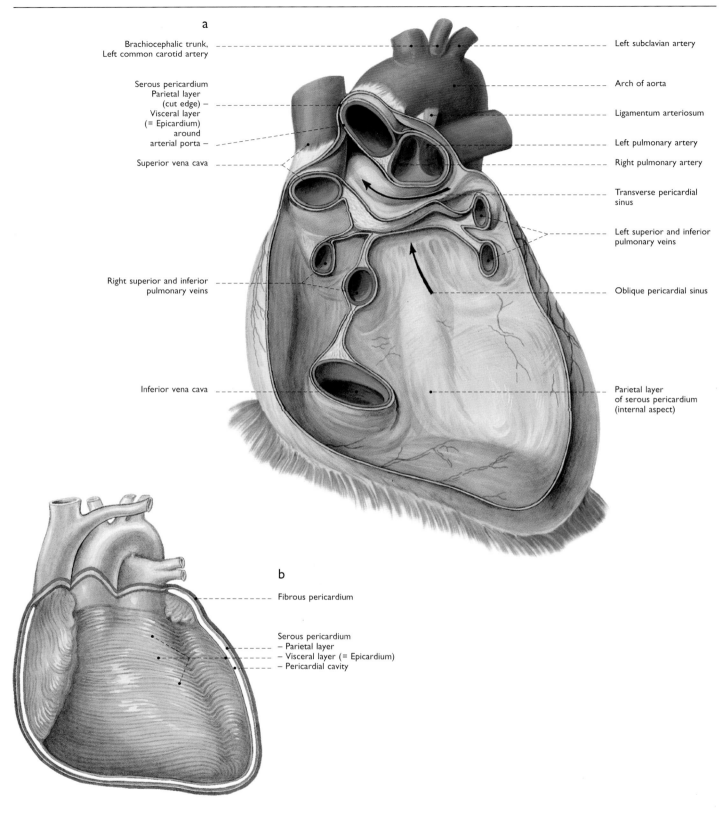

Brachiocephalic trunk,
Left common carotid artery

Serous pericardium
Parietal layer
(cut edge) –
Visceral layer
(= Epicardium)
around
arterial porta –

Superior vena cava

Right superior and inferior
pulmonary veins

Inferior vena cava

Left subclavian artery

Arch of aorta

Ligamentum arteriosum

Left pulmonary artery

Right pulmonary artery

Transverse pericardial
sinus

Left superior and inferior
pulmonary veins

Oblique pericardial sinus

Parietal layer
of serous pericardium
(internal aspect)

b

Fibrous pericardium

Serous pericardium
– Parietal layer
– Visceral layer (= Epicardium)
– Pericardial cavity

214 Pericardial sac
Ventral aspect
a Posterior wall of the pericardial cavity (75%)
b Construction of the pericardium (50%), schematic representation

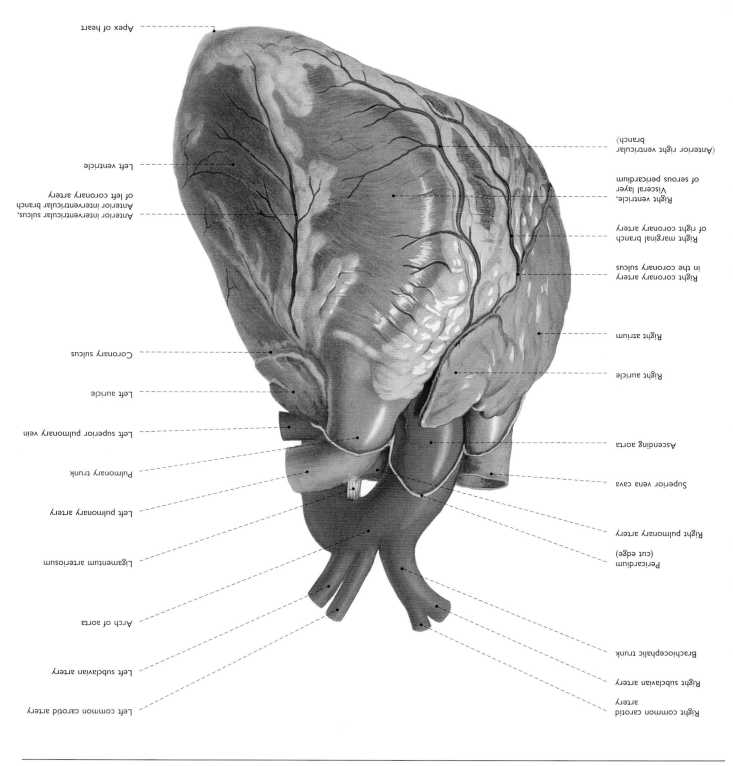

215 Heart and great blood vessels (100%)
The fibrous pericardium and the parietal layer
of the serous pericardium were removed. Ventral aspect

Apex of heart

Left ventricle

Anterior interventricular branch
of left coronary artery

Anterior interventricular sulcus,

Coronary sulcus

Left auricle

Left superior pulmonary vein

Pulmonary trunk

Left pulmonary artery

Ligamentum arteriosum

Arch of aorta

Left subclavian artery

Left common carotid artery

(Anterior right ventricular
branch)

Right ventricle,
Visceral layer
of serous pericardium

Right marginal branch
of right coronary artery

Right coronary artery
in the coronary sulcus

Right atrium

Right auricle

Ascending aorta

Superior vena cava

Right pulmonary artery

Pericardium
(cut edge)

Brachiocephalic trunk

Right subclavian artery

Right common carotid
artery

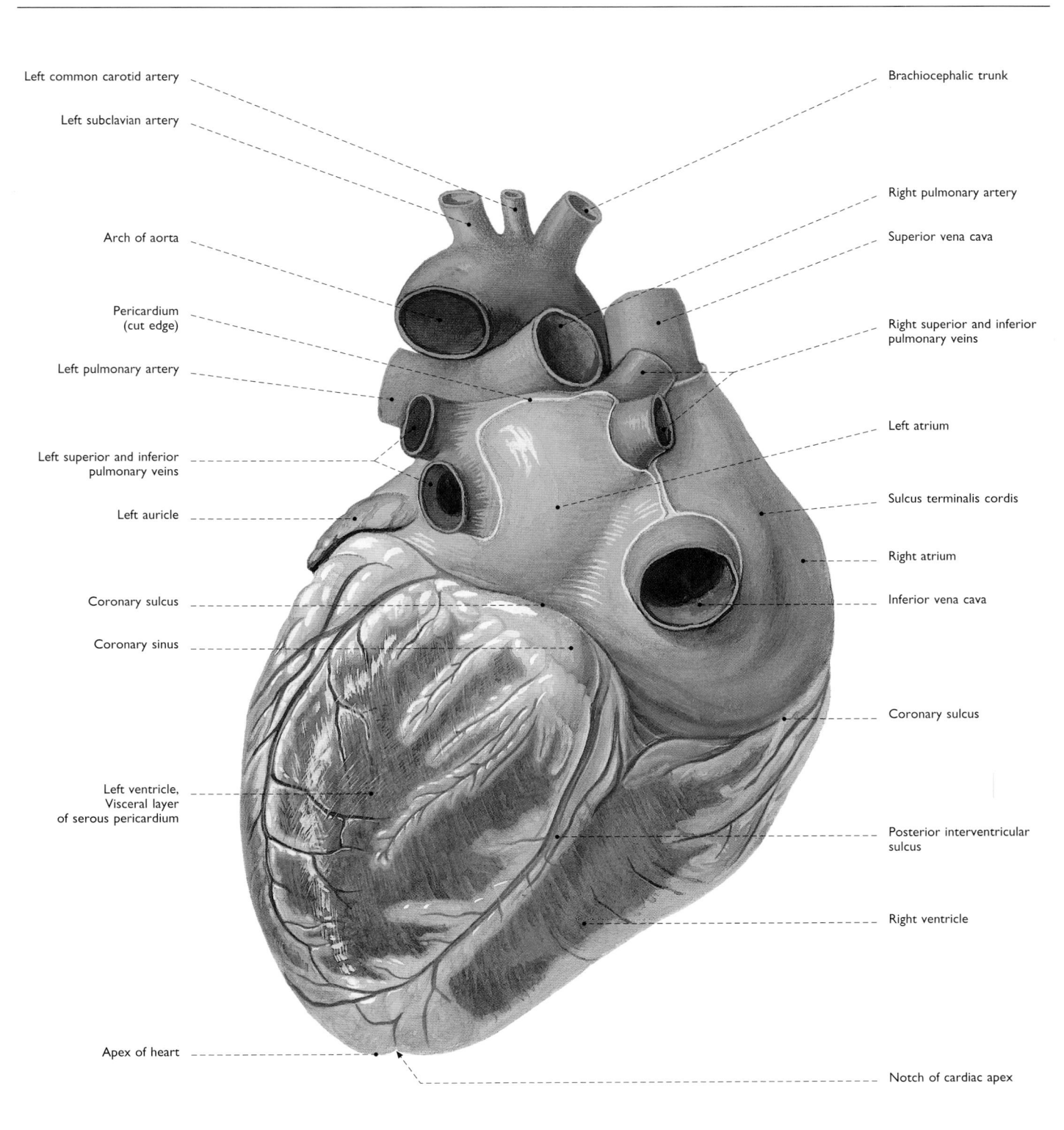

Left common carotid artery

Left subclavian artery

Arch of aorta

Pericardium
(cut edge)

Left pulmonary artery

Left superior and inferior
pulmonary veins

Left auricle

Coronary sulcus

Coronary sinus

Left ventricle,
Visceral layer
of serous pericardium

Apex of heart

Brachiocephalic trunk

Right pulmonary artery

Superior vena cava

Right superior and inferior
pulmonary veins

Left atrium

Sulcus terminalis cordis

Right atrium

Inferior vena cava

Coronary sulcus

Posterior interventricular
sulcus

Right ventricle

Notch of cardiac apex

216 Heart and great blood vessels (100%)
The fibrous pericardium and the parietal layer
of the serous pericardium were removed. Dorsal aspect

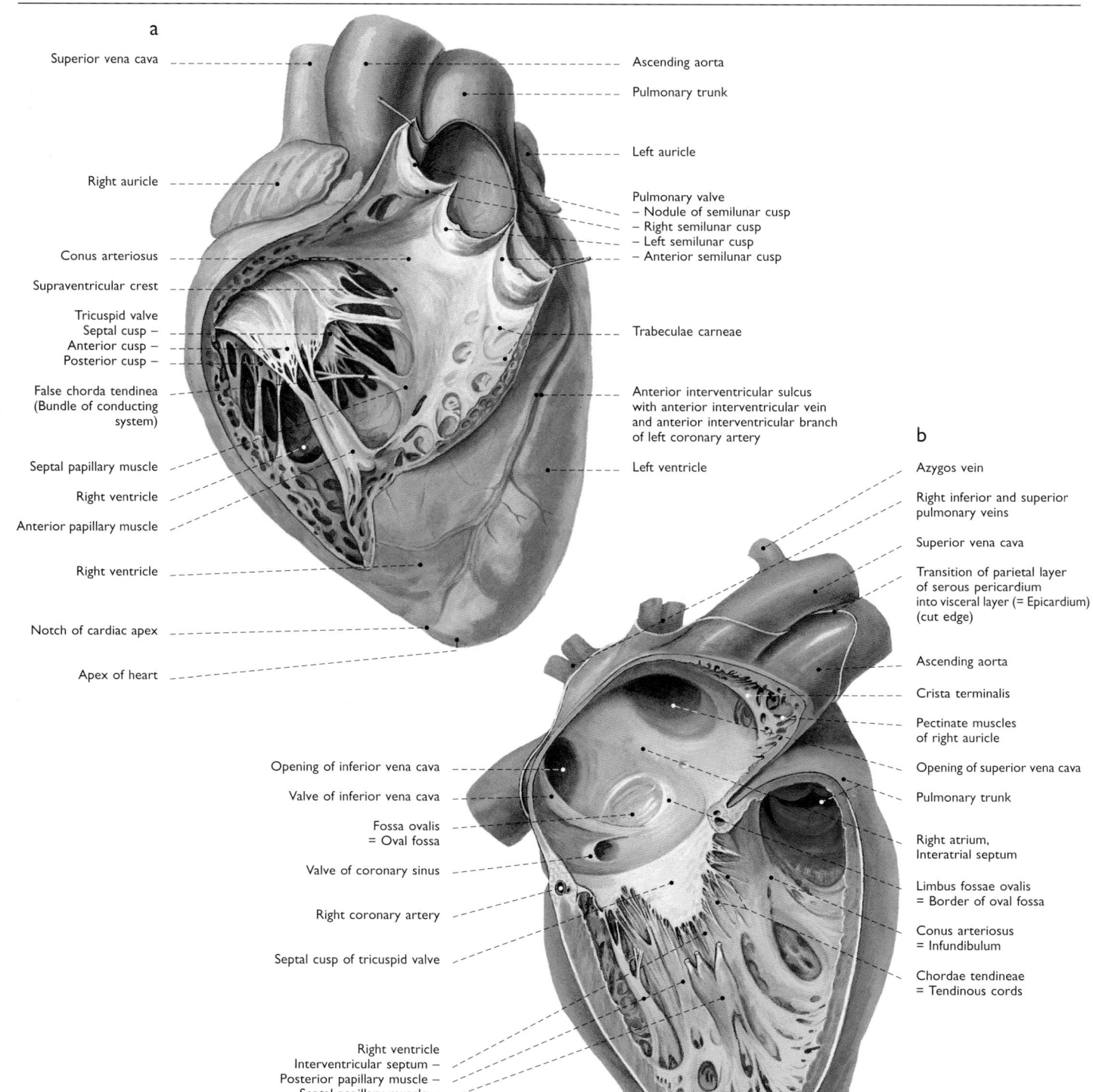

a

Superior vena cava

Right auricle

Conus arteriosus

Supraventricular crest

Tricuspid valve
Septal cusp –
Anterior cusp –
Posterior cusp –

False chorda tendinea
(Bundle of conducting
system)

Septal papillary muscle

Right ventricle

Anterior papillary muscle

Right ventricle

Notch of cardiac apex

Apex of heart

Ascending aorta

Pulmonary trunk

Left auricle

Pulmonary valve
– Nodule of semilunar cusp
– Right semilunar cusp
– Left semilunar cusp
– Anterior semilunar cusp

Trabeculae carneae

Anterior interventricular sulcus
with anterior interventricular vein
and anterior interventricular branch
of left coronary artery

Left ventricle

b

Azygos vein

Right inferior and superior
pulmonary veins

Superior vena cava

Transition of parietal layer
of serous pericardium
into visceral layer (= Epicardium)
(cut edge)

Ascending aorta

Crista terminalis

Pectinate muscles
of right auricle

Opening of superior vena cava

Pulmonary trunk

Right atrium,
Interatrial septum

Limbus fossae ovalis
= Border of oval fossa

Conus arteriosus
= Infundibulum

Chordae tendineae
= Tendinous cords

Opening of inferior vena cava

Valve of inferior vena cava

Fossa ovalis
= Oval fossa

Valve of coronary sinus

Right coronary artery

Septal cusp of tricuspid valve

Right ventricle
Interventricular septum –
Posterior papillary muscle –
Septal papillary muscle –

217 Heart (70%)
a Right ventricle and opening of the pulmonary trunk after
having cut a window in the anterior walls of the right ventricle
and the opening of the pulmonary trunk, ventral aspect
b Right atrium and right ventricle of the heart after removal
of the lateral walls of the right atrium and the right ventricle,
lateral aspect

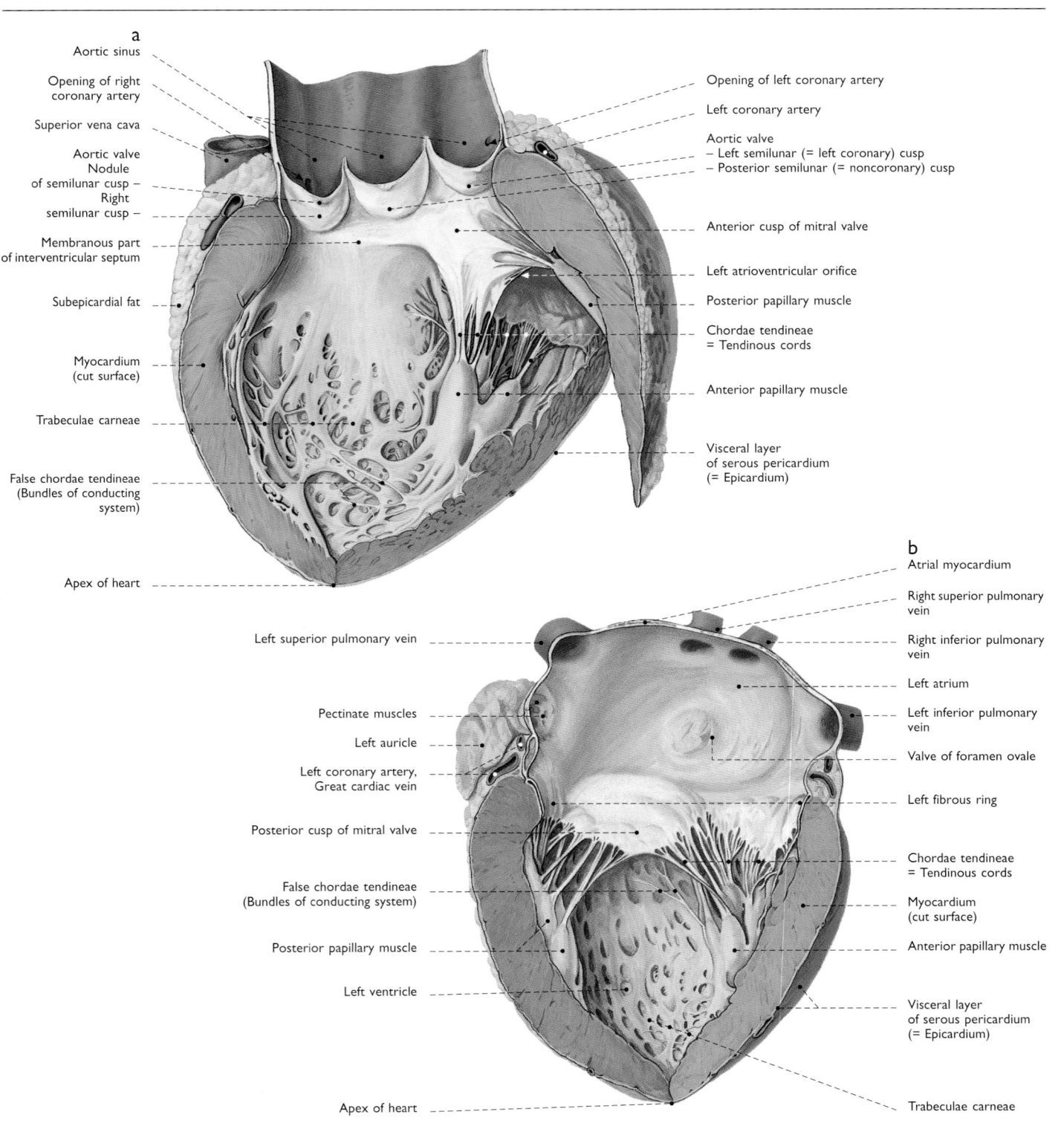

a

Aortic sinus

Opening of right coronary artery

Superior vena cava

Aortic valve
Nodule of semilunar cusp –
Right semilunar cusp –

Membranous part of interventricular septum

Subepicardial fat

Myocardium (cut surface)

Trabeculae carneae

False chordae tendineae (Bundles of conducting system)

Apex of heart

Opening of left coronary artery

Left coronary artery

Aortic valve
– Left semilunar (= left coronary) cusp
– Posterior semilunar (= noncoronary) cusp

Anterior cusp of mitral valve

Left atrioventricular orifice

Posterior papillary muscle

Chordae tendineae = Tendinous cords

Anterior papillary muscle

Visceral layer of serous pericardium (= Epicardium)

b

Atrial myocardium

Right superior pulmonary vein

Right inferior pulmonary vein

Left atrium

Left inferior pulmonary vein

Valve of foramen ovale

Left fibrous ring

Chordae tendineae = Tendinous cords

Myocardium (cut surface)

Anterior papillary muscle

Visceral layer of serous pericardium (= Epicardium)

Trabeculae carneae

Left superior pulmonary vein

Pectinate muscles

Left auricle

Left coronary artery, Great cardiac vein

Posterior cusp of mitral valve

False chordae tendineae (Bundles of conducting system)

Posterior papillary muscle

Left ventricle

Apex of heart

218 Heart (70%)

a Internal aspect of the left ventricle and the ascending aorta (effluent blood pathway). Incision from the apex of the heart to the point between the right and left semilunar cusps of the aortic valve. A second longitudinal section exposes the affluent blood pathway of the left ventricle.
b Internal aspect of the left atrium and the affluent blood pathway of the left ventricle. Longitudinal incision along the rounded (left) margin of the heart

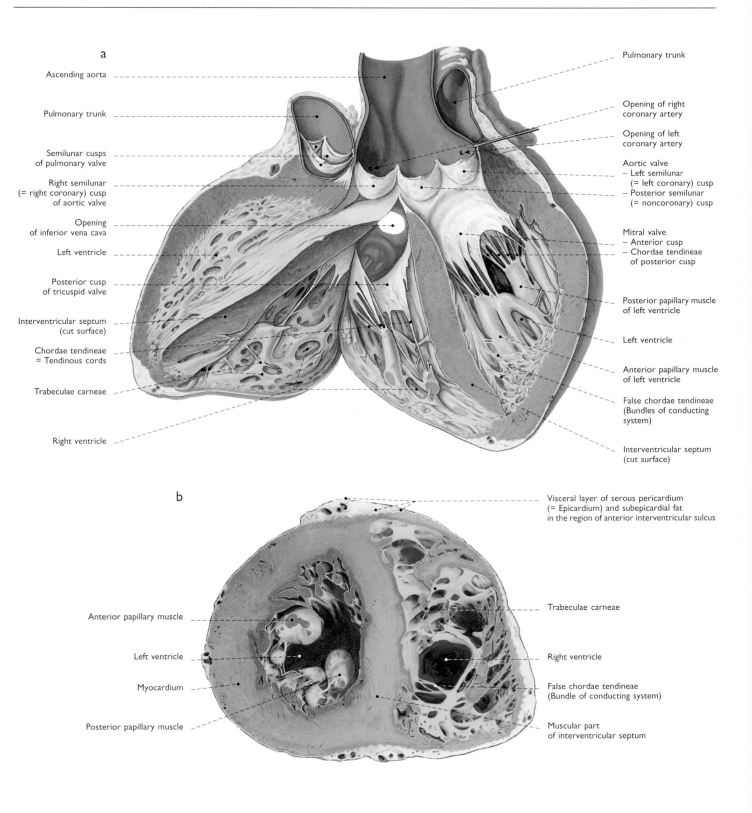

a

Ascending aorta

Pulmonary trunk

Semilunar cusps
of pulmonary valve

Right semilunar
(= right coronary) cusp
of aortic valve

Opening
of inferior vena cava

Left ventricle

Posterior cusp
of tricuspid valve

Interventricular septum
(cut surface)

Chordae tendineae
= Tendinous cords

Trabeculae carneae

Right ventricle

Pulmonary trunk

Opening of right
coronary artery

Opening of left
coronary artery

Aortic valve
– Left semilunar
(= left coronary) cusp
– Posterior semilunar
(= noncoronary) cusp

Mitral valve
– Anterior cusp
– Chordae tendineae
of posterior cusp

Posterior papillary muscle
of left ventricle

Left ventricle

Anterior papillary muscle
of left ventricle

False chordae tendineae
(Bundles of conducting
system)

Interventricular septum
(cut surface)

b

Anterior papillary muscle

Left ventricle

Myocardium

Posterior papillary muscle

Visceral layer of serous pericardium
(= Epicardium) and subepicardial fat
in the region of anterior interventricular sulcus

Trabeculae carneae

Right ventricle

False chordae tendineae
(Bundle of conducting system)

Muscular part
of interventricular septum

219   Heart

a   Internal aspect of both ventricles of the heart and
the left effluent blood pathway. Longitudinal section
perpendicular to the interventricular septum (70%)

b   Superior view into both ventricles of the heart,
opened by a transverse section (80%)

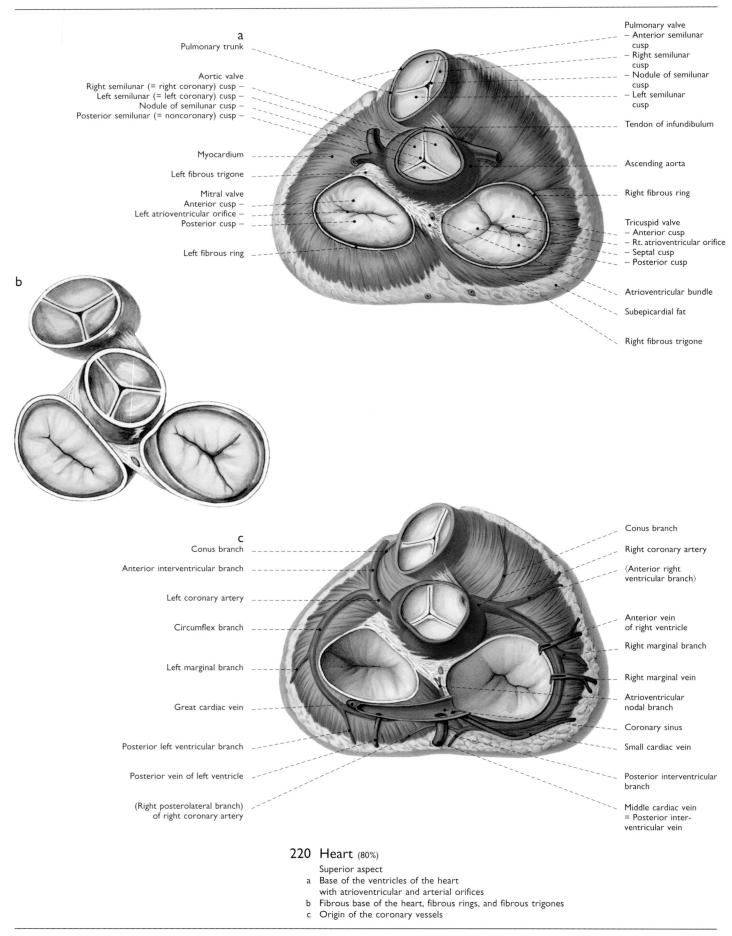

a

Pulmonary trunk

Aortic valve
Right semilunar (= right coronary) cusp –
Left semilunar (= left coronary) cusp –
Nodule of semilunar cusp –
Posterior semilunar (= noncoronary) cusp –

Myocardium

Left fibrous trigone

Mitral valve
Anterior cusp –
Left atrioventricular orifice –
Posterior cusp –

Left fibrous ring

Pulmonary valve
– Anterior semilunar cusp
– Right semilunar cusp
– Nodule of semilunar cusp
– Left semilunar cusp

Tendon of infundibulum

Ascending aorta

Right fibrous ring

Tricuspid valve
– Anterior cusp
– Rt. atrioventricular orifice
– Septal cusp
– Posterior cusp

Atrioventricular bundle

Subepicardial fat

Right fibrous trigone

b

c

Conus branch

Anterior interventricular branch

Left coronary artery

Circumflex branch

Left marginal branch

Great cardiac vein

Posterior left ventricular branch

Posterior vein of left ventricle

(Right posterolateral branch) of right coronary artery

Conus branch

Right coronary artery

⟨Anterior right ventricular branch⟩

Anterior vein of right ventricle

Right marginal branch

Right marginal vein

Atrioventricular nodal branch

Coronary sinus

Small cardiac vein

Posterior interventricular branch

Middle cardiac vein = Posterior interventricular vein

220 Heart (80%)

Superior aspect
a Base of the ventricles of the heart
  with atrioventricular and arterial orifices
b Fibrous base of the heart, fibrous rings, and fibrous trigones
c Origin of the coronary vessels

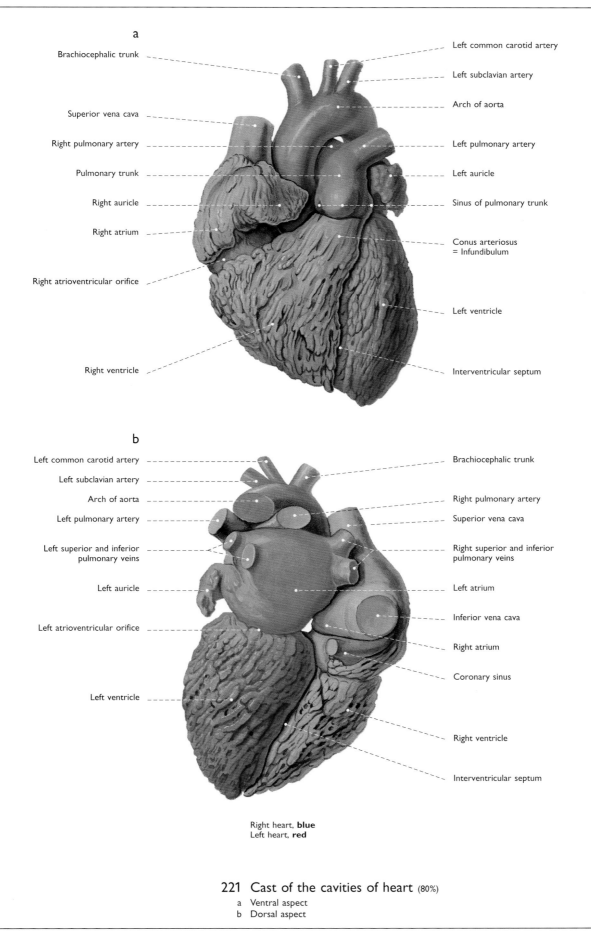

a

Brachiocephalic trunk

Superior vena cava

Right pulmonary artery

Pulmonary trunk

Right auricle

Right atrium

Right atrioventricular orifice

Right ventricle

Left common carotid artery

Left subclavian artery

Arch of aorta

Left pulmonary artery

Left auricle

Sinus of pulmonary trunk

Conus arteriosus
= Infundibulum

Left ventricle

Interventricular septum

b

Left common carotid artery

Left subclavian artery

Arch of aorta

Left pulmonary artery

Left superior and inferior
pulmonary veins

Left auricle

Left atrioventricular orifice

Left ventricle

Brachiocephalic trunk

Right pulmonary artery

Superior vena cava

Right superior and inferior
pulmonary veins

Left atrium

Inferior vena cava

Right atrium

Coronary sinus

Right ventricle

Interventricular septum

Right heart, **blue**
Left heart, **red**

221 Cast of the cavities of heart (80%)
a   Ventral aspect
b   Dorsal aspect

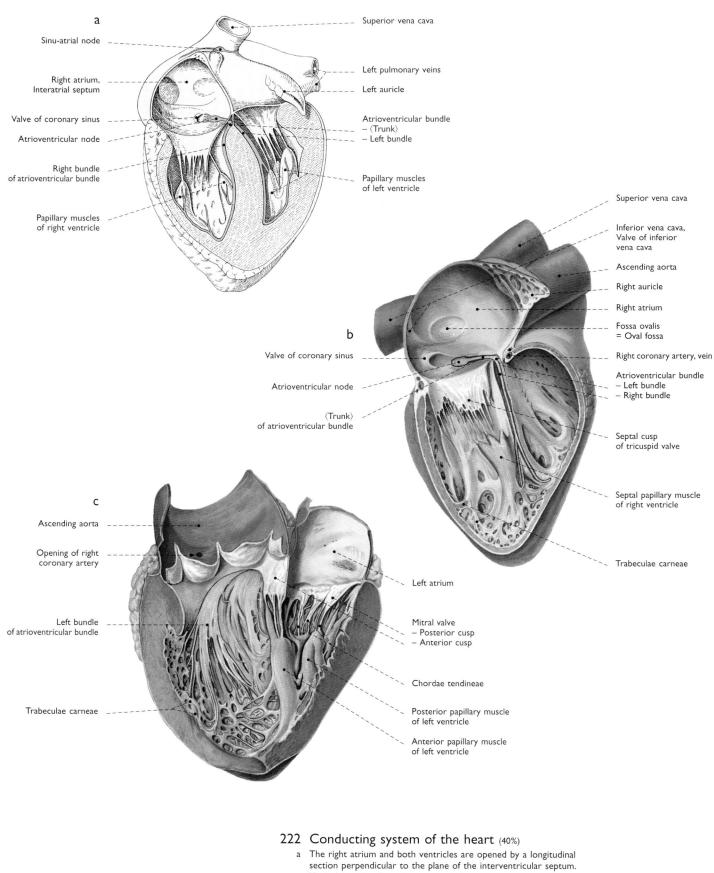

a

Sinu-atrial node

Right atrium,
Interatrial septum

Valve of coronary sinus

Atrioventricular node

Right bundle
of atrioventricular bundle

Papillary muscles
of right ventricle

Superior vena cava

Left pulmonary veins

Left auricle

Atrioventricular bundle
– ⟨Trunk⟩
– Left bundle

Papillary muscles
of left ventricle

b

Valve of coronary sinus

Atrioventricular node

⟨Trunk⟩
of atrioventricular bundle

Superior vena cava

Inferior vena cava,
Valve of inferior
vena cava

Ascending aorta

Right auricle

Right atrium

Fossa ovalis
= Oval fossa

Right coronary artery, vein

Atrioventricular bundle
– Left bundle
– Right bundle

Septal cusp
of tricuspid valve

Septal papillary muscle
of right ventricle

Trabeculae carneae

c

Ascending aorta

Opening of right
coronary artery

Left bundle
of atrioventricular bundle

Trabeculae carneae

Left atrium

Mitral valve
– Posterior cusp
– Anterior cusp

Chordae tendineae

Posterior papillary muscle
of left ventricle

Anterior papillary muscle
of left ventricle

222 Conducting system of the heart (40%)

a The right atrium and both ventricles are opened by a longitudinal
   section perpendicular to the plane of the interventricular septum.
   Ventral aspect, schematic representation
b View from the right to the interventricular septum
c View from the left to the interventricular septum

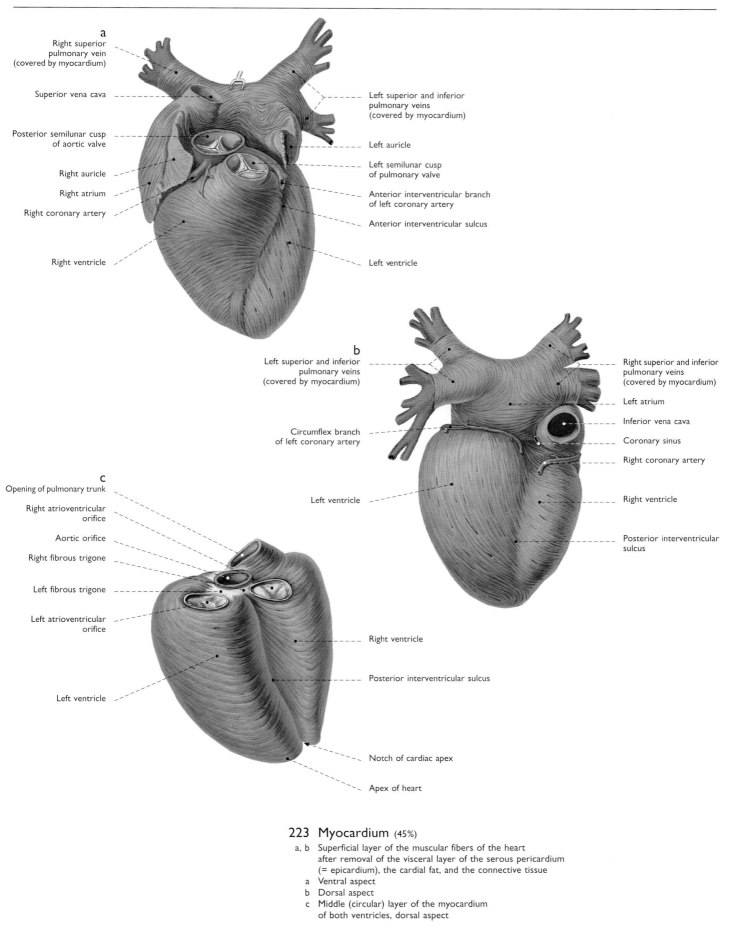

a

Right superior pulmonary vein (covered by myocardium)

Superior vena cava

Posterior semilunar cusp of aortic valve

Right auricle

Right atrium

Right coronary artery

Right ventricle

Left superior and inferior pulmonary veins (covered by myocardium)

Left auricle

Left semilunar cusp of pulmonary valve

Anterior interventricular branch of left coronary artery

Anterior interventricular sulcus

Left ventricle

b

Left superior and inferior pulmonary veins (covered by myocardium)

Circumflex branch of left coronary artery

Left ventricle

Right superior and inferior pulmonary veins (covered by myocardium)

Left atrium

Inferior vena cava

Coronary sinus

Right coronary artery

Right ventricle

Posterior interventricular sulcus

c

Opening of pulmonary trunk

Right atrioventricular orifice

Aortic orifice

Right fibrous trigone

Left fibrous trigone

Left atrioventricular orifice

Left ventricle

Right ventricle

Posterior interventricular sulcus

Notch of cardiac apex

Apex of heart

223 Myocardium (45%)

a, b Superficial layer of the muscular fibers of the heart
after removal of the visceral layer of the serous pericardium
(= epicardium), the cardial fat, and the connective tissue
a Ventral aspect
b Dorsal aspect
c Middle (circular) layer of the myocardium
of both ventricles, dorsal aspect

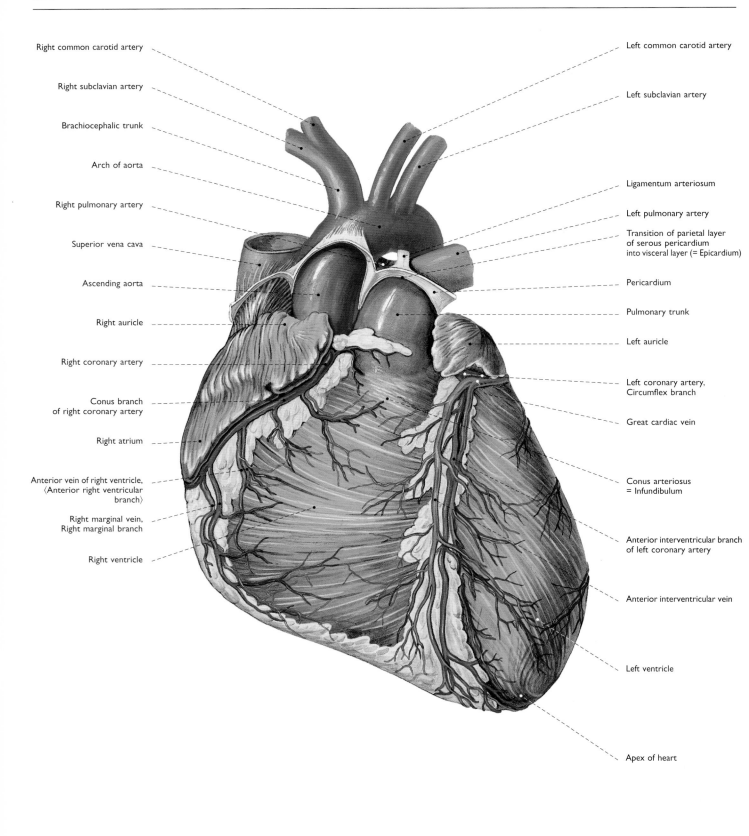

Right common carotid artery

Right subclavian artery

Brachiocephalic trunk

Arch of aorta

Right pulmonary artery

Superior vena cava

Ascending aorta

Right auricle

Right coronary artery

Conus branch
of right coronary artery

Right atrium

Anterior vein of right ventricle,
⟨Anterior right ventricular
branch⟩

Right marginal vein,
Right marginal branch

Right ventricle

Left common carotid artery

Left subclavian artery

Ligamentum arteriosum

Left pulmonary artery

Transition of parietal layer
of serous pericardium
into visceral layer (= Epicardium)

Pericardium

Pulmonary trunk

Left auricle

Left coronary artery,
Circumflex branch

Great cardiac vein

Conus arteriosus
= Infundibulum

Anterior interventricular branch
of left coronary artery

Anterior interventricular vein

Left ventricle

Apex of heart

224 Arteries and veins of the heart (100%)
The epicardium was removed.
Sternocostal (= anterior) surface, ventral aspect

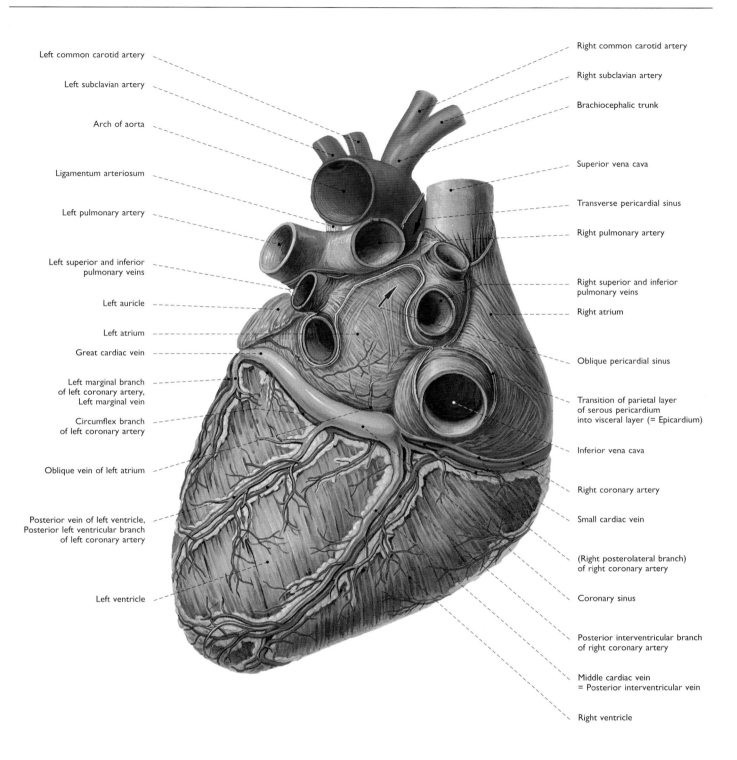

Left common carotid artery

Left subclavian artery

Arch of aorta

Ligamentum arteriosum

Left pulmonary artery

Left superior and inferior pulmonary veins

Left auricle

Left atrium

Great cardiac vein

Left marginal branch of left coronary artery, Left marginal vein

Circumflex branch of left coronary artery

Oblique vein of left atrium

Posterior vein of left ventricle, Posterior left ventricular branch of left coronary artery

Left ventricle

Right common carotid artery

Right subclavian artery

Brachiocephalic trunk

Superior vena cava

Transverse pericardial sinus

Right pulmonary artery

Right superior and inferior pulmonary veins

Right atrium

Oblique pericardial sinus

Transition of parietal layer of serous pericardium into visceral layer (= Epicardium)

Inferior vena cava

Right coronary artery

Small cardiac vein

(Right posterolateral branch) of right coronary artery

Coronary sinus

Posterior interventricular branch of right coronary artery

Middle cardiac vein = Posterior interventricular vein

Right ventricle

225 Arteries and veins of the heart (100%)
The epicardium was removed. Base of the heart and diaphragmatic (= inferior) surface, dorsal aspect

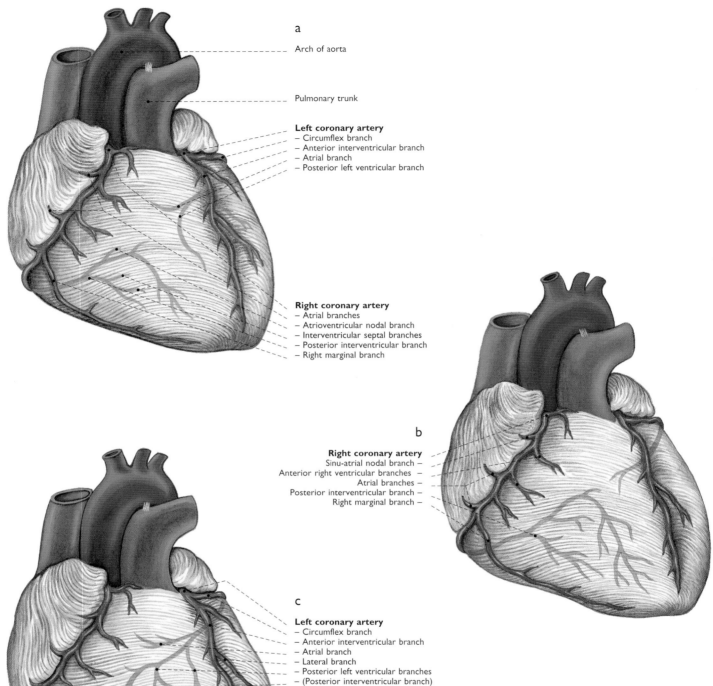

a

Arch of aorta

Pulmonary trunk

Left coronary artery
– Circumflex branch
– Anterior interventricular branch
– Atrial branch
– Posterior left ventricular branch

Right coronary artery
– Atrial branches
– Atrioventricular nodal branch
– Interventricular septal branches
– Posterior interventricular branch
– Right marginal branch

b

Right coronary artery
Sinu-atrial nodal branch –
Anterior right ventricular branches –
Atrial branches –
Posterior interventricular branch –
Right marginal branch –

c

Left coronary artery
– Circumflex branch
– Anterior interventricular branch
– Atrial branch
– Lateral branch
– Posterior left ventricular branches
– (Posterior interventricular branch)
– (Interventricular septal branches)

226 Coronary arteries of the heart (70%)
Schematic representations, ventral aspect
a Commonest arrangement
b Right dominance, the right coronary artery
 supplying the posterior walls of both ventricles
c Left dominance, the left coronary artery
 supplying the posterior walls of both ventricles

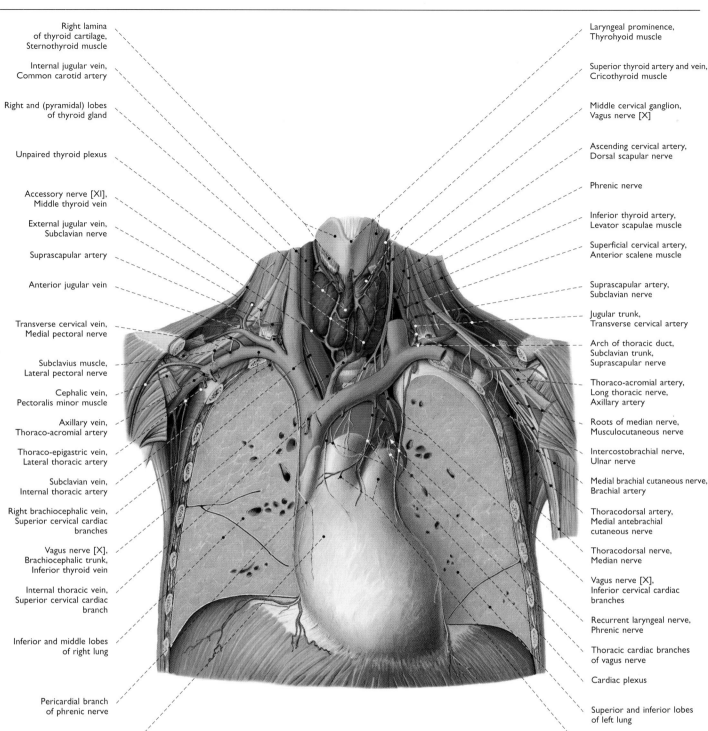

Right lamina of thyroid cartilage, Sternothyroid muscle

Internal jugular vein, Common carotid artery

Right and (pyramidal) lobes of thyroid gland

Unpaired thyroid plexus

Accessory nerve [XI], Middle thyroid vein

External jugular vein, Subclavian nerve

Suprascapular artery

Anterior jugular vein

Transverse cervical vein, Medial pectoral nerve

Subclavius muscle, Lateral pectoral nerve

Cephalic vein, Pectoralis minor muscle

Axillary vein, Thoraco-acromial artery

Thoraco-epigastric vein, Lateral thoracic artery

Subclavian vein, Internal thoracic artery

Right brachiocephalic vein, Superior cervical cardiac branches

Vagus nerve [X], Brachiocephalic trunk, Inferior thyroid vein

Internal thoracic vein, Superior cervical cardiac branch

Inferior and middle lobes of right lung

Pericardial branch of phrenic nerve

Fibrous pericardium

Laryngeal prominence, Thyrohyoid muscle

Superior thyroid artery and vein, Cricothyroid muscle

Middle cervical ganglion, Vagus nerve [X]

Ascending cervical artery, Dorsal scapular nerve

Phrenic nerve

Inferior thyroid artery, Levator scapulae muscle

Superficial cervical artery, Anterior scalene muscle

Suprascapular artery, Subclavian nerve

Jugular trunk, Transverse cervical artery

Arch of thoracic duct, Subclavian trunk, Suprascapular nerve

Thoraco-acromial artery, Long thoracic nerve, Axillary artery

Roots of median nerve, Musculocutaneous nerve

Intercostobrachial nerve, Ulnar nerve

Medial brachial cutaneous nerve, Brachial artery

Thoracodorsal artery, Medial antebrachial cutaneous nerve

Thoracodorsal nerve, Median nerve

Vagus nerve [X], Inferior cervical cardiac branches

Recurrent laryngeal nerve, Phrenic nerve

Thoracic cardiac branches of vagus nerve

Cardiac plexus

Superior and inferior lobes of left lung

Ascending aorta, Pulmonary trunk

227 Blood vessels and nerves of the neck, mediastinum, and axilla (40%)
The sternocleidomastoid and infrahyoid muscles, the anterior thoracic wall, and anterior parts of the lungs were removed. Ventral aspect

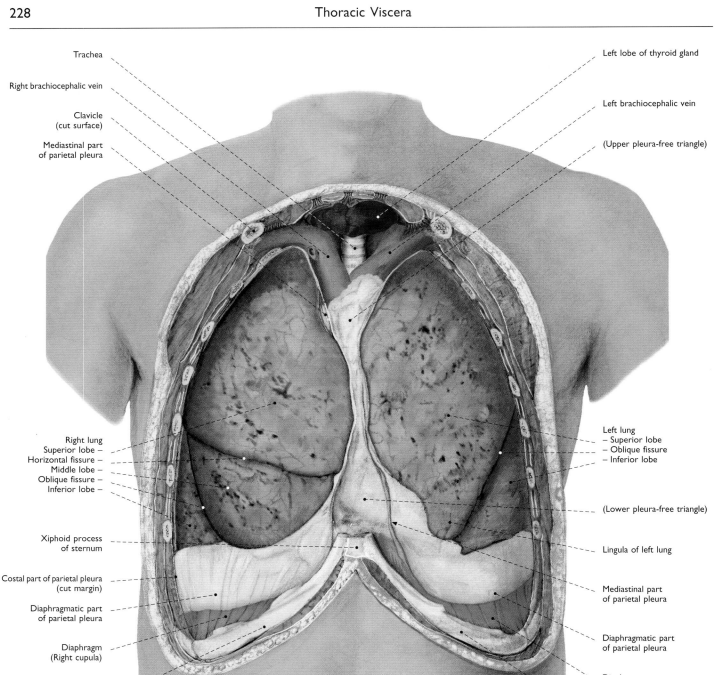

Trachea

Right brachiocephalic vein

Clavicle
(cut surface)

Mediastinal part
of parietal pleura

Left lobe of thyroid gland

Left brachiocephalic vein

(Upper pleura-free triangle)

Right lung
Superior lobe –
Horizontal fissure –
Middle lobe –
Oblique fissure –
Inferior lobe –

Left lung
– Superior lobe
– Oblique fissure
– Inferior lobe

(Lower pleura-free triangle)

Lingula of left lung

Xiphoid process
of sternum

Costal part of parietal pleura
(cut margin)

Diaphragmatic part
of parietal pleura

Diaphragm
(Right cupula)

Right costal arch

Mediastinal part
of parietal pleura

Diaphragmatic part
of parietal pleura

Diaphragm
(Left cupula)

Left costal arch

228   Thoracic viscera in situ (40%)
The anterior thoracic wall was removed.
Ventral aspect

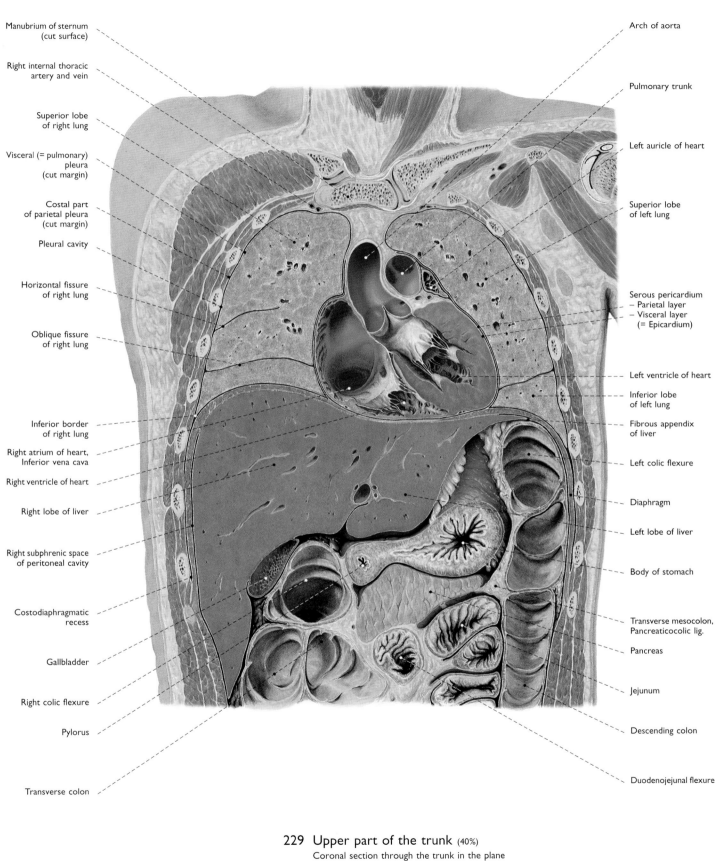

Manubrium of sternum
(cut surface)

Right internal thoracic
artery and vein

Superior lobe
of right lung

Visceral (= pulmonary)
pleura
(cut margin)

Costal part
of parietal pleura
(cut margin)

Pleural cavity

Horizontal fissure
of right lung

Oblique fissure
of right lung

Inferior border
of right lung

Right atrium of heart,
Inferior vena cava

Right ventricle of heart

Right lobe of liver

Right subphrenic space
of peritoneal cavity

Costodiaphragmatic
recess

Gallbladder

Right colic flexure

Pylorus

Transverse colon

Arch of aorta

Pulmonary trunk

Left auricle of heart

Superior lobe
of left lung

Serous pericardium
– Parietal layer
– Visceral layer
(= Epicardium)

Left ventricle of heart

Inferior lobe
of left lung

Fibrous appendix
of liver

Left colic flexure

Diaphragm

Left lobe of liver

Body of stomach

Transverse mesocolon,
Pancreaticocolic lig.

Pancreas

Jejunum

Descending colon

Duodenojejunal flexure

229 Upper part of the trunk (40%)
Coronal section through the trunk in the plane
of the two sternoclavicular joints,
drawing of an anatomical section, ventral aspect

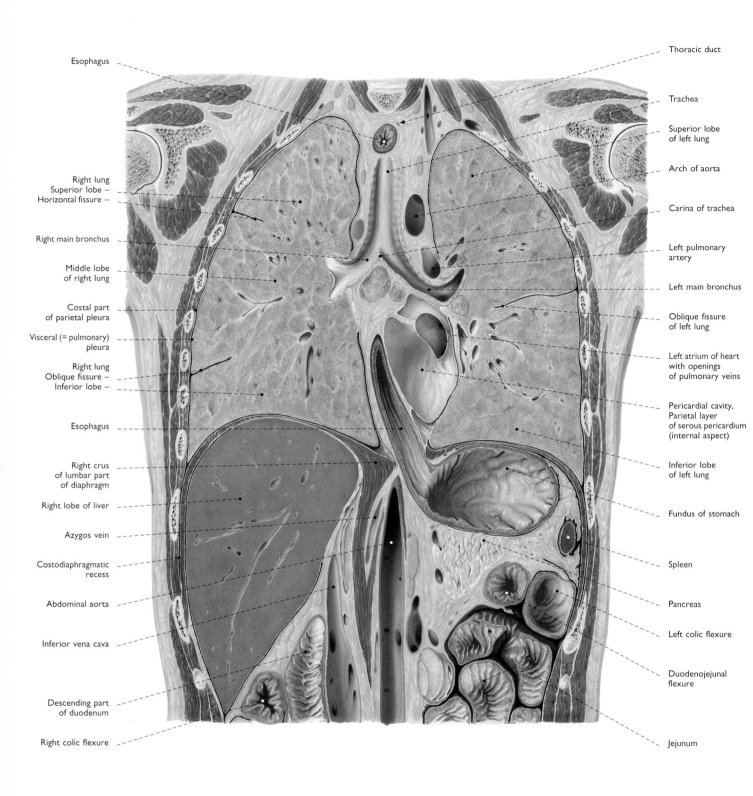

Esophagus

Right lung
Superior lobe –
Horizontal fissure –

Right main bronchus

Middle lobe
of right lung

Costal part
of parietal pleura

Visceral (= pulmonary)
pleura

Right lung
Oblique fissure –
Inferior lobe –

Esophagus

Right crus
of lumbar part
of diaphragm

Right lobe of liver

Azygos vein

Costodiaphragmatic
recess

Abdominal aorta

Inferior vena cava

Descending part
of duodenum

Right colic flexure

Thoracic duct

Trachea

Superior lobe
of left lung

Arch of aorta

Carina of trachea

Left pulmonary
artery

Left main bronchus

Oblique fissure
of left lung

Left atrium of heart
with openings
of pulmonary veins

Pericardial cavity,
Parietal layer
of serous pericardium
(internal aspect)

Inferior lobe
of left lung

Fundus of stomach

Spleen

Pancreas

Left colic flexure

Duodenojejunal
flexure

Jejunum

230 Upper part of the trunk (40%)
Coronal section through the trunk in the plane
of the tracheal bifurcation and the esophageal hiatus,
drawing of an anatomical section, ventral aspect

a

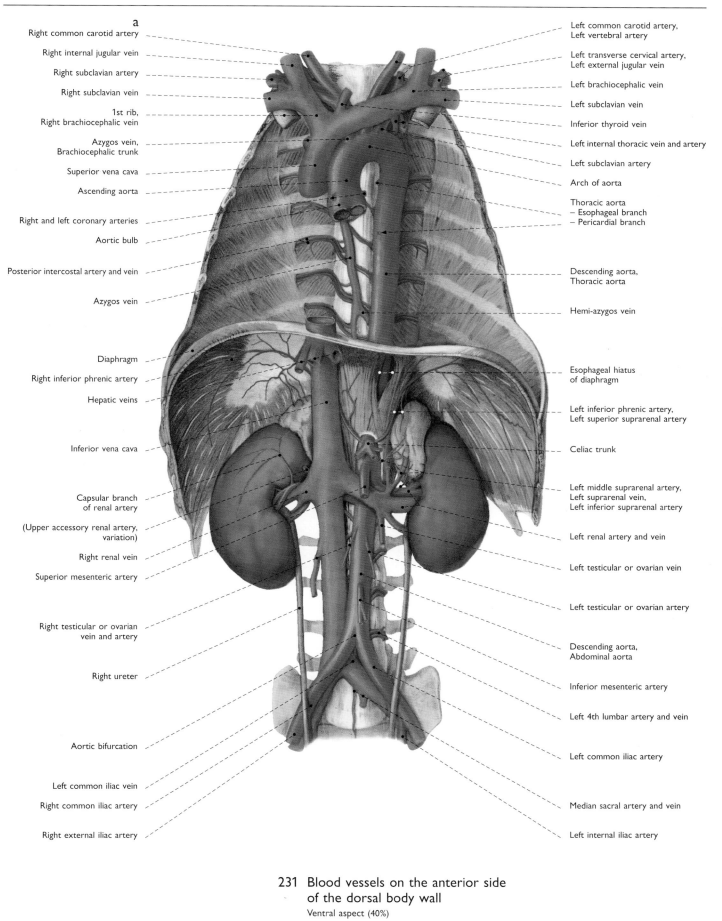

Right common carotid artery

Right internal jugular vein

Right subclavian artery

Right subclavian vein

1st rib,
Right brachiocephalic vein

Azygos vein,
Brachiocephalic trunk

Superior vena cava

Ascending aorta

Right and left coronary arteries

Aortic bulb

Posterior intercostal artery and vein

Azygos vein

Diaphragm

Right inferior phrenic artery

Hepatic veins

Inferior vena cava

Capsular branch
of renal artery

(Upper accessory renal artery,
variation)

Right renal vein

Superior mesenteric artery

Right testicular or ovarian
vein and artery

Right ureter

Aortic bifurcation

Left common iliac vein

Right common iliac artery

Right external iliac artery

Left common carotid artery,
Left vertebral artery

Left transverse cervical artery,
Left external jugular vein

Left brachiocephalic vein

Left subclavian vein

Inferior thyroid vein

Left internal thoracic vein and artery

Left subclavian artery

Arch of aorta

Thoracic aorta
− Esophageal branch
− Pericardial branch

Descending aorta,
Thoracic aorta

Hemi-azygos vein

Esophageal hiatus
of diaphragm

Left inferior phrenic artery,
Left superior suprarenal artery

Celiac trunk

Left middle suprarenal artery,
Left suprarenal vein,
Left inferior suprarenal artery

Left renal artery and vein

Left testicular or ovarian vein

Left testicular or ovarian artery

Descending aorta,
Abdominal aorta

Inferior mesenteric artery

Left 4th lumbar artery and vein

Left common iliac artery

Median sacral artery and vein

Left internal iliac artery

**231 Blood vessels on the anterior side
of the dorsal body wall**
Ventral aspect (40%)

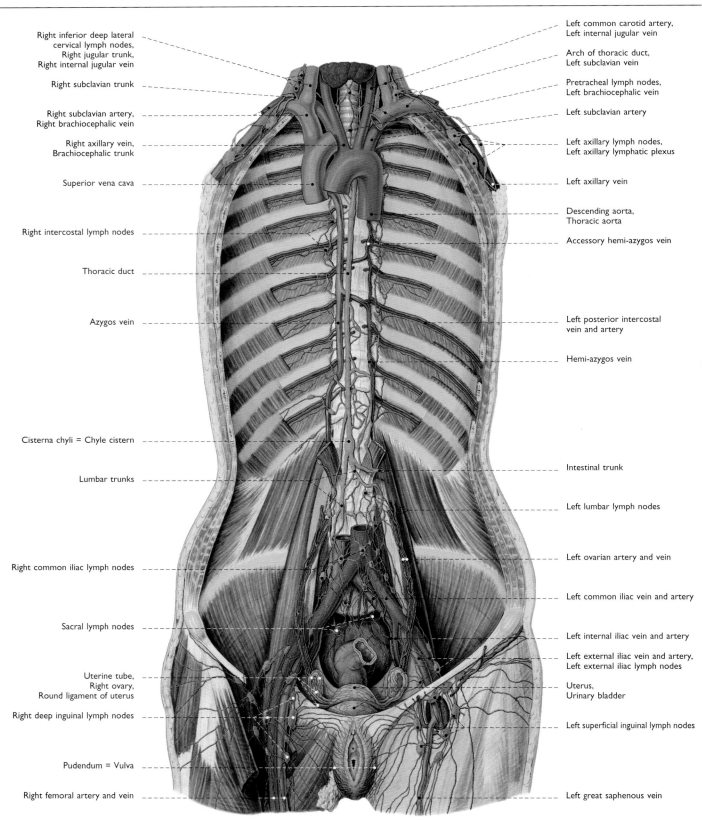

Right inferior deep lateral
cervical lymph nodes,
Right jugular trunk,
Right internal jugular vein

Right subclavian trunk

Right subclavian artery,
Right brachiocephalic vein

Right axillary vein,
Brachiocephalic trunk

Superior vena cava

Right intercostal lymph nodes

Thoracic duct

Azygos vein

Cisterna chyli = Chyle cistern

Lumbar trunks

Right common iliac lymph nodes

Sacral lymph nodes

Uterine tube,
Right ovary,
Round ligament of uterus

Right deep inguinal lymph nodes

Pudendum = Vulva

Right femoral artery and vein

Left common carotid artery,
Left internal jugular vein

Arch of thoracic duct,
Left subclavian vein

Pretracheal lymph nodes,
Left brachiocephalic vein

Left subclavian artery

Left axillary lymph nodes,
Left axillary lymphatic plexus

Left axillary vein

Descending aorta,
Thoracic aorta

Accessory hemi-azygos vein

Left posterior intercostal
vein and artery

Hemi-azygos vein

Intestinal trunk

Left lumbar lymph nodes

Left ovarian artery and vein

Left common iliac vein and artery

Left internal iliac vein and artery

Left external iliac vein and artery,
Left external iliac lymph nodes

Uterus,
Urinary bladder

Left superficial inguinal lymph nodes

Left great saphenous vein

232 Lymphatic vessels and lymph nodes of the thorax, abdomen, and pelvis of a female (30%)

Lymphatic vessels and lymph nodes on the anterior surface
of the dorsal body wall, pelvis, and inguinal region, ventral aspect

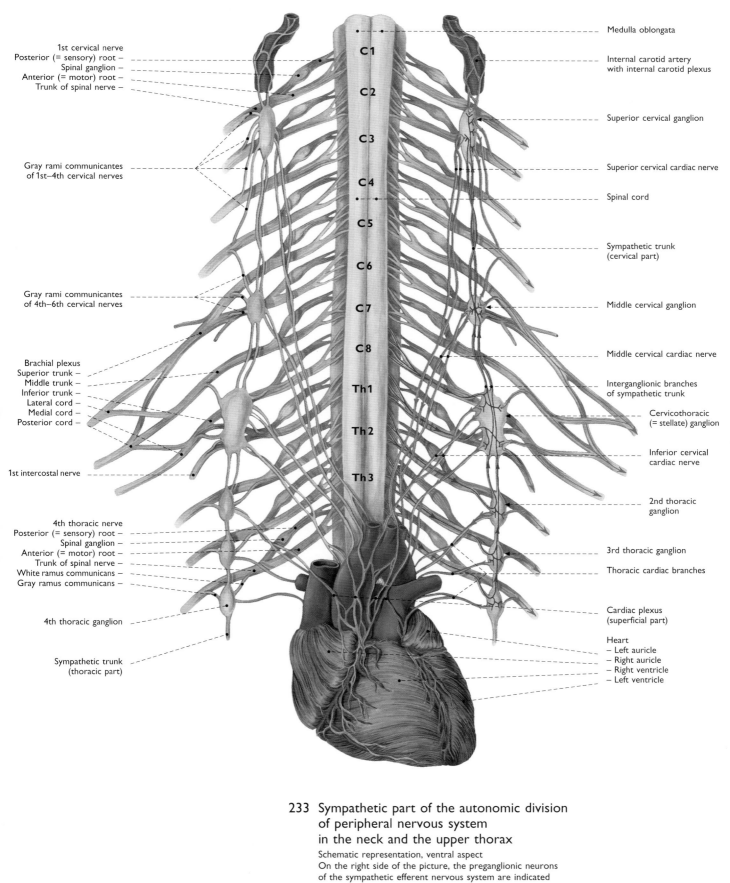

1st cervical nerve
Posterior (= sensory) root –
Spinal ganglion –
Anterior (= motor) root –
Trunk of spinal nerve –

Gray rami communicantes
of 1st–4th cervical nerves

Gray rami communicantes
of 4th–6th cervical nerves

Brachial plexus
Superior trunk –
Middle trunk –
Inferior trunk –
Lateral cord –
Medial cord –
Posterior cord –

1st intercostal nerve

4th thoracic nerve
Posterior (= sensory) root –
Spinal ganglion –
Anterior (= motor) root –
Trunk of spinal nerve –
White ramus communicans –
Gray ramus communicans –

4th thoracic ganglion

Sympathetic trunk
(thoracic part)

C1
C2
C3
C4
C5
C6
C7
C8
Th1
Th2
Th3

Medulla oblongata

Internal carotid artery
with internal carotid plexus

Superior cervical ganglion

Superior cervical cardiac nerve

Spinal cord

Sympathetic trunk
(cervical part)

Middle cervical ganglion

Middle cervical cardiac nerve

Interganglionic branches
of sympathetic trunk

Cervicothoracic
(= stellate) ganglion

Inferior cervical
cardiac nerve

2nd thoracic
ganglion

3rd thoracic ganglion

Thoracic cardiac branches

Cardiac plexus
(superficial part)

Heart
– Left auricle
– Right auricle
– Right ventricle
– Left ventricle

**233 Sympathetic part of the autonomic division
of peripheral nervous system
in the neck and the upper thorax**

Schematic representation, ventral aspect
On the right side of the picture, the preganglionic neurons
of the sympathetic efferent nervous system are indicated
by blue lines, the postganglionic neurons by green ones.

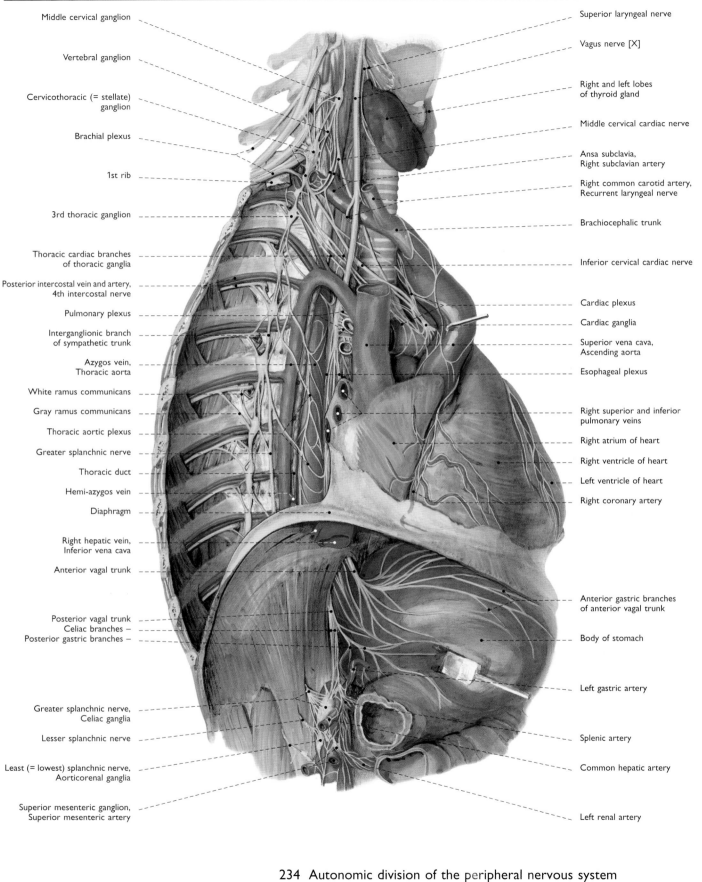

Middle cervical ganglion

Vertebral ganglion

Cervicothoracic (= stellate) ganglion

Brachial plexus

1st rib

3rd thoracic ganglion

Thoracic cardiac branches of thoracic ganglia

Posterior intercostal vein and artery, 4th intercostal nerve

Pulmonary plexus

Interganglionic branch of sympathetic trunk

Azygos vein, Thoracic aorta

White ramus communicans

Gray ramus communicans

Thoracic aortic plexus

Greater splanchnic nerve

Thoracic duct

Hemi-azygos vein

Diaphragm

Right hepatic vein, Inferior vena cava

Anterior vagal trunk

Posterior vagal trunk
Celiac branches –
Posterior gastric branches –

Greater splanchnic nerve, Celiac ganglia

Lesser splanchnic nerve

Least (= lowest) splanchnic nerve, Aorticorenal ganglia

Superior mesenteric ganglion, Superior mesenteric artery

Superior laryngeal nerve

Vagus nerve [X]

Right and left lobes of thyroid gland

Middle cervical cardiac nerve

Ansa subclavia, Right subclavian artery

Right common carotid artery, Recurrent laryngeal nerve

Brachiocephalic trunk

Inferior cervical cardiac nerve

Cardiac plexus

Cardiac ganglia

Superior vena cava, Ascending aorta

Esophageal plexus

Right superior and inferior pulmonary veins

Right atrium of heart

Right ventricle of heart

Left ventricle of heart

Right coronary artery

Anterior gastric branches of anterior vagal trunk

Body of stomach

Left gastric artery

Splenic artery

Common hepatic artery

Left renal artery

234 Autonomic division of the peripheral nervous system in the thorax and the upper abdomen (50%)
Anterolateral aspect

Abdominal and Pelvic Viscera

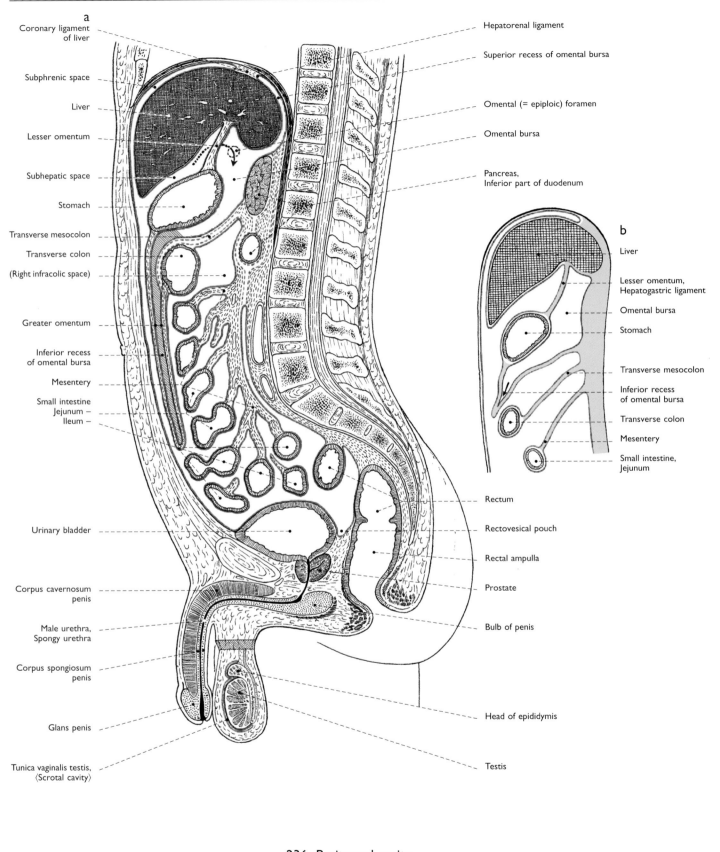

a

Coronary ligament of liver

Subphrenic space

Liver

Lesser omentum

Subhepatic space

Stomach

Transverse mesocolon

Transverse colon

(Right infracolic space)

Greater omentum

Inferior recess of omental bursa

Mesentery

Small intestine
Jejunum –
Ileum –

Urinary bladder

Corpus cavernosum penis

Male urethra, Spongy urethra

Corpus spongiosum penis

Glans penis

Tunica vaginalis testis, ⟨Scrotal cavity⟩

Hepatorenal ligament

Superior recess of omental bursa

Omental (= epiploic) foramen

Omental bursa

Pancreas, Inferior part of duodenum

Rectum

Rectovesical pouch

Rectal ampulla

Prostate

Bulb of penis

Head of epididymis

Testis

b

Liver

Lesser omentum, Hepatogastric ligament

Omental bursa

Stomach

Transverse mesocolon

Inferior recess of omental bursa

Transverse colon

Mesentery

Small intestine, Jejunum

236 Peritoneal cavity
Schematized median sections
a Adult stage of a male. Step-cut in the scrotum in order
to expose the right cavity of the tunica vaginalis
b Fetal stage

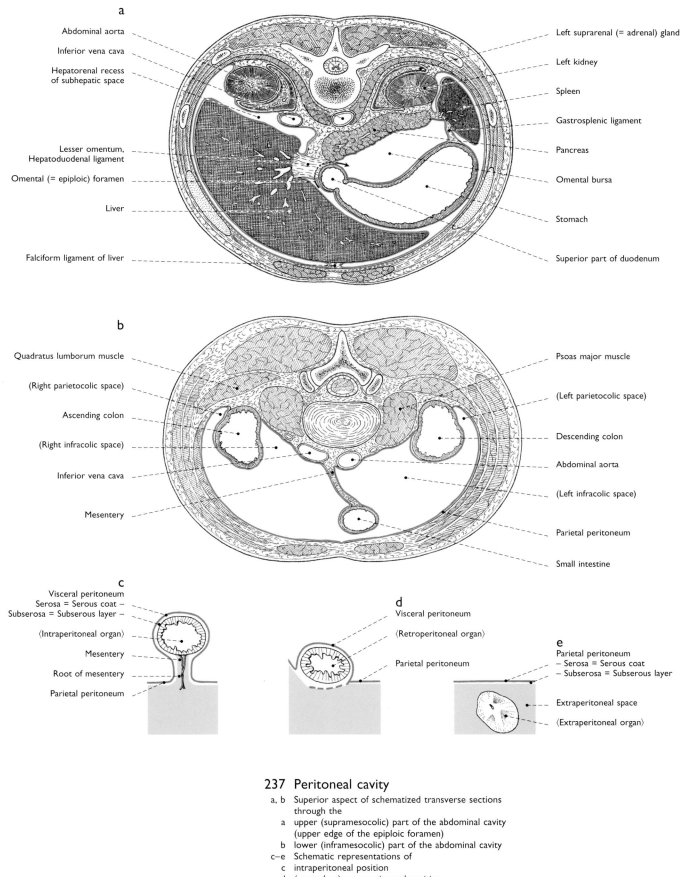

a

Abdominal aorta

Inferior vena cava

Hepatorenal recess
of subhepatic space

Lesser omentum,
Hepatoduodenal ligament

Omental (= epiploic) foramen

Liver

Falciform ligament of liver

Left suprarenal (= adrenal) gland

Left kidney

Spleen

Gastrosplenic ligament

Pancreas

Omental bursa

Stomach

Superior part of duodenum

b

Quadratus lumborum muscle

(Right parietocolic space)

Ascending colon

(Right infracolic space)

Inferior vena cava

Mesentery

Psoas major muscle

(Left parietocolic space)

Descending colon

Abdominal aorta

(Left infracolic space)

Parietal peritoneum

Small intestine

c

Visceral peritoneum
Serosa = Serous coat –
Subserosa = Subserous layer –

⟨Intraperitoneal organ⟩

Mesentery

Root of mesentery

Parietal peritoneum

d

Visceral peritoneum

⟨Retroperitoneal organ⟩

Parietal peritoneum

e

Parietal peritoneum
– Serosa = Serous coat
– Subserosa = Subserous layer

Extraperitoneal space

⟨Extraperitoneal organ⟩

237 Peritoneal cavity

a, b Superior aspect of schematized transverse sections
through the
a upper (supramesocolic) part of the abdominal cavity
(upper edge of the epiploic foramen)
b lower (inframesocolic) part of the abdominal cavity
c–e Schematic representations of
c intraperitoneal position
d (secondary) retroperitoneal position
e extraperitoneal position

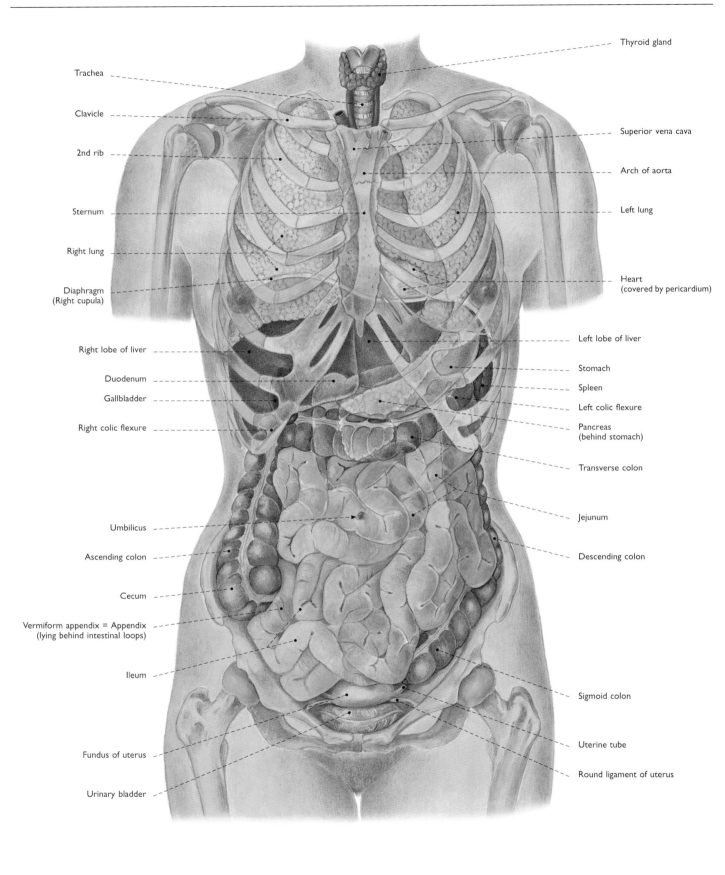

Thyroid gland

Trachea

Clavicle

Superior vena cava

2nd rib

Arch of aorta

Sternum

Left lung

Right lung

Diaphragm
(Right cupula)

Heart
(covered by pericardium)

Right lobe of liver

Left lobe of liver

Duodenum

Stomach

Gallbladder

Spleen

Right colic flexure

Left colic flexure

Pancreas
(behind stomach)

Transverse colon

Jejunum

Umbilicus

Ascending colon

Descending colon

Cecum

Vermiform appendix = Appendix
(lying behind intestinal loops)

Ileum

Sigmoid colon

Uterine tube

Fundus of uterus

Round ligament of uterus

Urinary bladder

238 Thoracic and abdominal viscera (30%)
Surface projections onto the chest and
anterior abdominal walls, ventral aspect

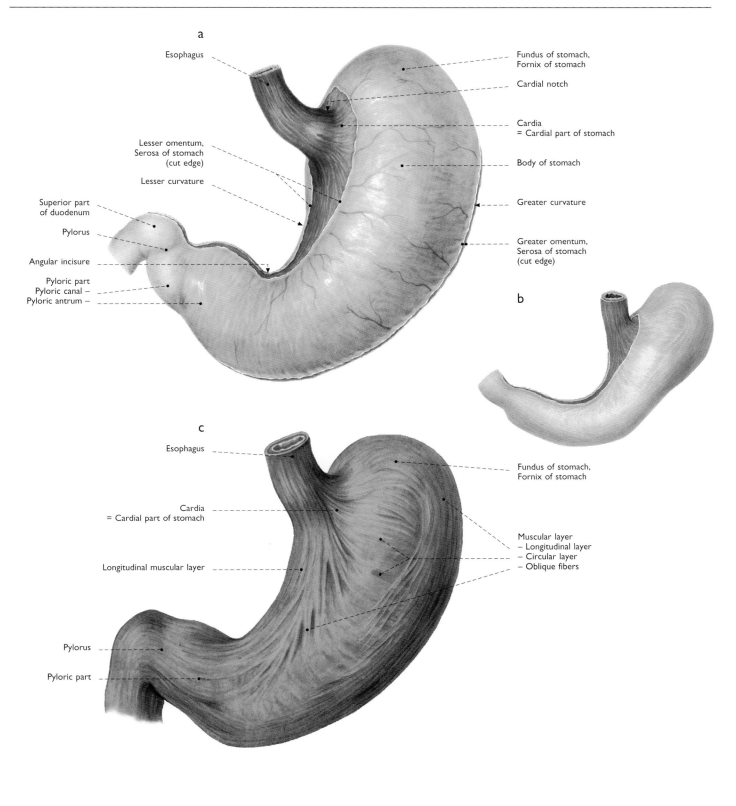

a

Esophagus

Fundus of stomach,
Fornix of stomach

Cardial notch

Cardia
= Cardial part of stomach

Lesser omentum,
Serosa of stomach
(cut edge)

Body of stomach

Lesser curvature

Superior part
of duodenum

Greater curvature

Pylorus

Greater omentum,
Serosa of stomach
(cut edge)

Angular incisure

Pyloric part
Pyloric canal
Pyloric antrum

b

c

Esophagus

Fundus of stomach,
Fornix of stomach

Cardia
= Cardial part of stomach

Muscular layer
– Longitudinal layer
– Circular layer
– Oblique fibers

Longitudinal muscular layer

Pylorus

Pyloric part

239 Stomach (55%)

Ventral aspect
a   External aspect of a completely filled stomach with distended walls
b   Empty stomach with extremely contracted walls
c   Muscle tracts in the anterior wall of the stomach after removal
of the serosa and subserosa

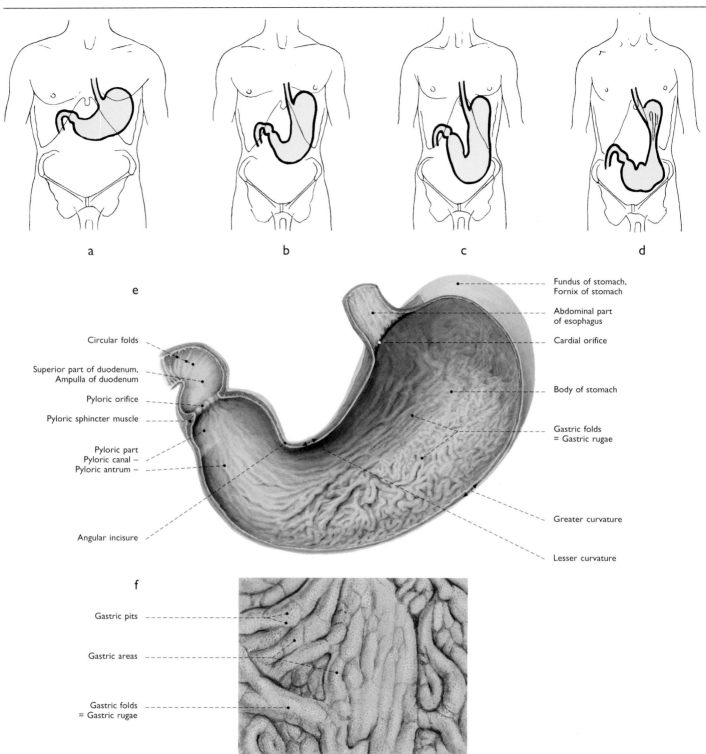

a

b

c

d

e

Fundus of stomach,
Fornix of stomach

Abdominal part
of esophagus

Cardial orifice

Circular folds

Superior part of duodenum,
Ampulla of duodenum

Pyloric orifice

Pyloric sphincter muscle

Body of stomach

Pyloric part
Pyloric canal
Pyloric antrum

Gastric folds
= Gastric rugae

Angular incisure

Greater curvature

Lesser curvature

f

Gastric pits

Gastric areas

Gastric folds
= Gastric rugae

240 Stomach

a–d Schematic representations of some important functional forms
of the stomach
a Hypertonic stomach
b Orthotonic stomach
c Hypotonic stomach
d Atonic stomach
e, f Internal aspect of the stomach
e Internal aspect of the posterior wall after removal
of the anterior wall (50%)
f Mucosa of the body of stomach (300%)

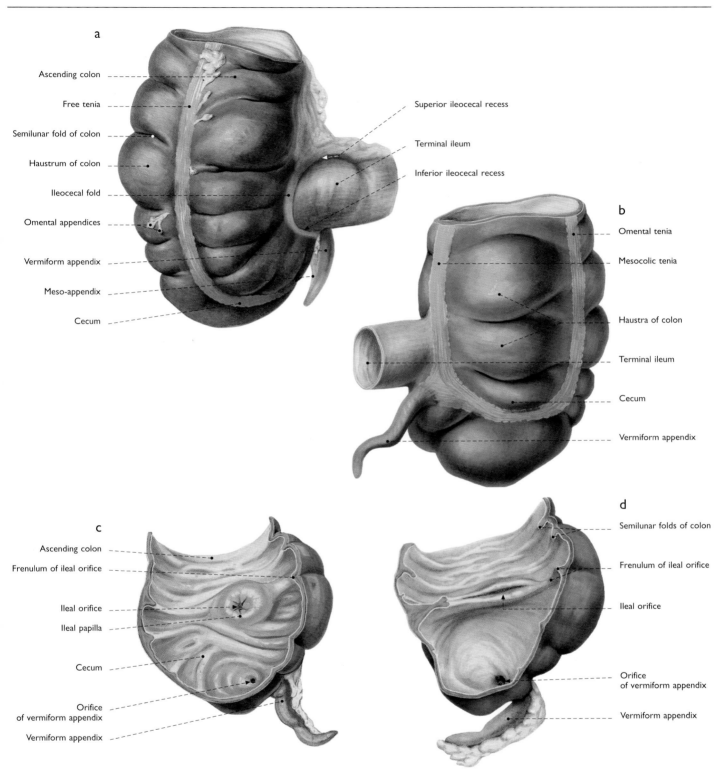

a

Ascending colon

Free tenia

Semilunar fold of colon

Haustrum of colon

Ileocecal fold

Omental appendices

Vermiform appendix

Meso-appendix

Cecum

Superior ileocecal recess

Terminal ileum

Inferior ileocecal recess

b

Omental tenia

Mesocolic tenia

Haustra of colon

Terminal ileum

Cecum

Vermiform appendix

c

Ascending colon

Frenulum of ileal orifice

Ileal orifice

Ileal papilla

Cecum

Orifice
of vermiform appendix

Vermiform appendix

d

Semilunar folds of colon

Frenulum of ileal orifice

Ileal orifice

Orifice
of vermiform appendix

Vermiform appendix

241 Cecum and vermiform appendix (70%)

   a Ventral aspect
   b Dorsal aspect
c, d Ileal orifice and orifice of vermiform appendix after removal
      of the ventrolateral wall of the cecum and the ascending colon
   c View during life
   d View of the relaxed intestine after death

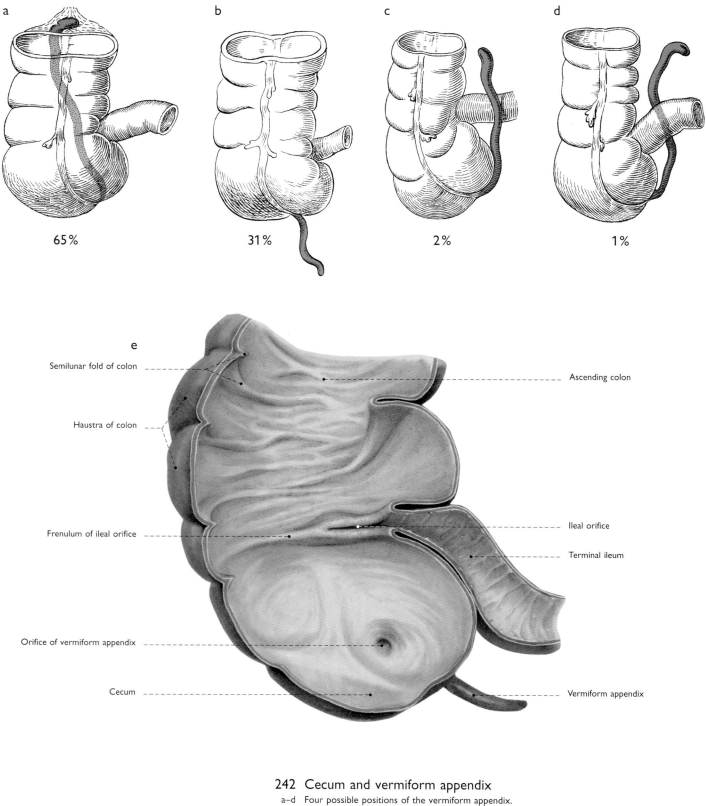

a   b   c   d

65%   31%   2%   1%

e

Semilunar fold of colon

Haustra of colon

Frenulum of ileal orifice

Orifice of vermiform appendix

Cecum

Ascending colon

Ileal orifice

Terminal ileum

Vermiform appendix

242 Cecum and vermiform appendix

a–d Four possible positions of the vermiform appendix.
The percentile numbers beneath the pictures indicate
the approximate frequency.
a Retrocecal and retrocolic position
b Pendulous position into the lesser pelvis
c Pre-ileal position
d Retro-ileal position
e Internal aspect of the dorsal wall of the cecum
after removal of the ventral intestinal wall (80%)

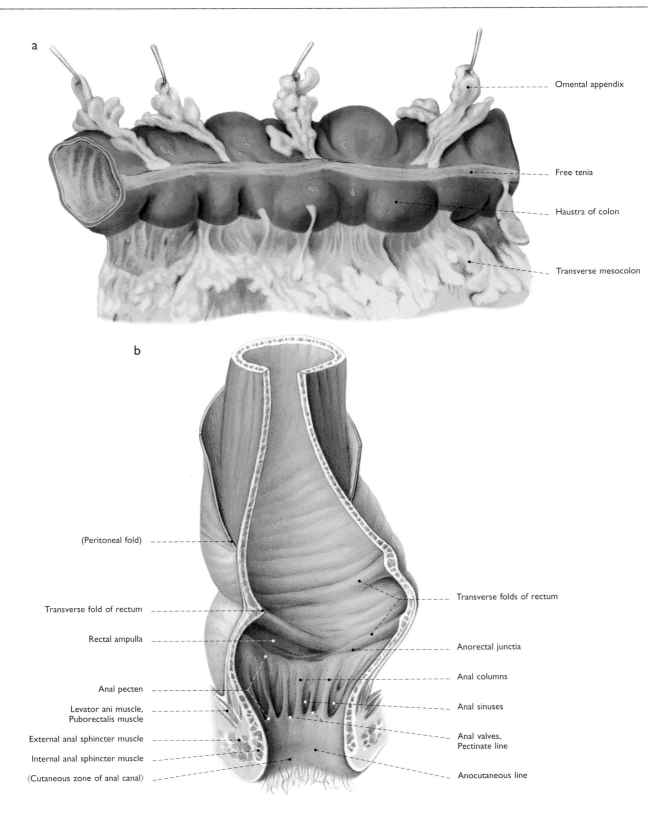

a

Omental appendix

Free tenia

Haustra of colon

Transverse mesocolon

b

(Peritoneal fold)

Transverse fold of rectum

Rectal ampulla

Anal pecten

Levator ani muscle,
Puborectalis muscle

External anal sphincter muscle

Internal anal sphincter muscle

(Cutaneous zone of anal canal)

Transverse folds of rectum

Anorectal junctia

Anal columns

Anal sinuses

Anal valves,
Pectinate line

Anocutaneous line

243 Transverse colon and rectum (70%)
a Dorsal aspect of the middle part of the transverse colon
b Interior of the rectum revealed by incising along the midventral line

a

Coronary ligament
(superior layer)

Right triangular ligament

Diaphragmatic surface
Superior part –
Right part –
Anterior part –

Right lobe of liver

Gallbladder

Inferior border

Inferior vena cava

Fibrous appendix of liver

Left triangular ligament

Bare area

Left lobe of liver

Falciform ligament of liver

Round ligament of liver

b

Falciform ligament of liver

Left lobe of liver

Bare area

Coronary ligament

Caudate lobe

Fibrous appendix of liver

Right lobe of liver

Hepatic veins

Inferior vena cava

Coronary ligament
(superior layer)

Bare area

Right triangular ligament

Coronary ligament
(inferior layer)

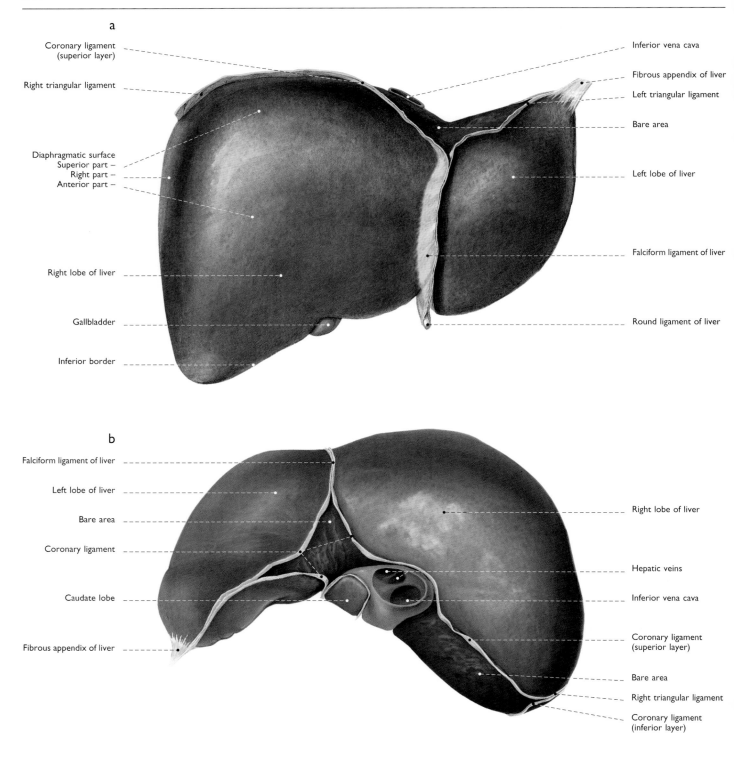

244 Liver (45%)
Diaphragmatic surface
a Ventral aspect
b Superior aspect

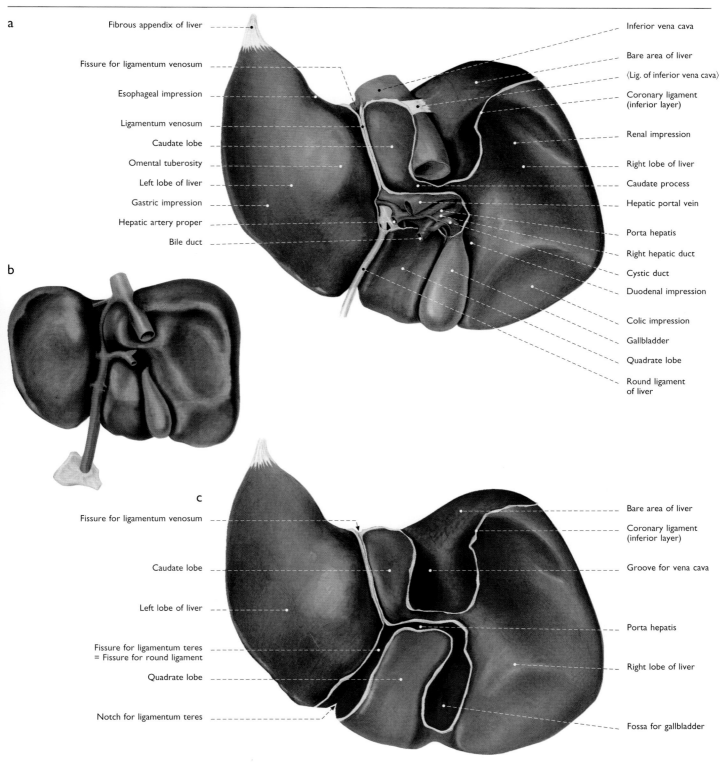

a

Fibrous appendix of liver

Fissure for ligamentum venosum

Esophageal impression

Ligamentum venosum

Caudate lobe

Omental tuberosity

Left lobe of liver

Gastric impression

Hepatic artery proper

Bile duct

Inferior vena cava

Bare area of liver

⟨Lig. of inferior vena cava⟩

Coronary ligament
(inferior layer)

Renal impression

Right lobe of liver

Caudate process

Hepatic portal vein

Porta hepatis

Right hepatic duct

Cystic duct

Duodenal impression

Colic impression

Gallbladder

Quadrate lobe

Round ligament
of liver

b

c

Fissure for ligamentum venosum

Caudate lobe

Left lobe of liver

Fissure for ligamentum teres
= Fissure for round ligament

Quadrate lobe

Notch for ligamentum teres

Bare area of liver

Coronary ligament
(inferior layer)

Groove for vena cava

Porta hepatis

Right lobe of liver

Fossa for gallbladder

245 Liver (45%)

Visceral surface, dorsal aspect
a Liver of an adult
b Liver of a newborn child
c Liver of an adult after removal of the gallbladder,
   the inferior vena cava, the hepatic portal vein,
   and the remnants of embryonal vessels

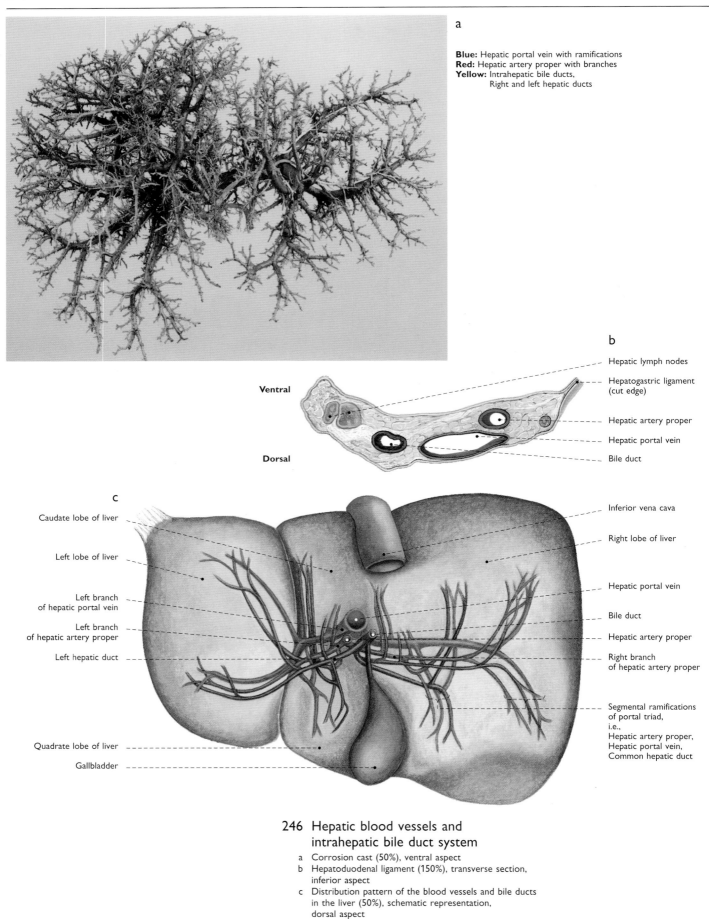

a

Blue: Hepatic portal vein with ramifications
Red: Hepatic artery proper with branches
Yellow: Intrahepatic bile ducts,
                  Right and left hepatic ducts

b

Ventral

Hepatic lymph nodes

Hepatogastric ligament
(cut edge)

Hepatic artery proper

Hepatic portal vein

Dorsal

Bile duct

c

Caudate lobe of liver

Left lobe of liver

Left branch
of hepatic portal vein

Left branch
of hepatic artery proper

Left hepatic duct

Quadrate lobe of liver

Gallbladder

Inferior vena cava

Right lobe of liver

Hepatic portal vein

Bile duct

Hepatic artery proper

Right branch
of hepatic artery proper

Segmental ramifications
of portal triad,
i.e.,
Hepatic artery proper,
Hepatic portal vein,
Common hepatic duct

246 Hepatic blood vessels and
intrahepatic bile duct system

a Corrosion cast (50%), ventral aspect
b Hepatoduodenal ligament (150%), transverse section,
   inferior aspect
c Distribution pattern of the blood vessels and bile ducts
   in the liver (50%), schematic representation,
   dorsal aspect

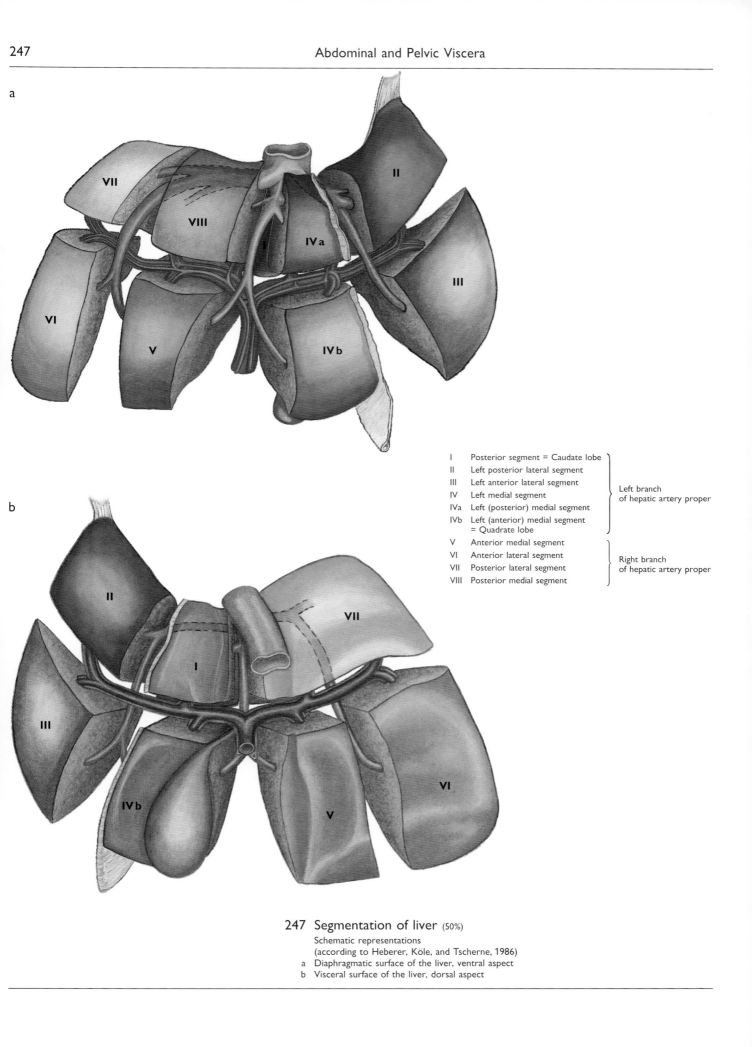

a

b

I Posterior segment = Caudate lobe
II Left posterior lateral segment
III Left anterior lateral segment
IV Left medial segment
IVa Left (posterior) medial segment
IVb Left (anterior) medial segment
= Quadrate lobe

Left branch
of hepatic artery proper

V Anterior medial segment
VI Anterior lateral segment
VII Posterior lateral segment
VIII Posterior medial segment

Right branch
of hepatic artery proper

247 Segmentation of liver (50%)
Schematic representations
(according to Heberer, Köle, and Tscherne, 1986)
a Diaphragmatic surface of the liver, ventral aspect
b Visceral surface of the liver, dorsal aspect

a

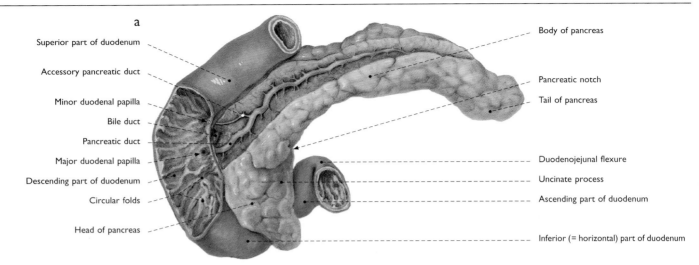

Superior part of duodenum

Accessory pancreatic duct

Minor duodenal papilla

Bile duct

Pancreatic duct

Major duodenal papilla

Descending part of duodenum

Circular folds

Head of pancreas

Body of pancreas

Pancreatic notch

Tail of pancreas

Duodenojejunal flexure

Uncinate process

Ascending part of duodenum

Inferior (= horizontal) part of duodenum

b

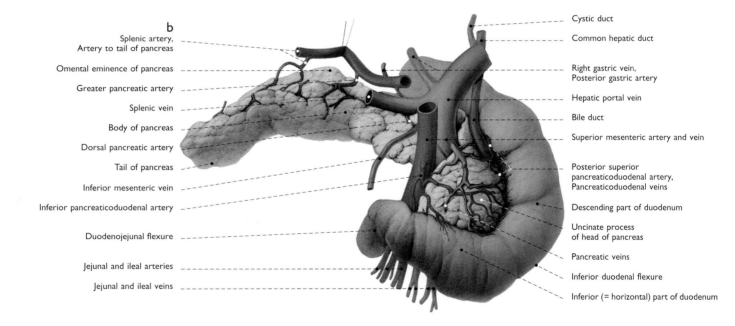

Splenic artery,
Artery to tail of pancreas

Omental eminence of pancreas

Greater pancreatic artery

Splenic vein

Body of pancreas

Dorsal pancreatic artery

Tail of pancreas

Inferior mesenteric vein

Inferior pancreaticoduodenal artery

Duodenojejunal flexure

Jejunal and ileal arteries

Jejunal and ileal veins

Cystic duct

Common hepatic duct

Right gastric vein,
Posterior gastric artery

Hepatic portal vein

Bile duct

Superior mesenteric artery and vein

Posterior superior
pancreaticoduodenal artery,
Pancreaticoduodenal veins

Descending part of duodenum

Uncinate process
of head of pancreas

Pancreatic veins

Inferior duodenal flexure

Inferior (= horizontal) part of duodenum

248 Pancreas and duodenum (60%)

a The anterior wall of the descending part of the duodenum
   was removed. The pancreatic duct and the accessory pancreatic duct
   are exposed. Ventral aspect
b Pancreas, duodenum, bile duct, and adjacent blood vessels,
   dorsal aspect

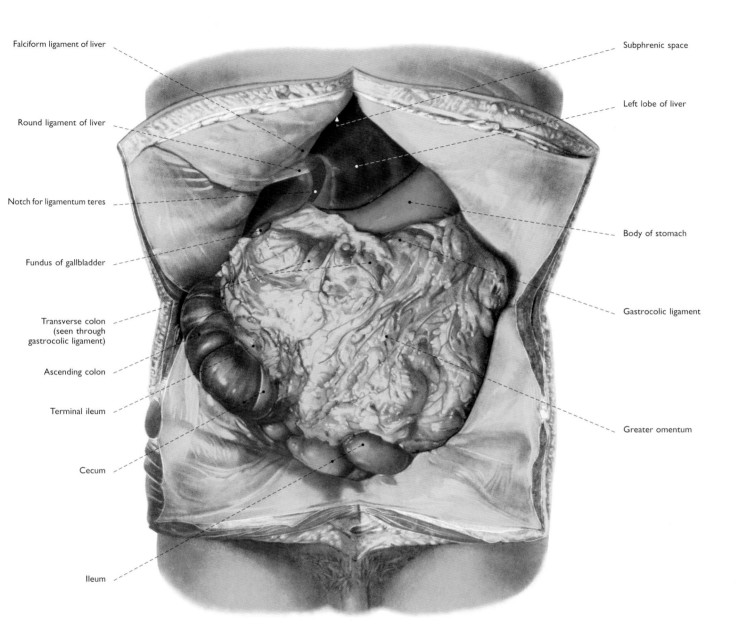

Falciform ligament of liver

Round ligament of liver

Notch for ligamentum teres

Fundus of gallbladder

Transverse colon
(seen through
gastrocolic ligament)

Ascending colon

Terminal ileum

Cecum

Ileum

Subphrenic space

Left lobe of liver

Body of stomach

Gastrocolic ligament

Greater omentum

249 Superficial abdominal viscera (30%)
The abdominal wall was opened by a cruciate incision
and retracted. Ventral aspect

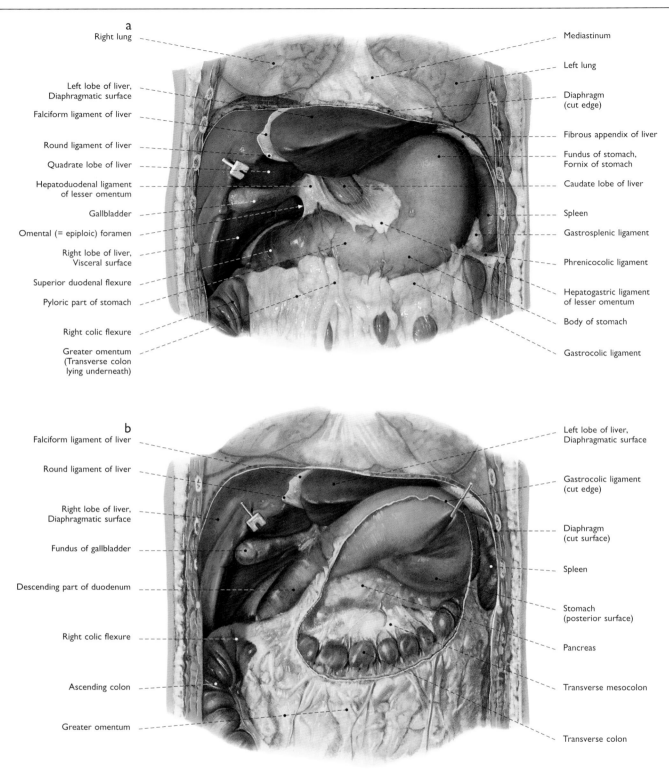

a

Right lung

Left lobe of liver,
Diaphragmatic surface

Falciform ligament of liver

Round ligament of liver

Quadrate lobe of liver

Hepatoduodenal ligament
of lesser omentum

Gallbladder

Omental (= epiploic) foramen

Right lobe of liver,
Visceral surface

Superior duodenal flexure

Pyloric part of stomach

Right colic flexure

Greater omentum
(Transverse colon
lying underneath)

Mediastinum

Left lung

Diaphragm
(cut edge)

Fibrous appendix of liver

Fundus of stomach,
Fornix of stomach

Caudate lobe of liver

Spleen

Gastrosplenic ligament

Phrenicocolic ligament

Hepatogastric ligament
of lesser omentum

Body of stomach

Gastrocolic ligament

b

Falciform ligament of liver

Round ligament of liver

Right lobe of liver,
Diaphragmatic surface

Fundus of gallbladder

Descending part of duodenum

Right colic flexure

Ascending colon

Greater omentum

Left lobe of liver,
Diaphragmatic surface

Gastrocolic ligament
(cut edge)

Diaphragm
(cut surface)

Spleen

Stomach
(posterior surface)

Pancreas

Transverse mesocolon

Transverse colon

250 Viscera of the upper abdomen (35%)

The ventral body wall was removed, the liver is raised by hooks.
a   Ventral aspect
b   The gastrocolic ligament was additionally cut transversely
and the stomach retracted cranially. By this, the dorsal wall
of the omental bursa is exposed. Ventral aspect

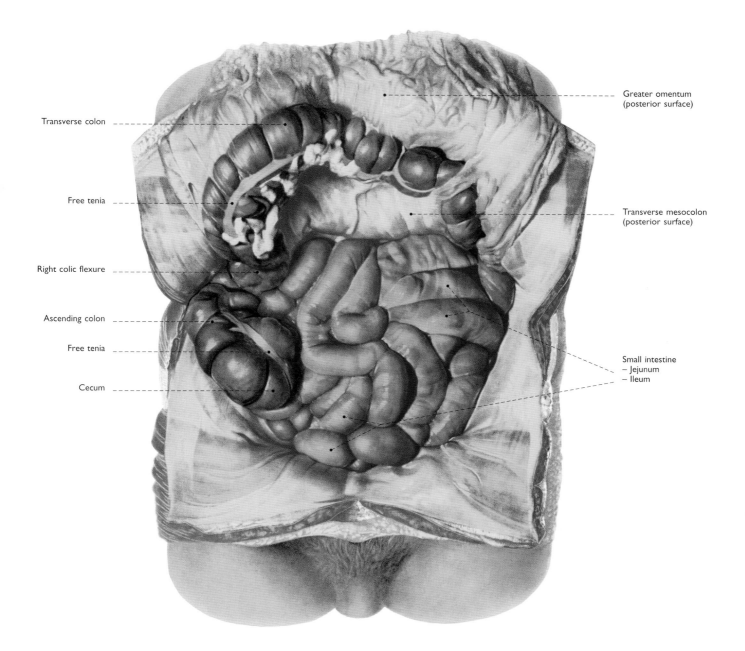

Transverse colon

Free tenia

Right colic flexure

Ascending colon

Free tenia

Cecum

Greater omentum
(posterior surface)

Transverse mesocolon
(posterior surface)

Small intestine
– Jejunum
– Ileum

251 Intraperitoneal viscera of the lower abdomen (30%)
The greater omentum and the transverse colon
are retracted upwards. Ventral aspect

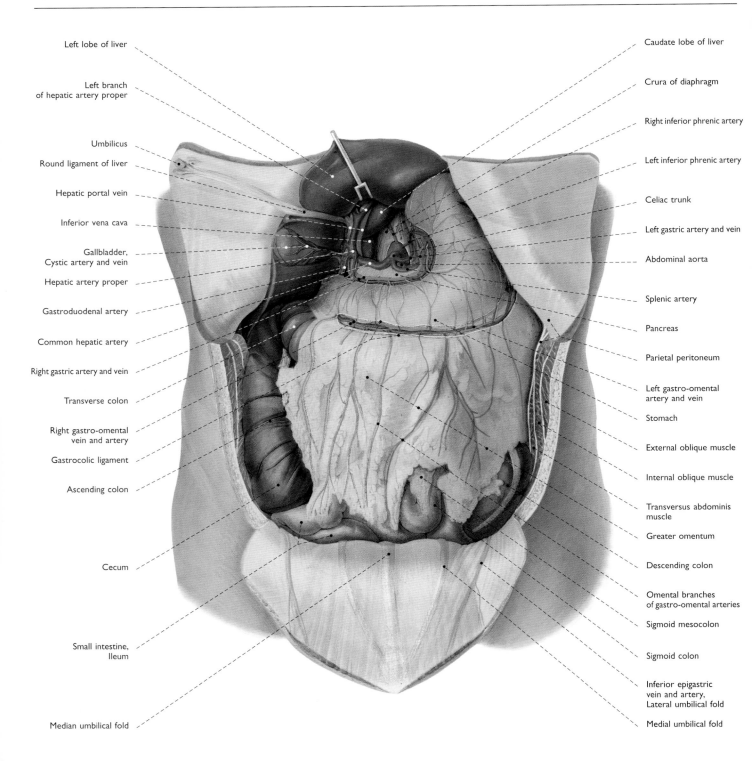

Left lobe of liver

Left branch
of hepatic artery proper

Umbilicus

Round ligament of liver

Hepatic portal vein

Inferior vena cava

Gallbladder,
Cystic artery and vein

Hepatic artery proper

Gastroduodenal artery

Common hepatic artery

Right gastric artery and vein

Transverse colon

Right gastro-omental
vein and artery

Gastrocolic ligament

Ascending colon

Cecum

Small intestine,
Ileum

Median umbilical fold

Caudate lobe of liver

Crura of diaphragm

Right inferior phrenic artery

Left inferior phrenic artery

Celiac trunk

Left gastric artery and vein

Abdominal aorta

Splenic artery

Pancreas

Parietal peritoneum

Left gastro-omental
artery and vein

Stomach

External oblique muscle

Internal oblique muscle

Transversus abdominis
muscle

Greater omentum

Descending colon

Omental branches
of gastro-omental arteries

Sigmoid mesocolon

Sigmoid colon

Inferior epigastric
vein and artery,
Lateral umbilical fold

Medial umbilical fold

252 Celiac trunk and its branches (30%)
The lesser omentum was removed,
the liver is retracted upwards. Ventral aspect

Left lobe of liver

Right gastro-omental
vein and artery

Round ligament of liver

Quadrate lobe of liver

Common hepatic artery

Gallbladder

Cystic vein and artery

Pyloric part of stomach

Gastroduodenal artery

Splenic artery

Anterior superior
pancreaticoduodenal artery

Transverse colon

Middle colic vein and artery

Inferior pancreatico-
duodenal artery

Head of pancreas

Parietal peritoneum
(cut edge)

Descending part
of duodenum

Superior mesenteric vein

Ascending colon

Right colic artery

Ileocolic artery

Terminal ileum

Inferior epigastric
veins and artery,
Lateral umbilical fold

Body of stomach

Left gastric artery and vein

Left crus of diaphragm

Celiac trunk

Inferior phrenic artery

Splenic artery

Splenic vein

Left colic flexure

Tail of pancreas

Superior mesenteric artery

Root
of transverse mesocolon

Inferior mesenteric vein

Jejunal and ileal
veins and arteries

Small intestine

Jejunal and ileal
arteries and veins

Median umbilical fold

Medial umbilical fold

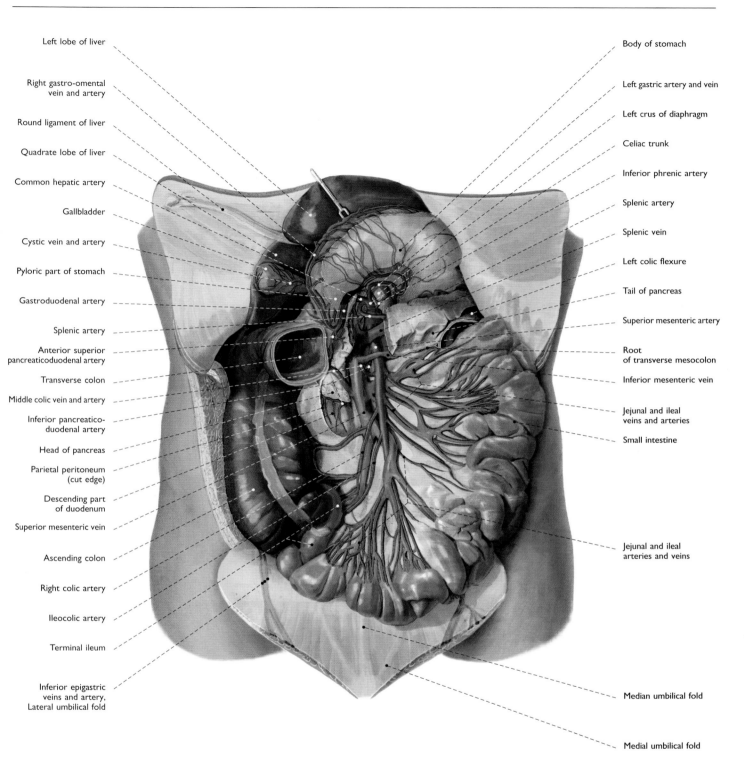

**253   Celiac trunk and superior mesenteric vessels
with their branches** (30%)
The greater omentum and the transverse colon were removed,
the stomach is retracted upwards and the small intestine
drawn to the left. Ventral aspect

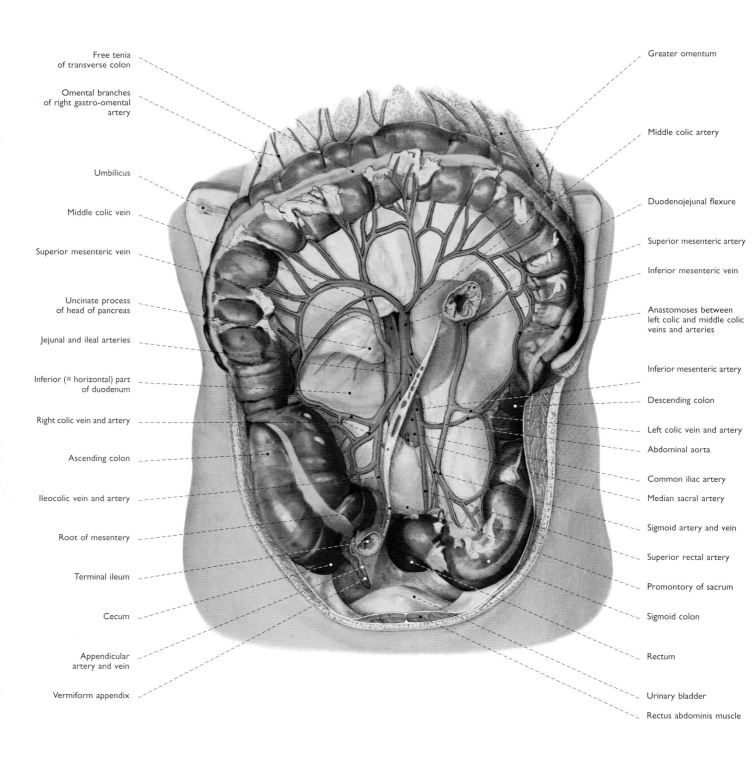

Free tenia
of transverse colon

Omental branches
of right gastro-omental
artery

Umbilicus

Middle colic vein

Superior mesenteric vein

Uncinate process
of head of pancreas

Jejunal and ileal arteries

Inferior (= horizontal) part
of duodenum

Right colic vein and artery

Ascending colon

Ileocolic vein and artery

Root of mesentery

Terminal ileum

Cecum

Appendicular
artery and vein

Vermiform appendix

Greater omentum

Middle colic artery

Duodenojejunal flexure

Superior mesenteric artery

Inferior mesenteric vein

Anastomoses between
left colic and middle colic
veins and arteries

Inferior mesenteric artery

Descending colon

Left colic vein and artery

Abdominal aorta

Common iliac artery

Median sacral artery

Sigmoid artery and vein

Superior rectal artery

Promontory of sacrum

Sigmoid colon

Rectum

Urinary bladder

Rectus abdominis muscle

254 Vascular supply of the large intestine (30%)
The greater omentum and the transverse colon are retracted upwards.
The small intestine was transected both at the duodenojejunal flexure
and the ileocecal transition, and removed along with its mesentery
which was severed at the root of mesentery. Ventral aspect

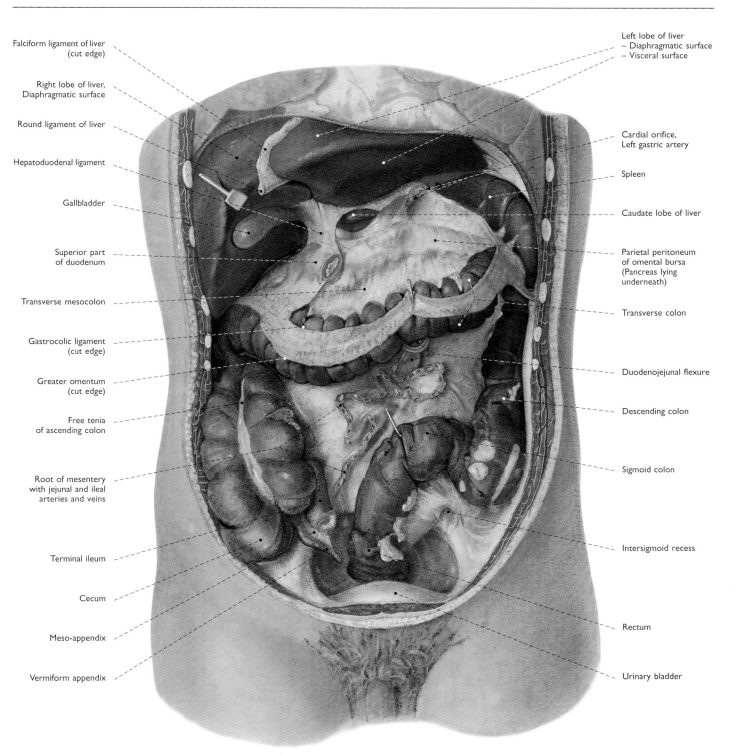

Falciform ligament of liver (cut edge)

Right lobe of liver, Diaphragmatic surface

Round ligament of liver

Hepatoduodenal ligament

Gallbladder

Superior part of duodenum

Transverse mesocolon

Gastrocolic ligament (cut edge)

Greater omentum (cut edge)

Free tenia of ascending colon

Root of mesentery with jejunal and ileal arteries and veins

Terminal ileum

Cecum

Meso-appendix

Vermiform appendix

Left lobe of liver
– Diaphragmatic surface
– Visceral surface

Cardial orifice, Left gastric artery

Spleen

Caudate lobe of liver

Parietal peritoneum of omental bursa (Pancreas lying underneath)

Transverse colon

Duodenojejunal flexure

Descending colon

Sigmoid colon

Intersigmoid recess

Rectum

Urinary bladder

255 Large intestine and mesenteries (30%)
The anterior body wall was removed. The stomach was taken away
by severing the hepatogastric, gastrosplenic, and gastrocolic ligaments.
The greater omentum, the jejunum, and the ileum with their mesenteries,
excepting the terminal ileum, were also removed. Ventral aspect

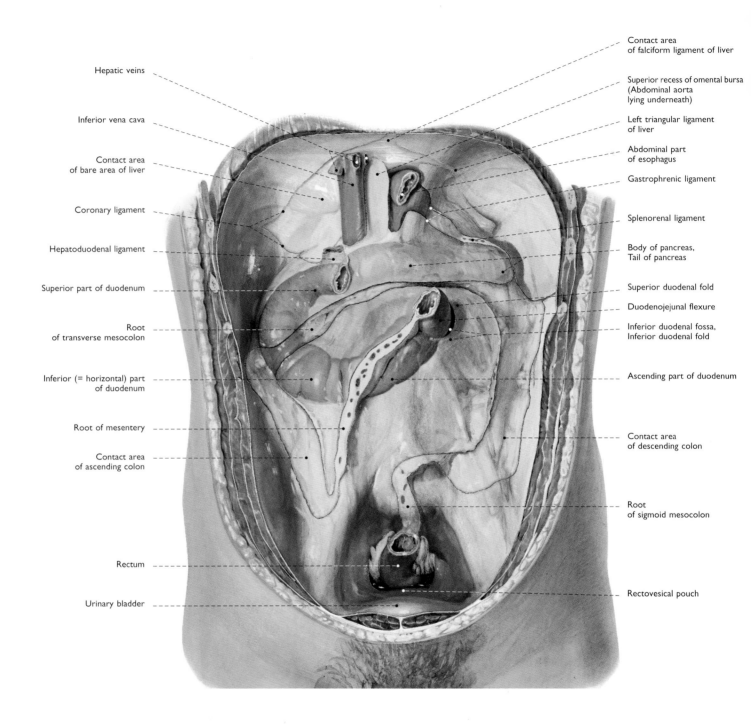

Hepatic veins

Inferior vena cava

Contact area
of bare area of liver

Coronary ligament

Hepatoduodenal ligament

Superior part of duodenum

Root
of transverse mesocolon

Inferior (= horizontal) part
of duodenum

Root of mesentery

Contact area
of ascending colon

Rectum

Urinary bladder

Contact area
of falciform ligament of liver

Superior recess of omental bursa
(Abdominal aorta
lying underneath)

Left triangular ligament
of liver

Abdominal part
of esophagus

Gastrophrenic ligament

Splenorenal ligament

Body of pancreas,
Tail of pancreas

Superior duodenal fold

Duodenojejunal flexure

Inferior duodenal fossa,
Inferior duodenal fold

Ascending part of duodenum

Contact area
of descending colon

Root
of sigmoid mesocolon

Rectovesical pouch

256 Posterior abdominal wall (30%)
Duodenum and pancreas remaining in situ.
The contact areas of the retroperitoneal parts of colon
and the liver as well as the roots of mesenteries are shown.
Ventral aspect

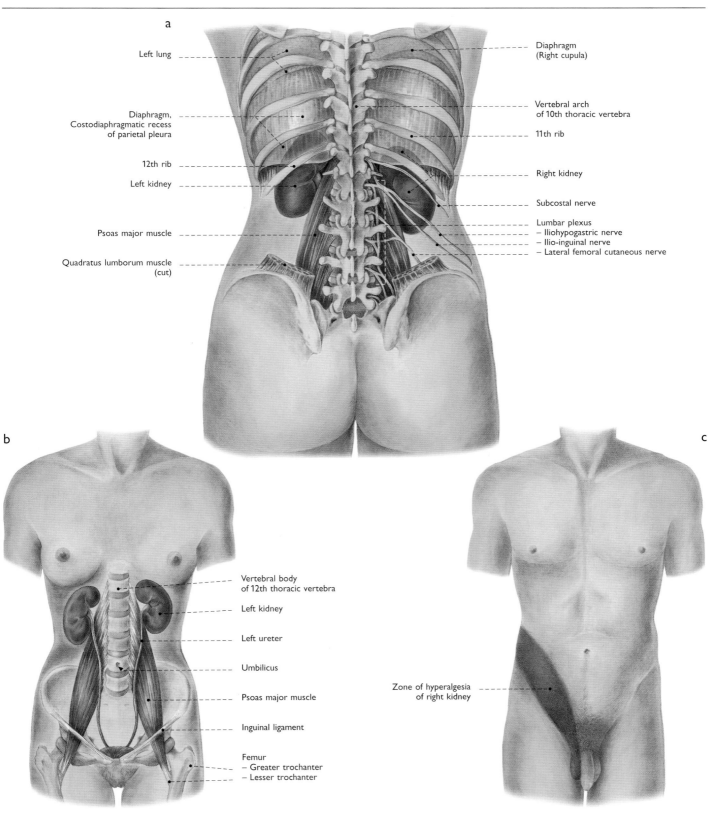

a

Left lung

Diaphragm,
Costodiaphragmatic recess
of parietal pleura

12th rib

Left kidney

Psoas major muscle

Quadratus lumborum muscle
(cut)

Diaphragm
(Right cupula)

Vertebral arch
of 10th thoracic vertebra

11th rib

Right kidney

Subcostal nerve

Lumbar plexus
– Iliohypogastric nerve
– Ilio-inguinal nerve
– Lateral femoral cutaneous nerve

b

Vertebral body
of 12th thoracic vertebra

Left kidney

Left ureter

Umbilicus

Psoas major muscle

Inguinal ligament

Femur
– Greater trochanter
– Lesser trochanter

c

Zone of hyperalgesia
of right kidney

257 Kidneys

a Kidneys and adjacent structures projected to
  the posterior abdominal wall (25%), dorsal aspect
b Projection of the kidneys and both psoas major muscles
  onto the anterior abdominal wall (15%), ventral aspect
c Zone of hyperalgesia of the right kidney (15%), ventral aspect

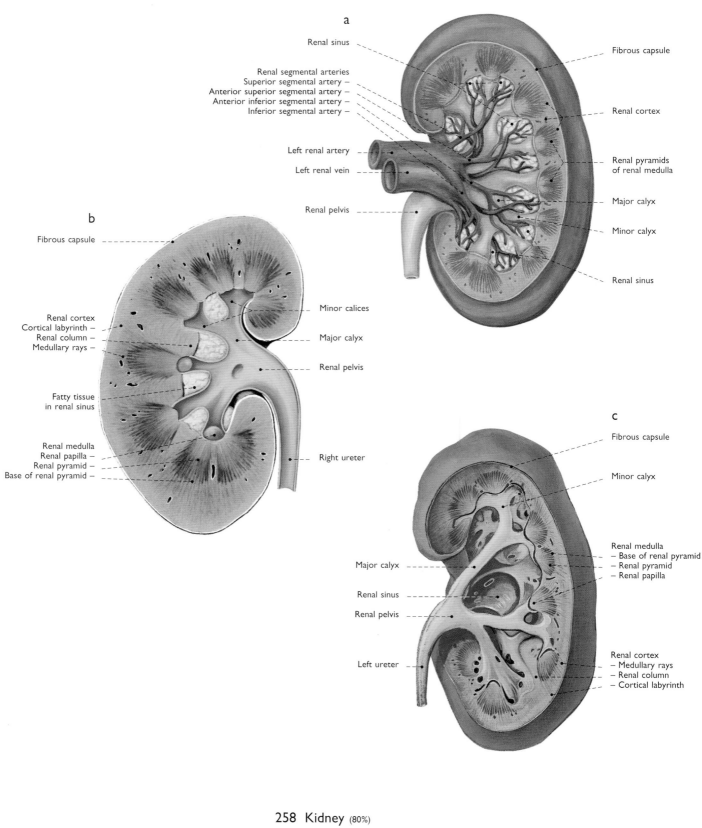

a

Renal sinus

Renal segmental arteries
Superior segmental artery –
Anterior superior segmental artery –
Anterior inferior segmental artery –
Inferior segmental artery –

Left renal artery

Left renal vein

Renal pelvis

Fibrous capsule

Renal cortex

Renal pyramids
of renal medulla

Major calyx

Minor calyx

Renal sinus

b

Fibrous capsule

Renal cortex
Cortical labyrinth –
Renal column –
Medullary rays –

Fatty tissue
in renal sinus

Renal medulla
Renal papilla –
Renal pyramid –
Base of renal pyramid –

Minor calices

Major calyx

Renal pelvis

Right ureter

c

Fibrous capsule

Minor calyx

Major calyx

Renal sinus

Renal pelvis

Left ureter

Renal medulla
– Base of renal pyramid
– Renal pyramid
– Renal papilla

Renal cortex
– Medullary rays
– Renal column
– Cortical labyrinth

258 Kidney (80%)

Ventral aspect

a   Left renal sinus with renal pelvis and renal blood vessels.
The renal parenchyma of the anterior part of kidney was removed.

b   Longitudinal section through the right kidney, cut surface
of the posterior part

c   Left renal sinus with the renal pelvis and its division into calices.
The fatty tissue and blood vessels in the renal sinus were removed.

a

11th rib

Vertebral body
of 12th thoracic
vertebra

Inferior vena cava

Renal segmental
arteries
Superior –
Anterior superior –
Anterior inferior –
Inferior –

Right renal pelvis

Vertebral body
of 2nd lumbar
vertebra

Right ureter

Abdominal aorta

Superior mesenteric
artery

Right renal artery

Left renal artery

Left renal vein

Left renal pelvis

Left ureter

b

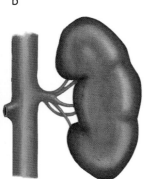

60%

c

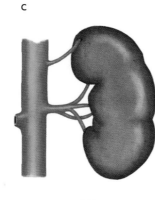

8%

d

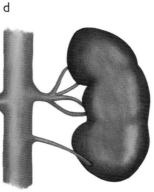

6%

e

5%

259 Kidneys and renal blood vessels

Ventral aspect

a   Corrosion cast of the renal blood vessels and the ureters
    of both kidneys of a 15-year-old girl (60%)
    (Anatomical Collection, Basel)

b–e Variations of the renal artery. The percentile numbers
     beneath the pictures indicate the approximate frequency.

b   'Normal pattern' with only one renal artery from the aorta
    on the left side of the body

c   Additional upper accessory renal artery from the aorta

d   Additional lower accessory renal artery from the aorta

c   Multiple (more than two) renal arteries from the aorta

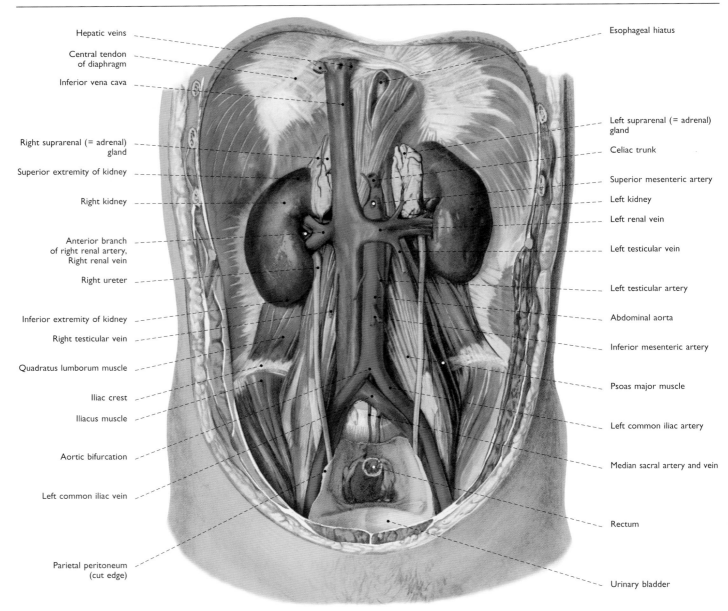

Hepatic veins

Central tendon
of diaphragm

Inferior vena cava

Right suprarenal (= adrenal)
gland

Superior extremity of kidney

Right kidney

Anterior branch
of right renal artery,
Right renal vein

Right ureter

Inferior extremity of kidney

Right testicular vein

Quadratus lumborum muscle

Iliac crest

Iliacus muscle

Aortic bifurcation

Left common iliac vein

Parietal peritoneum
(cut edge)

Esophageal hiatus

Left suprarenal (= adrenal)
gland

Celiac trunk

Superior mesenteric artery

Left kidney

Left renal vein

Left testicular vein

Left testicular artery

Abdominal aorta

Inferior mesenteric artery

Psoas major muscle

Left common iliac artery

Median sacral artery and vein

Rectum

Urinary bladder

260 Urinary system and great abdominal blood vessels
Ventral aspect (30%)

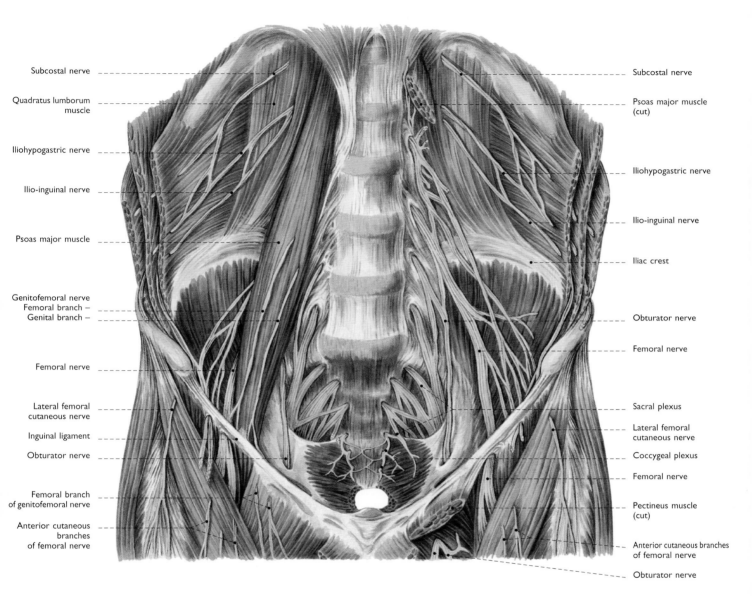

Subcostal nerve

Quadratus lumborum muscle

Iliohypogastric nerve

Ilio-inguinal nerve

Psoas major muscle

Genitofemoral nerve
Femoral branch –
Genital branch –

Femoral nerve

Lateral femoral cutaneous nerve

Inguinal ligament

Obturator nerve

Femoral branch of genitofemoral nerve

Anterior cutaneous branches of femoral nerve

Subcostal nerve

Psoas major muscle (cut)

Iliohypogastric nerve

Ilio-inguinal nerve

Iliac crest

Obturator nerve

Femoral nerve

Sacral plexus

Lateral femoral cutaneous nerve

Coccygeal plexus

Femoral nerve

Pectineus muscle (cut)

Anterior cutaneous branches of femoral nerve

Obturator nerve

261 Lumbosacral plexus (40%)
On the left side of the body, the psoas major muscle was removed and the pectineus muscle cut near to its origin. Ventral aspect

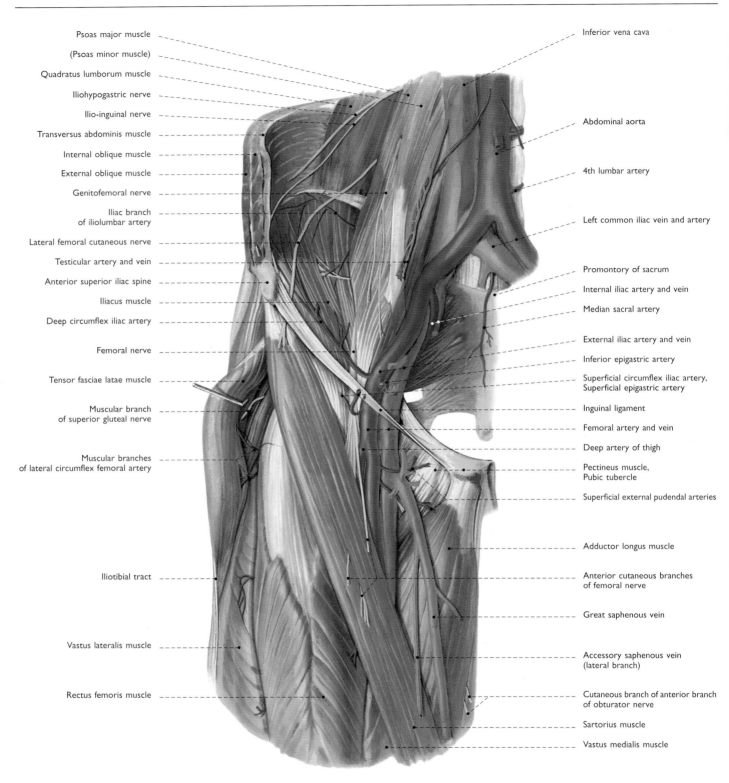

Psoas major muscle

(Psoas minor muscle)

Quadratus lumborum muscle

Iliohypogastric nerve

Ilio-inguinal nerve

Transversus abdominis muscle

Internal oblique muscle

External oblique muscle

Genitofemoral nerve

Iliac branch
of iliolumbar artery

Lateral femoral cutaneous nerve

Testicular artery and vein

Anterior superior iliac spine

Iliacus muscle

Deep circumflex iliac artery

Femoral nerve

Tensor fasciae latae muscle

Muscular branch
of superior gluteal nerve

Muscular branches
of lateral circumflex femoral artery

Iliotibial tract

Vastus lateralis muscle

Rectus femoris muscle

Inferior vena cava

Abdominal aorta

4th lumbar artery

Left common iliac vein and artery

Promontory of sacrum

Internal iliac artery and vein

Median sacral artery

External iliac artery and vein

Inferior epigastric artery

Superficial circumflex iliac artery,
Superficial epigastric artery

Inguinal ligament

Femoral artery and vein

Deep artery of thigh

Pectineus muscle,
Pubic tubercle

Superficial external pudendal arteries

Adductor longus muscle

Anterior cutaneous branches
of femoral nerve

Great saphenous vein

Accessory saphenous vein
(lateral branch)

Cutaneous branch of anterior branch
of obturator nerve

Sartorius muscle

Vastus medialis muscle

**262 Blood vessels and nerves
of the posterior abdominal wall
and the thigh of a male** (50%)
Ventral aspect

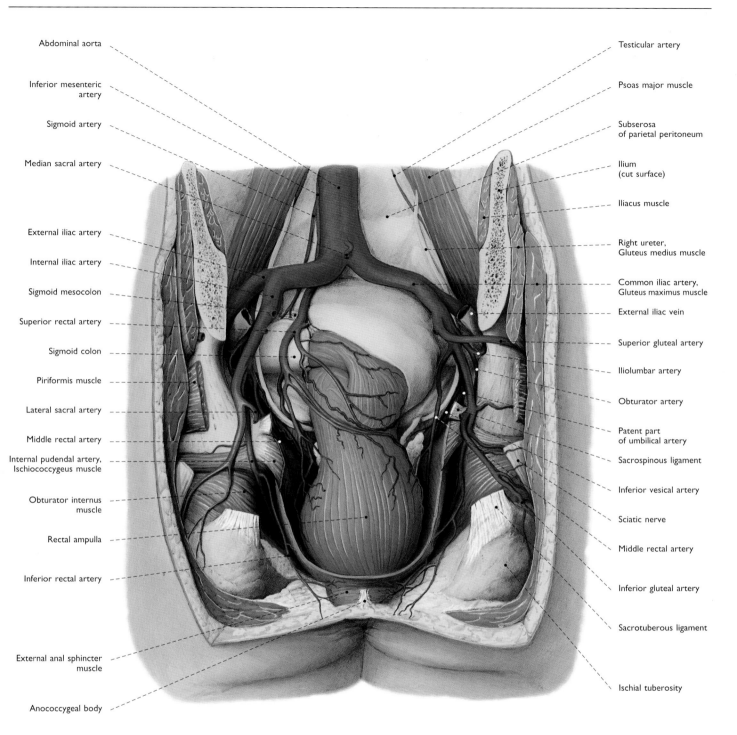

Abdominal aorta

Inferior mesenteric artery

Sigmoid artery

Median sacral artery

External iliac artery

Internal iliac artery

Sigmoid mesocolon

Superior rectal artery

Sigmoid colon

Piriformis muscle

Lateral sacral artery

Middle rectal artery

Internal pudendal artery, Ischiococcygeus muscle

Obturator internus muscle

Rectal ampulla

Inferior rectal artery

External anal sphincter muscle

Anococcygeal body

Testicular artery

Psoas major muscle

Subserosa of parietal peritoneum

Ilium (cut surface)

Iliacus muscle

Right ureter, Gluteus medius muscle

Common iliac artery, Gluteus maximus muscle

External iliac vein

Superior gluteal artery

Iliolumbar artery

Obturator artery

Patent part of umbilical artery

Sacrospinous ligament

Inferior vesical artery

Sciatic nerve

Middle rectal artery

Inferior gluteal artery

Sacrotuberous ligament

Ischial tuberosity

263 Arterial supply of the rectum of a male (60%)
The sacrum was removed, the coccygeus muscle,
the muscles of the gluteal region, the sacrospinous and
sacrotuberous ligaments were cut off and partially excised.
Dorsal aspect

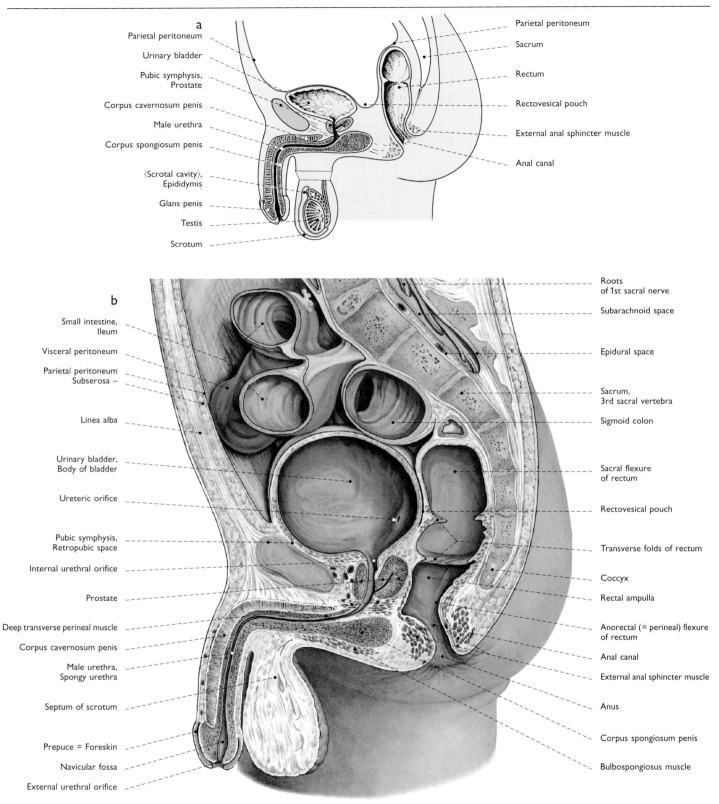

a

Parietal peritoneum

Urinary bladder

Pubic symphysis,
Prostate

Corpus cavernosum penis

Male urethra

Corpus spongiosum penis

⟨Scrotal cavity⟩,
Epididymis

Glans penis

Testis

Scrotum

Parietal peritoneum

Sacrum

Rectum

Rectovesical pouch

External anal sphincter muscle

Anal canal

b

Small intestine,
Ileum

Visceral peritoneum

Parietal peritoneum
Subserosa —

Linea alba

Urinary bladder,
Body of bladder

Ureteric orifice

Pubic symphysis,
Retropubic space

Internal urethral orifice

Prostate

Deep transverse perineal muscle

Corpus cavernosum penis

Male urethra,
Spongy urethra

Septum of scrotum

Prepuce = Foreskin

Navicular fossa

External urethral orifice

Roots
of 1st sacral nerve

Subarachnoid space

Epidural space

Sacrum,
3rd sacral vertebra

Sigmoid colon

Sacral flexure
of rectum

Rectovesical pouch

Transverse folds of rectum

Coccyx

Rectal ampulla

Anorectal (= perineal) flexure
of rectum

Anal canal

External anal sphincter muscle

Anus

Corpus spongiosum penis

Bulbospongiosus muscle

264 Male pelvis and urogenital system
a   Male pelvis, schematized median section, medial aspect
b   Pelvic viscera of an 18-year-old male,
      median section (55%), medial aspect of the right half
      (Anatomical Collection, Basel)

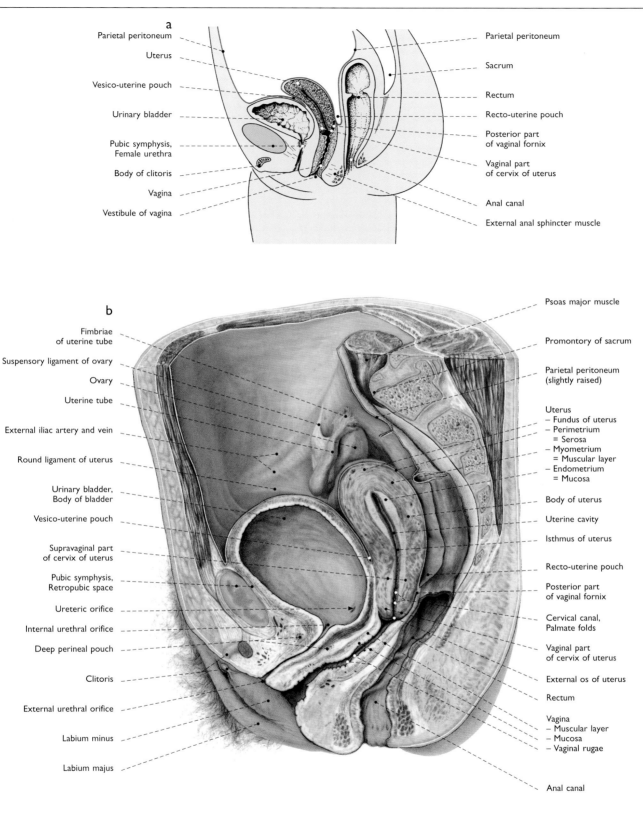

a

Parietal peritoneum

Uterus

Vesico-uterine pouch

Urinary bladder

Pubic symphysis,
Female urethra

Body of clitoris

Vagina

Vestibule of vagina

Parietal peritoneum

Sacrum

Rectum

Recto-uterine pouch

Posterior part
of vaginal fornix

Vaginal part
of cervix of uterus

Anal canal

External anal sphincter muscle

b

Fimbriae
of uterine tube

Suspensory ligament of ovary

Ovary

Uterine tube

External iliac artery and vein

Round ligament of uterus

Urinary bladder,
Body of bladder

Vesico-uterine pouch

Supravaginal part
of cervix of uterus

Pubic symphysis,
Retropubic space

Ureteric orifice

Internal urethral orifice

Deep perineal pouch

Clitoris

External urethral orifice

Labium minus

Labium majus

Psoas major muscle

Promontory of sacrum

Parietal peritoneum
(slightly raised)

Uterus
– Fundus of uterus
– Perimetrium
  = Serosa
– Myometrium
  = Muscular layer
– Endometrium
  = Mucosa

Body of uterus

Uterine cavity

Isthmus of uterus

Recto-uterine pouch

Posterior part
of vaginal fornix

Cervical canal,
Palmate folds

Vaginal part
of cervix of uterus

External os of uterus

Rectum

Vagina
– Muscular layer
– Mucosa
– Vaginal rugae

Anal canal

265 Female pelvis and urogenital apparatus

a Female pelvis, schematized median section, medial aspect
b Pelvic viscera of a 23-year-old female, median section (55%),
medial aspect of the right half (Anatomical Collection, Basel)

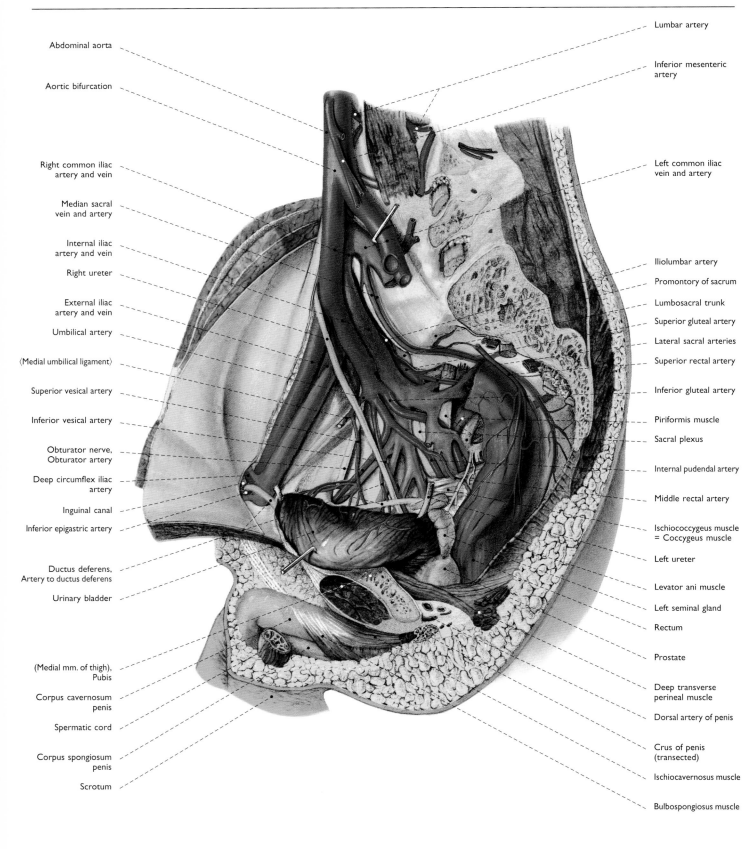

Lumbar artery

Inferior mesenteric artery

Abdominal aorta

Aortic bifurcation

Left common iliac vein and artery

Right common iliac artery and vein

Median sacral vein and artery

Internal iliac artery and vein

Right ureter

Iliolumbar artery

Promontory of sacrum

Lumbosacral trunk

External iliac artery and vein

Superior gluteal artery

Lateral sacral arteries

Umbilical artery

Superior rectal artery

〈Medial umbilical ligament〉

Inferior gluteal artery

Superior vesical artery

Piriformis muscle

Inferior vesical artery

Sacral plexus

Obturator nerve, Obturator artery

Internal pudendal artery

Deep circumflex iliac artery

Middle rectal artery

Inguinal canal

Ischiococcygeus muscle = Coccygeus muscle

Inferior epigastric artery

Left ureter

Levator ani muscle

Ductus deferens, Artery to ductus deferens

Left seminal gland

Urinary bladder

Rectum

Prostate

〈Medial mm. of thigh〉, Pubis

Deep transverse perineal muscle

Corpus cavernosum penis

Dorsal artery of penis

Spermatic cord

Crus of penis (transected)

Corpus spongiosum penis

Ischiocavernosus muscle

Scrotum

Bulbospongiosus muscle

266 Blood vessels and nerves of the male pelvis (70%)
Sagittal section to the left of the median plane,
medial aspect of the right part

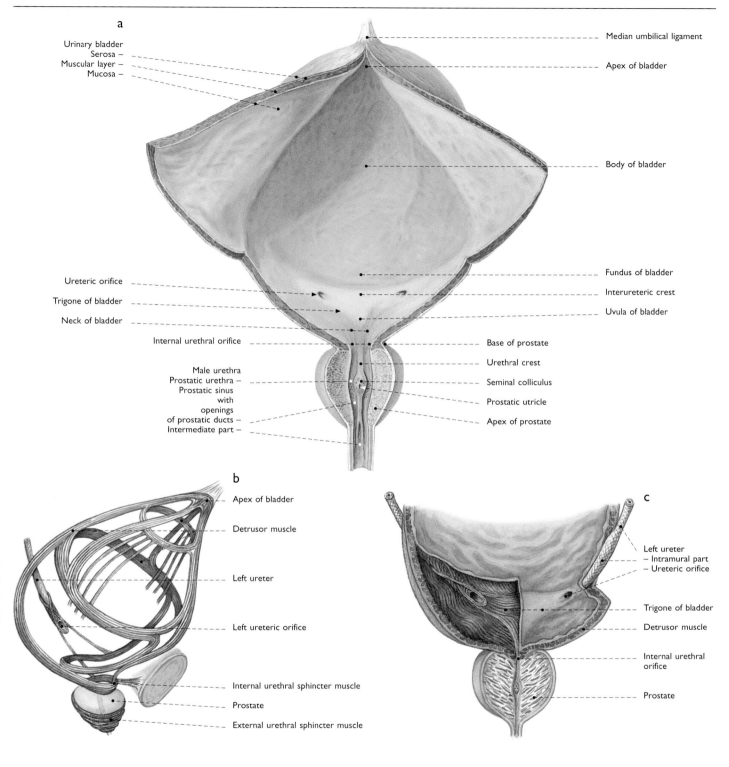

a

Urinary bladder
　Serosa –
　Muscular layer –
　Mucosa –

Median umbilical ligament

Apex of bladder

Body of bladder

Fundus of bladder

Interureteric crest

Uvula of bladder

Ureteric orifice

Trigone of bladder

Neck of bladder

Internal urethral orifice

Base of prostate

Urethral crest

Male urethra
Prostatic urethra –
Prostatic sinus
with
openings
of prostatic ducts –
Intermediate part –

Seminal colliculus

Prostatic utricle

Apex of prostate

b

Apex of bladder

Detrusor muscle

Left ureter

Left ureteric orifice

Internal urethral sphincter muscle

Prostate

External urethral sphincter muscle

c

Left ureter
– Intramural part
– Ureteric orifice

Trigone of bladder

Detrusor muscle

Internal urethral
orifice

Prostate

267　Urinary bladder

a　Urinary bladder and urethra of a male. The bladder and prostate
　were incised along the midsagittal plane and opened (80%). Ventral aspect
b　Arrangement of the muscles in the bladder wall (according to Ferner,
　1975) (60%), schematic representation, right lateral aspect
c　Ureteric orifices and trigone of bladder (according to Ferner, 1975,
　and Leonhardt, 1987). In the left half of the bladder, a step-cut
　in the bladder wall for showing the intramural part of ureter.
　On the right side of the bladder, the mucosa was removed
　in order to demonstrate the muscle arrangement in the trigone (80%),
　ventral aspect

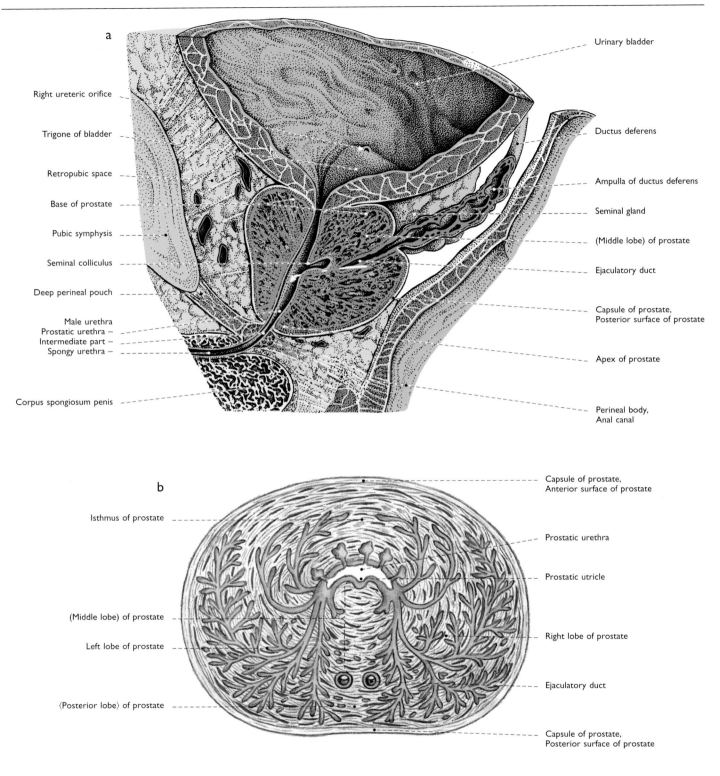

a

Urinary bladder

Right ureteric orifice

Trigone of bladder

Retropubic space

Base of prostate

Pubic symphysis

Seminal colliculus

Deep perineal pouch

Male urethra
Prostatic urethra –
Intermediate part –
Spongy urethra –

Corpus spongiosum penis

Ductus deferens

Ampulla of ductus deferens

Seminal gland

(Middle lobe) of prostate

Ejaculatory duct

Capsule of prostate,
Posterior surface of prostate

Apex of prostate

Perineal body,
Anal canal

b

Isthmus of prostate

(Middle lobe) of prostate

Left lobe of prostate

〈Posterior lobe〉 of prostate

Capsule of prostate,
Anterior surface of prostate

Prostatic urethra

Prostatic utricle

Right lobe of prostate

Ejaculatory duct

Capsule of prostate,
Posterior surface of prostate

268 Urinary bladder, deferent duct,
seminal gland, and prostate

a Median section, medial aspect of the right half (120%)
b Transverse section through the prostate (250%),
   schematic representation, superior aspect

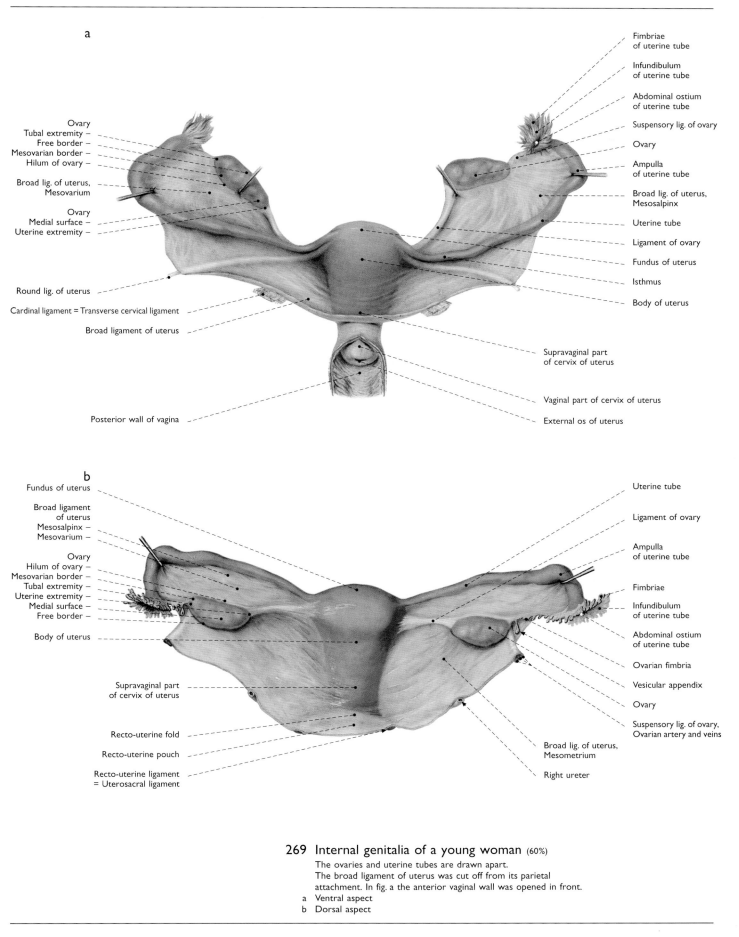

a

Fimbriae
of uterine tube

Infundibulum
of uterine tube

Abdominal ostium
of uterine tube

Suspensory lig. of ovary

Ovary

Ampulla
of uterine tube

Broad lig. of uterus,
Mesosalpinx

Uterine tube

Ligament of ovary

Fundus of uterus

Isthmus

Body of uterus

Ovary
Tubal extremity –
Free border –
Mesovarian border –
Hilum of ovary –

Broad lig. of uterus,
Mesovarium

Ovary
Medial surface –
Uterine extremity –

Round lig. of uterus

Cardinal ligament = Transverse cervical ligament

Broad ligament of uterus

Supravaginal part
of cervix of uterus

Vaginal part of cervix of uterus

Posterior wall of vagina

External os of uterus

b

Fundus of uterus

Broad ligament
of uterus
Mesosalpinx –
Mesovarium –

Ovary
Hilum of ovary –
Mesovarian border –
Tubal extremity –
Uterine extremity –
Medial surface –
Free border –

Body of uterus

Supravaginal part
of cervix of uterus

Recto-uterine fold

Recto-uterine pouch

Recto-uterine ligament
= Uterosacral ligament

Uterine tube

Ligament of ovary

Ampulla
of uterine tube

Fimbriae

Infundibulum
of uterine tube

Abdominal ostium
of uterine tube

Ovarian fimbria

Vesicular appendix

Ovary

Suspensory lig. of ovary,
Ovarian artery and veins

Broad lig. of uterus,
Mesometrium

Right ureter

269 Internal genitalia of a young woman (60%)
The ovaries and uterine tubes are drawn apart.
The broad ligament of uterus was cut off from its parietal
attachment. In fig. a the anterior vaginal wall was opened in front.
a   Ventral aspect
b   Dorsal aspect

a

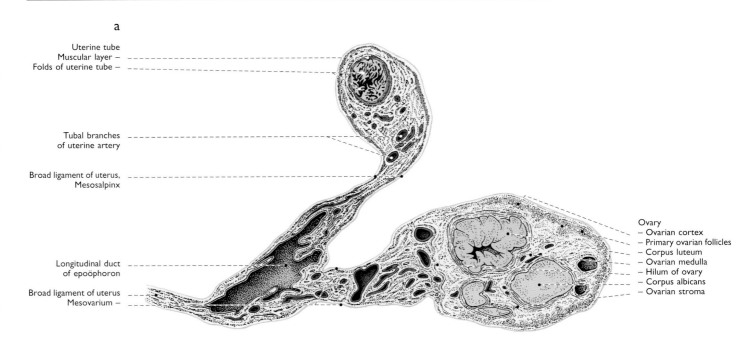

Uterine tube
Muscular layer –
Folds of uterine tube –

Tubal branches
of uterine artery

Broad ligament of uterus,
Mesosalpinx

Longitudinal duct
of epoöphoron

Broad ligament of uterus
Mesovarium –

Ovary
– Ovarian cortex
– Primary ovarian follicles
– Corpus luteum
– Ovarian medulla
– Hilum of ovary
– Corpus albicans
– Ovarian stroma

b

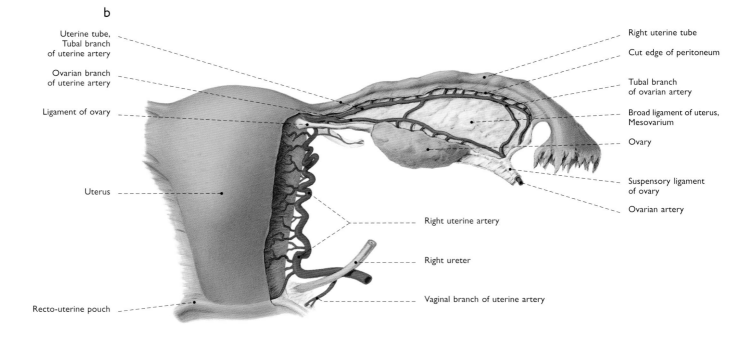

Uterine tube,
Tubal branch
of uterine artery

Ovarian branch
of uterine artery

Ligament of ovary

Uterus

Recto-uterine pouch

Right uterine tube

Cut edge of peritoneum

Tubal branch
of ovarian artery

Broad ligament of uterus,
Mesovarium

Ovary

Suspensory ligament
of ovary

Ovarian artery

Right uterine artery

Right ureter

Vaginal branch of uterine artery

270 Female internal genitalia

a Sagittal section through the ovary and the uterine tube (230%)
b Blood supply of the uterus, the uterine tubes, and the ovary (70%),
    dorsal aspect

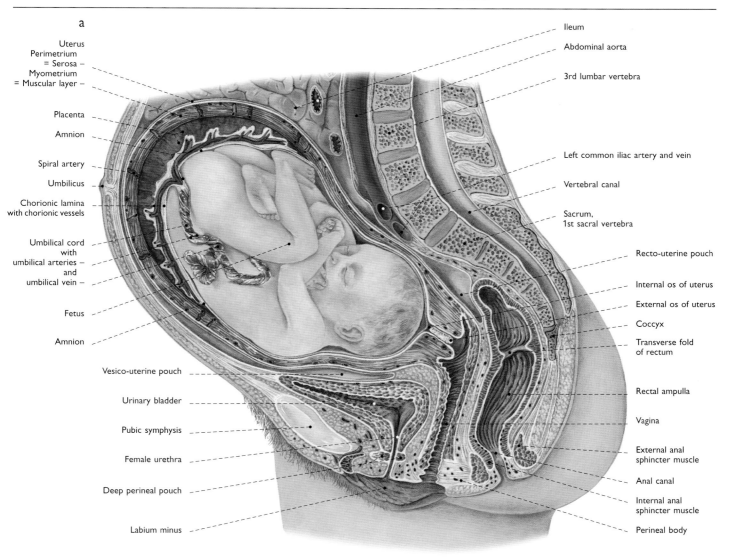

a

Uterus
Perimetrium
= Serosa
Myometrium
= Muscular layer

Placenta

Amnion

Spiral artery

Umbilicus

Chorionic lamina
with chorionic vessels

Umbilical cord
with
umbilical arteries
and
umbilical vein

Fetus

Amnion

Vesico-uterine pouch

Urinary bladder

Pubic symphysis

Female urethra

Deep perineal pouch

Labium minus

Ileum

Abdominal aorta

3rd lumbar vertebra

Left common iliac artery and vein

Vertebral canal

Sacrum,
1st sacral vertebra

Recto-uterine pouch

Internal os of uterus

External os of uterus

Coccyx

Transverse fold
of rectum

Rectal ampulla

Vagina

External anal
sphincter muscle

Anal canal

Internal anal
sphincter muscle

Perineal body

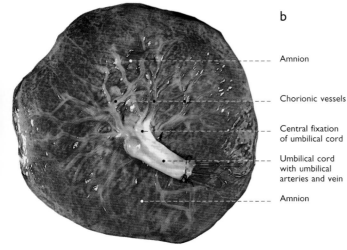

b

Amnion

Chorionic vessels

Central fixation
of umbilical cord

Umbilical cord
with umbilical
arteries and vein

Amnion

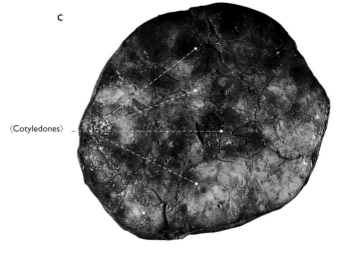

c

⟨Cotyledones⟩

271 Female internal genitalia and placenta (40%)

   a   Female pelvis with an uterus advanced in pregnancy,
       medial aspect of a median section (fetus and umbilical cord
       were left intact)
b, c   Recently delivered placenta
   b   Fetal surface with umbilical cord
   c   Maternal surface

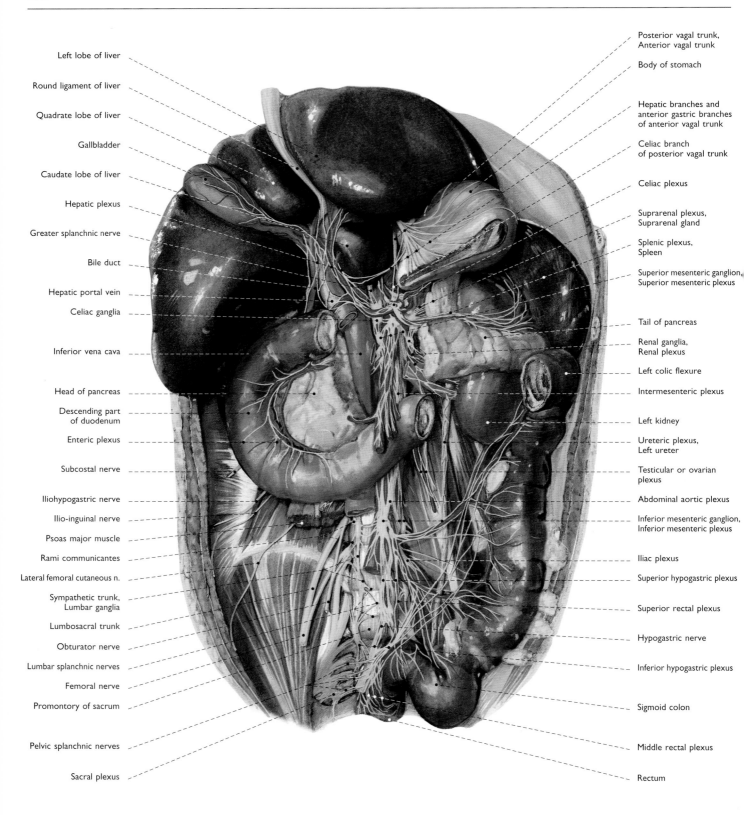

Left lobe of liver

Round ligament of liver

Quadrate lobe of liver

Gallbladder

Caudate lobe of liver

Hepatic plexus

Greater splanchnic nerve

Bile duct

Hepatic portal vein

Celiac ganglia

Inferior vena cava

Head of pancreas

Descending part of duodenum

Enteric plexus

Subcostal nerve

Iliohypogastric nerve

Ilio-inguinal nerve

Psoas major muscle

Rami communicantes

Lateral femoral cutaneous n.

Sympathetic trunk, Lumbar ganglia

Lumbosacral trunk

Obturator nerve

Lumbar splanchnic nerves

Femoral nerve

Promontory of sacrum

Pelvic splanchnic nerves

Sacral plexus

Posterior vagal trunk, Anterior vagal trunk

Body of stomach

Hepatic branches and anterior gastric branches of anterior vagal trunk

Celiac branch of posterior vagal trunk

Celiac plexus

Suprarenal plexus, Suprarenal gland

Splenic plexus, Spleen

Superior mesenteric ganglion, Superior mesenteric plexus

Tail of pancreas

Renal ganglia, Renal plexus

Left colic flexure

Intermesenteric plexus

Left kidney

Ureteric plexus, Left ureter

Testicular or ovarian plexus

Abdominal aortic plexus

Inferior mesenteric ganglion, Inferior mesenteric plexus

Iliac plexus

Superior hypogastric plexus

Superior rectal plexus

Hypogastric nerve

Inferior hypogastric plexus

Sigmoid colon

Middle rectal plexus

Rectum

272 Autonomic division of peripheral nervous system in the retroperitoneal space (40%)
Ventral aspect

Pelvic Diaphragm and External Genitalia

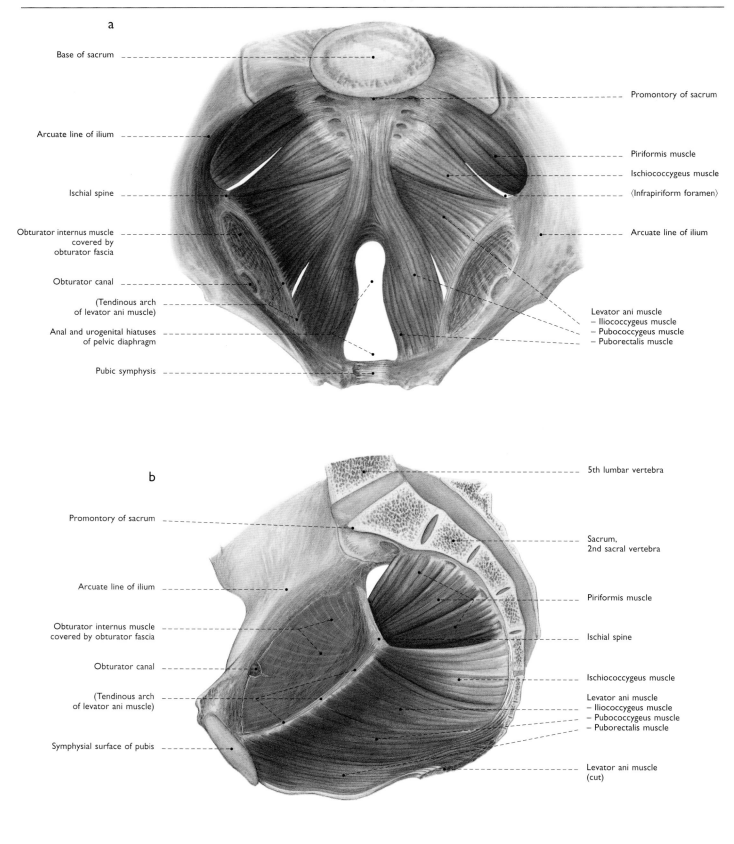

a

Base of sacrum

Arcuate line of ilium

Ischial spine

Obturator internus muscle covered by obturator fascia

Obturator canal

(Tendinous arch of levator ani muscle)

Anal and urogenital hiatuses of pelvic diaphragm

Pubic symphysis

Promontory of sacrum

Piriformis muscle

Ischiococcygeus muscle

(Infrapiriform foramen)

Arcuate line of ilium

Levator ani muscle
– Iliococcygeus muscle
– Pubococcygeus muscle
– Puborectalis muscle

b

Promontory of sacrum

Arcuate line of ilium

Obturator internus muscle covered by obturator fascia

Obturator canal

(Tendinous arch of levator ani muscle)

Symphysial surface of pubis

5th lumbar vertebra

Sacrum,
2nd sacral vertebra

Piriformis muscle

Ischial spine

Ischiococcygeus muscle

Levator ani muscle
– Iliococcygeus muscle
– Pubococcygeus muscle
– Puborectalis muscle

Levator ani muscle
(cut)

274 Pelvic diaphragm (60%)

Muscles of the pelvic diaphragm (= pelvic floor)
a   Superior aspect
b   Medial aspect of the right half of the pelvis

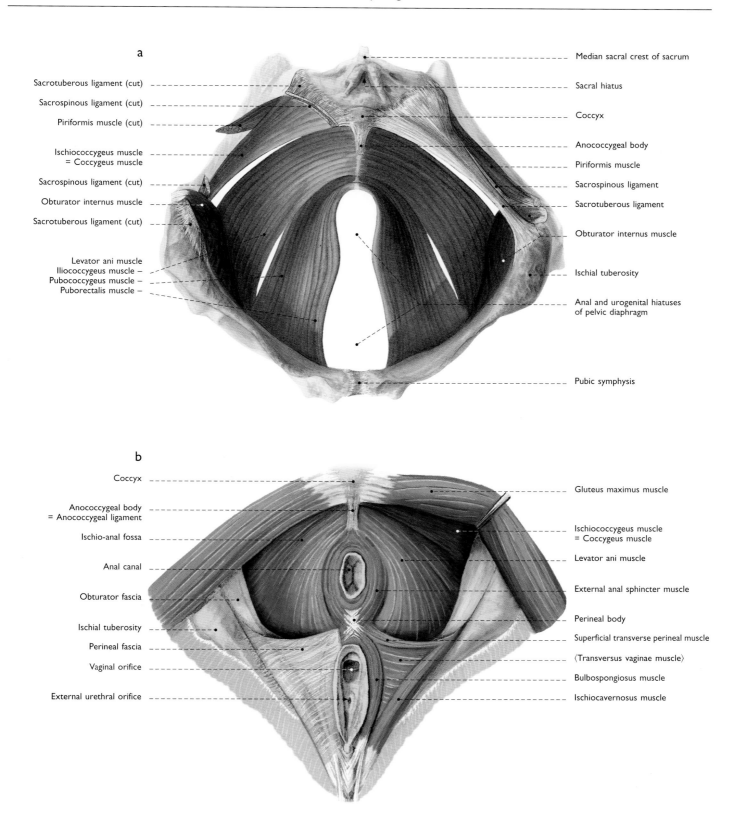

a

Sacrotuberous ligament (cut)

Sacrospinous ligament (cut)

Piriformis muscle (cut)

Ischiococcygeus muscle
= Coccygeus muscle

Sacrospinous ligament (cut)

Obturator internus muscle

Sacrotuberous ligament (cut)

Levator ani muscle
Iliococcygeus muscle –
Pubococcygeus muscle –
Puborectalis muscle –

Median sacral crest of sacrum

Sacral hiatus

Coccyx

Anococcygeal body

Piriformis muscle

Sacrospinous ligament

Sacrotuberous ligament

Obturator internus muscle

Ischial tuberosity

Anal and urogenital hiatuses
of pelvic diaphragm

Pubic symphysis

b

Coccyx

Anococcygeal body
= Anococcygeal ligament

Ischio-anal fossa

Anal canal

Obturator fascia

Ischial tuberosity

Perineal fascia

Vaginal orifice

External urethral orifice

Gluteus maximus muscle

Ischiococcygeus muscle
= Coccygeus muscle

Levator ani muscle

External anal sphincter muscle

Perineal body

Superficial transverse perineal muscle

⟨Transversus vaginae muscle⟩

Bulbospongiosus muscle

Ischiocavernosus muscle

275 Pelvic diaphragm and perineum of a female (60%)
Inferior aspect
a Muscles of the pelvic diaphragm (= pelvic floor)
b Muscles of the pelvic diaphragm and perineal muscles

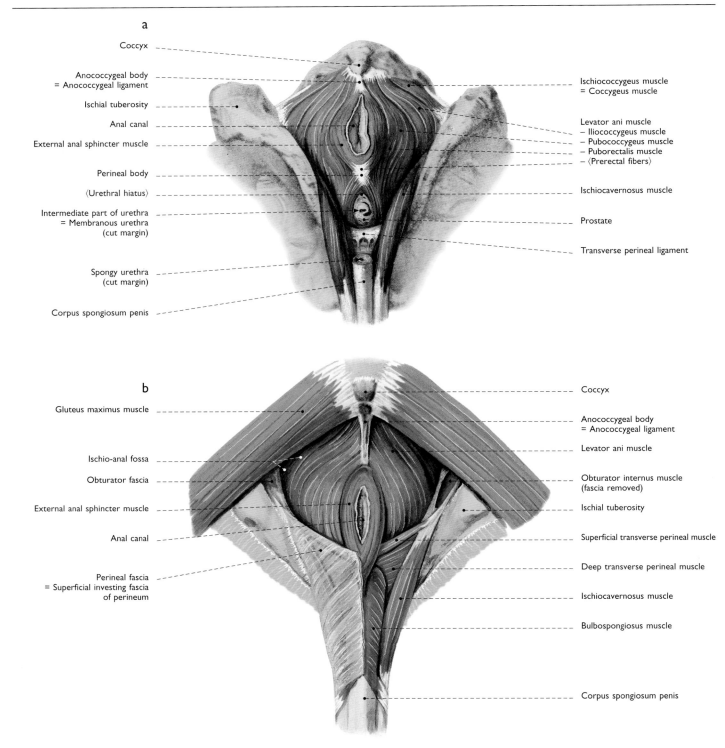

a

Coccyx

Anococcygeal body
= Anococcygeal ligament

Ischial tuberosity

Anal canal

External anal sphincter muscle

Perineal body

⟨Urethral hiatus⟩

Intermediate part of urethra
= Membranous urethra
(cut margin)

Spongy urethra
(cut margin)

Corpus spongiosum penis

Ischiococcygeus muscle
= Coccygeus muscle

Levator ani muscle
– Iliococcygeus muscle
– Pubococcygeus muscle
– Puborectalis muscle
– ⟨Prerectal fibers⟩

Ischiocavernosus muscle

Prostate

Transverse perineal ligament

b

Gluteus maximus muscle

Ischio-anal fossa

Obturator fascia

External anal sphincter muscle

Anal canal

Perineal fascia
= Superficial investing fascia
of perineum

Coccyx

Anococcygeal body
= Anococcygeal ligament

Levator ani muscle

Obturator internus muscle
(fascia removed)

Ischial tuberosity

Superficial transverse perineal muscle

Deep transverse perineal muscle

Ischiocavernosus muscle

Bulbospongiosus muscle

Corpus spongiosum penis

276 Pelvic diaphragm and perineum of a male (60%)
Inferior aspect
a   Muscles of the pelvic diaphragm (= pelvic floor)
b   Muscles of the pelvic diaphragm and perineal muscles

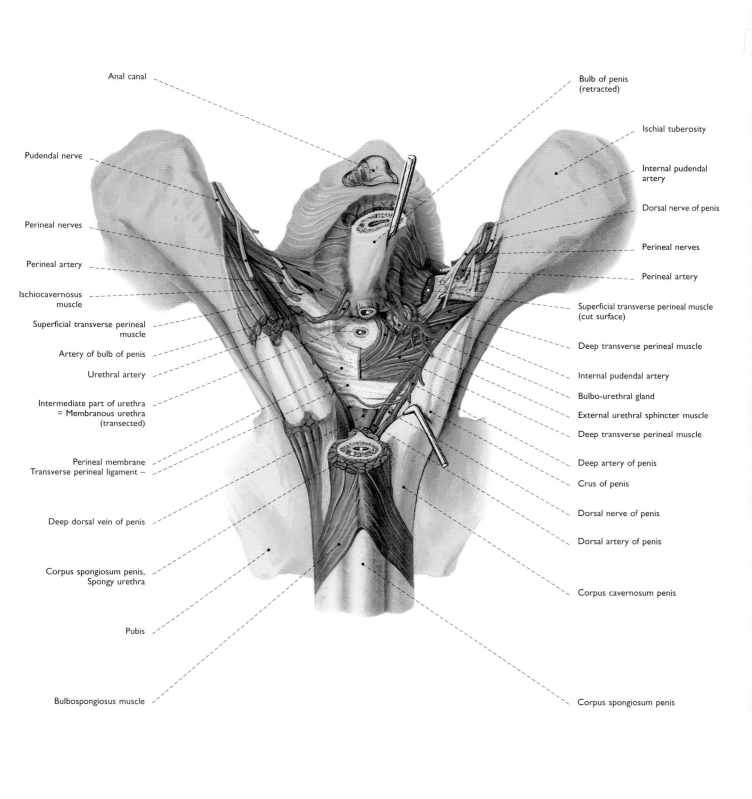

Anal canal

Pudendal nerve

Perineal nerves

Perineal artery

Ischiocavernosus
muscle

Superficial transverse perineal
muscle

Artery of bulb of penis

Urethral artery

Intermediate part of urethra
= Membranous urethra
(transected)

Perineal membrane
Transverse perineal ligament

Deep dorsal vein of penis

Corpus spongiosum penis,
Spongy urethra

Pubis

Bulbospongiosus muscle

Bulb of penis
(retracted)

Ischial tuberosity

Internal pudendal
artery

Dorsal nerve of penis

Perineal nerves

Perineal artery

Superficial transverse perineal muscle
(cut surface)

Deep transverse perineal muscle

Internal pudendal artery

Bulbo-urethral gland

External urethral sphincter muscle

Deep transverse perineal muscle

Deep artery of penis

Crus of penis

Dorsal nerve of penis

Dorsal artery of penis

Corpus cavernosum penis

Corpus spongiosum penis

277 Penis and deep perineal space (90%)
Inferior aspect
The corpus spongiosum penis was transected
and the bulb of penis retracted.

a

Mons pubis

Anterior commissure of labia majora

Glans of clitoris

Prepuce of clitoris

Frenulum of clitoris

(Bulge caused
by underlying bulb of vestibule)

Labium majus

External urethral orifice

Labium minus

Hymen

Vaginal rugae

Vestibule of vagina

Vaginal orifice

Frenulum of labia minora

Posterior commissure of labia majora

Vestibular fossa of vagina

b

Mons pubis

Glans of clitoris

Anterior commissure of labia majora

Prepuce of clitoris

Frenulum of clitoris

Labium majus

External urethral orifice

Labium minus

Hymenal caruncles

Opening
of greater vestibular gland

Posterior commissure of labia majora

Vaginal orifice

Perineal raphe

Anus

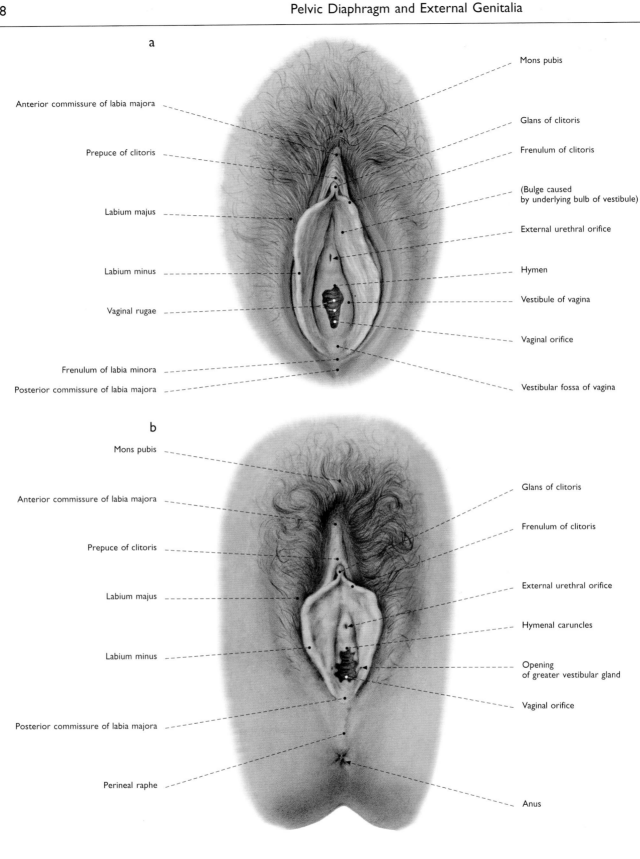

278 Female external genitalia (80%)

Inferior aspect
a Female genitals of a virgin
b Female genitals after defloration

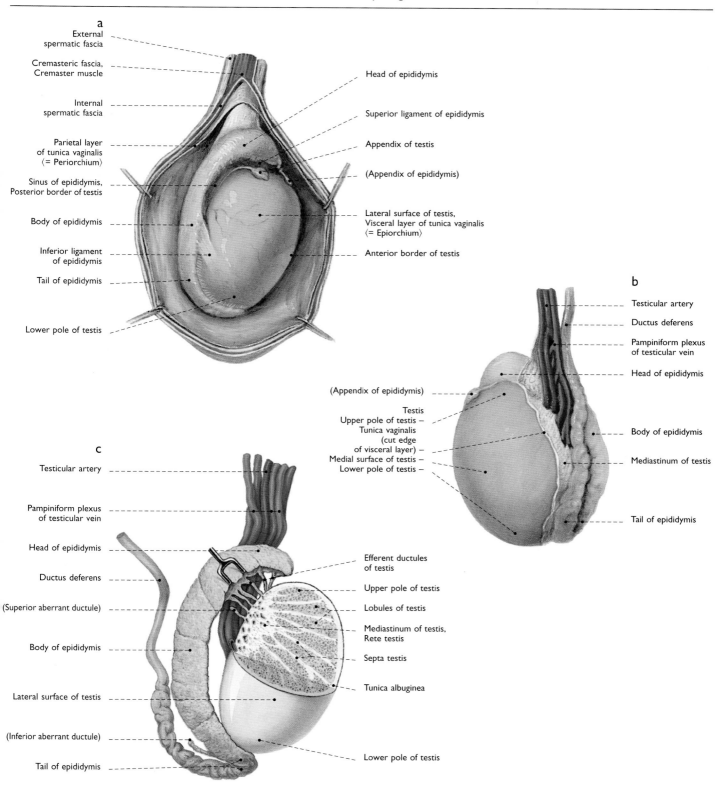

a

External spermatic fascia

Cremasteric fascia, Cremaster muscle

Internal spermatic fascia

Parietal layer of tunica vaginalis ⟨= Periorchium⟩

Sinus of epididymis, Posterior border of testis

Body of epididymis

Inferior ligament of epididymis

Tail of epididymis

Lower pole of testis

Head of epididymis

Superior ligament of epididymis

Appendix of testis

(Appendix of epididymis)

Lateral surface of testis, Visceral layer of tunica vaginalis ⟨= Epiorchium⟩

Anterior border of testis

b

Testicular artery

Ductus deferens

Pampiniform plexus of testicular vein

Head of epididymis

Body of epididymis

Mediastinum of testis

Tail of epididymis

(Appendix of epididymis)

Testis
Upper pole of testis —
Tunica vaginalis (cut edge of visceral layer) —
Medial surface of testis —
Lower pole of testis —

c

Testicular artery

Pampiniform plexus of testicular vein

Head of epididymis

Ductus deferens

(Superior aberrant ductule)

Body of epididymis

Lateral surface of testis

(Inferior aberrant ductule)

Tail of epididymis

Efferent ductules of testis

Upper pole of testis

Lobules of testis

Mediastinum of testis, Rete testis

Septa testis

Tunica albuginea

Lower pole of testis

279 Testis and epididymis (100%)
a, b The coverings of the right testis were opened (a) and removed (b).
a Right lateral aspect
b Medial aspect
c The lateral superior quadrant of the right testis was excised by a rectangular incision. Right lateral aspect

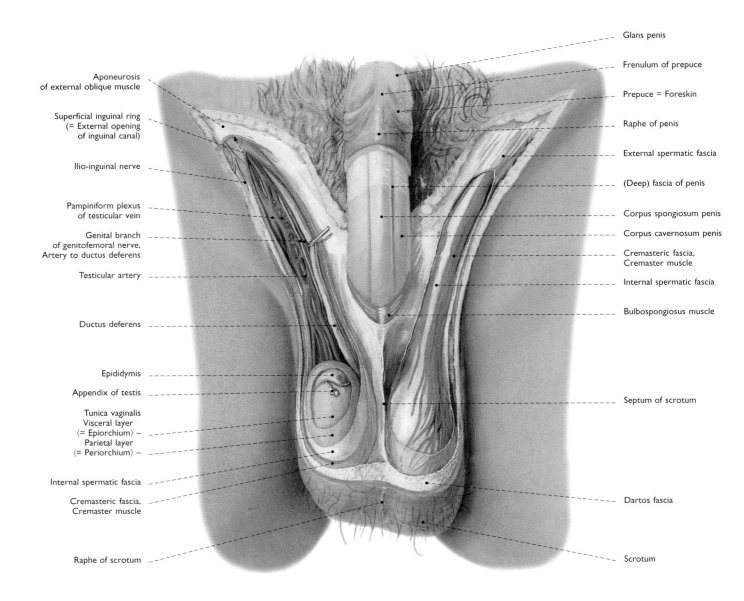

Glans penis

Frenulum of prepuce

Prepuce = Foreskin

Raphe of penis

External spermatic fascia

(Deep) fascia of penis

Corpus spongiosum penis

Corpus cavernosum penis

Cremasteric fascia, Cremaster muscle

Internal spermatic fascia

Bulbospongiosus muscle

Septum of scrotum

Dartos fascia

Scrotum

Aponeurosis of external oblique muscle

Superficial inguinal ring (= External opening of inguinal canal)

Ilio-inguinal nerve

Pampiniform plexus of testicular vein

Genital branch of genitofemoral nerve, Artery to ductus deferens

Testicular artery

Ductus deferens

Epididymis

Appendix of testis

Tunica vaginalis Visceral layer ⟨= Epiorchium⟩ – Parietal layer ⟨= Periorchium⟩ –

Internal spermatic fascia

Cremasteric fascia, Cremaster muscle

Raphe of scrotum

280 Male external genitalia (70%)

The penis was turned upwards, the skin and the deep fascia of penis
were largely removed from the body of penis. The skin of the scrotum
was excised in the region of both spermatic cords. On the right side
of the body, the structures of the spermatic cord are exposed by
opening the external and internal spermatic fasciae, the interposed
cremasteric fascia, and the cremaster muscle. On the left side of the body,
the cremaster muscle is shown after having removed the external spermatic
and the cremasteric fasciae. Ventral aspect

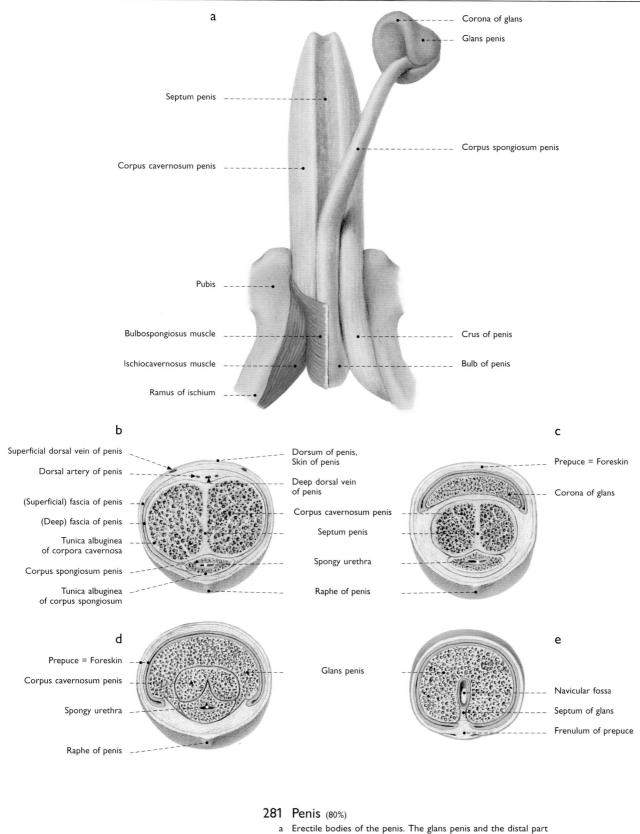

a

Corona of glans

Glans penis

Septum penis

Corpus spongiosum penis

Corpus cavernosum penis

Pubis

Bulbospongiosus muscle — Crus of penis

Ischiocavernosus muscle — Bulb of penis

Ramus of ischium

b

Superficial dorsal vein of penis — Dorsum of penis, Skin of penis

Dorsal artery of penis — Deep dorsal vein of penis

(Superficial) fascia of penis — Corpus cavernosum penis

(Deep) fascia of penis — Septum penis

Tunica albuginea of corpora cavernosa — Spongy urethra

Corpus spongiosum penis — Raphe of penis

Tunica albuginea of corpus spongiosum

c

Prepuce = Foreskin

Corona of glans

d

Prepuce = Foreskin

Corpus cavernosum penis — Glans penis

Spongy urethra

Raphe of penis

e

Navicular fossa

Septum of glans

Frenulum of prepuce

281 Penis (80%)

   a   Erectile bodies of the penis. The glans penis and the distal part of the corpus spongiosum penis were detached from the corpora cavernosa penis and displaced to the left. Inferior aspect

b–e   Distal aspect of transverse sections

   b   through the body of penis

   c   at the level of the neck of glans penis

   d   through the posterior part of the glans penis

   e   through the anterior part of the glans penis

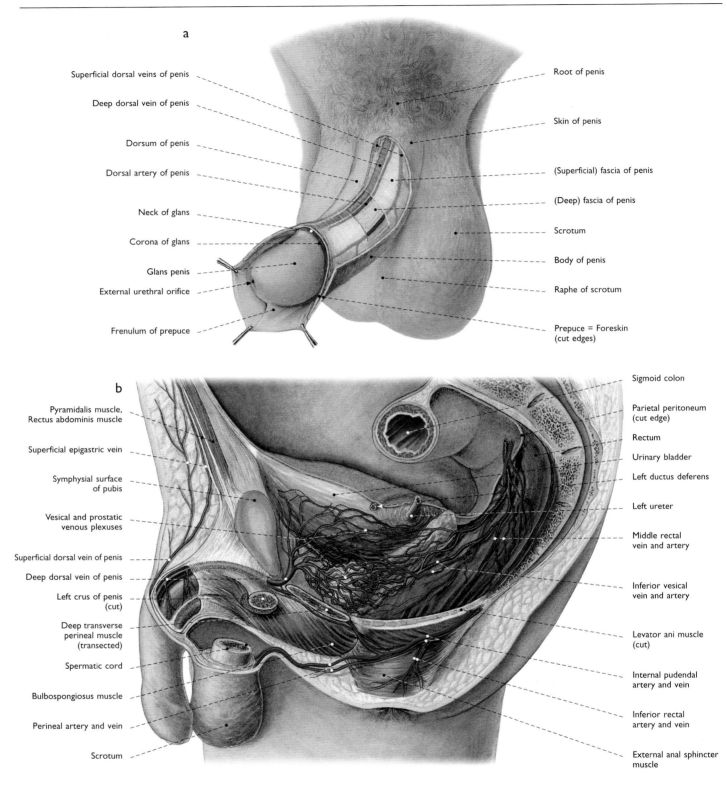

a

Superficial dorsal veins of penis

Deep dorsal vein of penis

Dorsum of penis

Dorsal artery of penis

Neck of glans

Corona of glans

Glans penis

External urethral orifice

Frenulum of prepuce

Root of penis

Skin of penis

(Superficial) fascia of penis

(Deep) fascia of penis

Scrotum

Body of penis

Raphe of scrotum

Prepuce = Foreskin
(cut edges)

b

Pyramidalis muscle,
Rectus abdominis muscle

Superficial epigastric vein

Symphysial surface
of pubis

Vesical and prostatic
venous plexuses

Superficial dorsal vein of penis

Deep dorsal vein of penis

Left crus of penis
(cut)

Deep transverse
perineal muscle
(transected)

Spermatic cord

Bulbospongiosus muscle

Perineal artery and vein

Scrotum

Sigmoid colon

Parietal peritoneum
(cut edge)

Rectum

Urinary bladder

Left ductus deferens

Left ureter

Middle rectal
vein and artery

Inferior vesical
vein and artery

Levator ani muscle
(cut)

Internal pudendal
artery and vein

Inferior rectal
artery and vein

External anal sphincter
muscle

282 Male external genitalia

a The preputial sac was opened and a longitudinal strip of skin
removed from the left side of the prepuce and the body of penis.
A small rectangular window was cut into the (superficial) fascia
of penis (80%). Ventral aspect

b Drainage of superficial and deep dorsal veins of penis (60%).
Left paramedian section through the lesser pelvis, medial aspect
of the right part

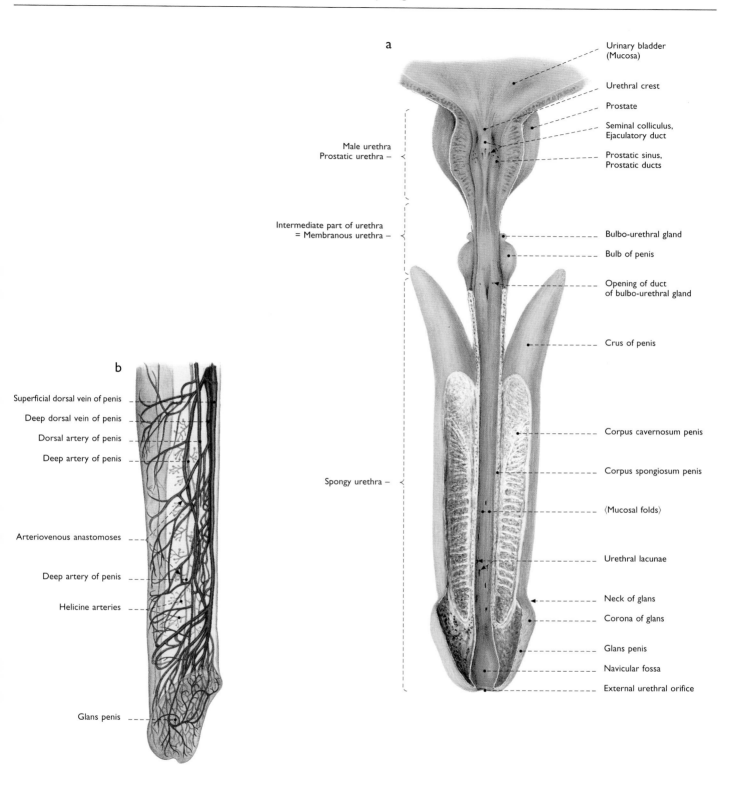

a

Urinary bladder
(Mucosa)

Urethral crest

Prostate

Seminal colliculus,
Ejaculatory duct

Prostatic sinus,
Prostatic ducts

Male urethra
Prostatic urethra —

Intermediate part of urethra
= Membranous urethra —

Bulbo-urethral gland

Bulb of penis

Opening of duct
of bulbo-urethral gland

Crus of penis

Corpus cavernosum penis

Corpus spongiosum penis

⟨Mucosal folds⟩

Urethral lacunae

Neck of glans

Corona of glans

Glans penis

Navicular fossa

External urethral orifice

Spongy urethra —

b

Superficial dorsal vein of penis

Deep dorsal vein of penis

Dorsal artery of penis

Deep artery of penis

Arteriovenous anastomoses

Deep artery of penis

Helicine arteries

Glans penis

283 Male urethra and arteries of the penis (80%)

a The urethra was opened by a median section
from the internal to the external urethral orifices.
The cut surfaces were turned outwards. Ventral aspect

b Arteries of the penis including the helicine arteries
(according to Ferner, 1975, and Lierse, 1984),
schematic representation, right lateral aspect

Anococcygeal nerves

Anococcygeal body
= Anococcygeal ligament

Levator ani muscle

Inferior rectal arteries

Inferior clunial nerves

Obturator fascia

Pudendal canal

Sacrotuberous ligament

Inferior anal nerves
(of pudendal nerve)

Internal pudendal
vein and artery

External anal sphincter
muscle

Perineal branches
of posterior femoral
cutaneous nerve

Perineal nerves
(of pudendal nerve)

Perineal veins and artery

Ramus of ischium

Perineal fascia

Posterior scrotal branches
of perineal artery,
Perineal veins

Ischiococcygeus muscle
= Coccygeus muscle

Sacrotuberous ligament

Sacrospinous ligament

Pudendal nerve

Inferior rectal artery,
Gemellus superior muscle

Inferior gluteal nerve

Internal pudendal artery

Obturator internus muscle

Sciatic nerve

Posterior femoral
cutaneous nerve

Gemellus inferior muscle

Sacrotuberous ligament

Perineal nerves

Dorsal artery, nerve of penis

Ischial tuberosity

Perineal branches
of post. fem. cut. nerve

Perineal membrane,
Superficial transverse
perineal muscle
(transected)

Perineal artery

Ischiocavernosus muscle

Bulbospongiosus muscle

Posterior scrotal nerves

Corpus cavernosum penis

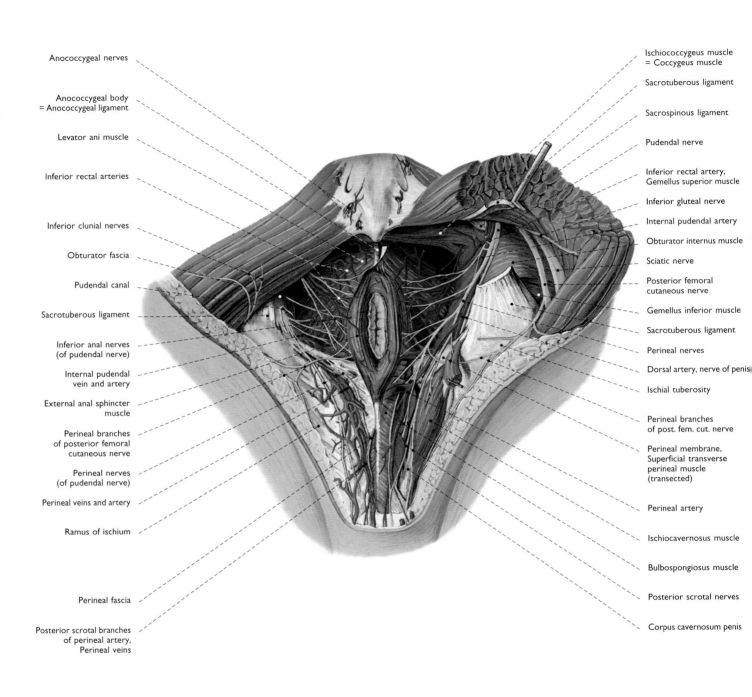

**284 Blood vessels and nerves
of the perineal region of a male** (70%)
Inferior aspect

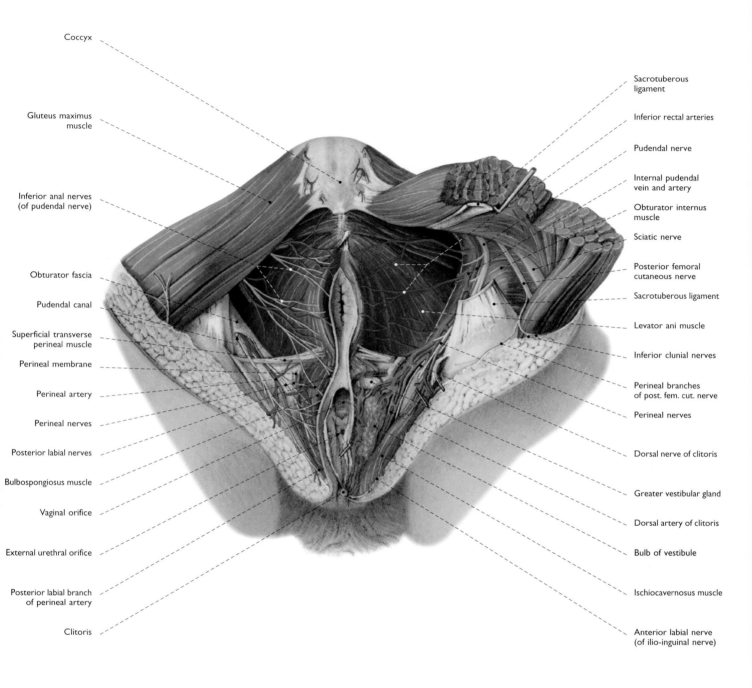

Coccyx

Gluteus maximus muscle

Inferior anal nerves (of pudendal nerve)

Obturator fascia

Pudendal canal

Superficial transverse perineal muscle

Perineal membrane

Perineal artery

Perineal nerves

Posterior labial nerves

Bulbospongiosus muscle

Vaginal orifice

External urethral orifice

Posterior labial branch of perineal artery

Clitoris

Sacrotuberous ligament

Inferior rectal arteries

Pudendal nerve

Internal pudendal vein and artery

Obturator internus muscle

Sciatic nerve

Posterior femoral cutaneous nerve

Sacrotuberous ligament

Levator ani muscle

Inferior clunial nerves

Perineal branches of post. fem. cut. nerve

Perineal nerves

Dorsal nerve of clitoris

Greater vestibular gland

Dorsal artery of clitoris

Bulb of vestibule

Ischiocavernosus muscle

Anterior labial nerve (of ilio-inguinal nerve)

285 Blood vessels and nerves of the perineal region of a female (70%)
Inferior aspect

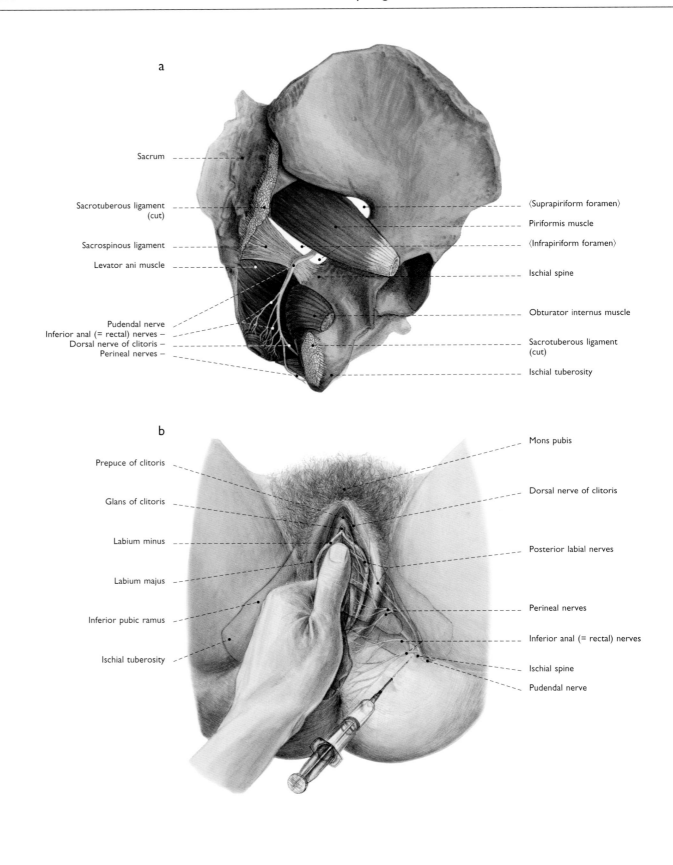

a

Sacrum

Sacrotuberous ligament (cut)

Sacrospinous ligament

Levator ani muscle

Pudendal nerve
Inferior anal (= rectal) nerves –
Dorsal nerve of clitoris –
Perineal nerves –

⟨Suprapiriform foramen⟩

Piriformis muscle

⟨Infrapiriform foramen⟩

Ischial spine

Obturator internus muscle

Sacrotuberous ligament (cut)

Ischial tuberosity

b

Prepuce of clitoris

Glans of clitoris

Labium minus

Labium majus

Inferior pubic ramus

Ischial tuberosity

Mons pubis

Dorsal nerve of clitoris

Posterior labial nerves

Perineal nerves

Inferior anal (= rectal) nerves

Ischial spine

Pudendal nerve

286 Nerves of the perineal region of a female (40%)

    a　Course of the right pudendal nerve, right dorsolateral aspect
    b　Anesthesia of the pudendal nerve on the left side of the body, schematic representation, inferior aspect

Central Nervous System

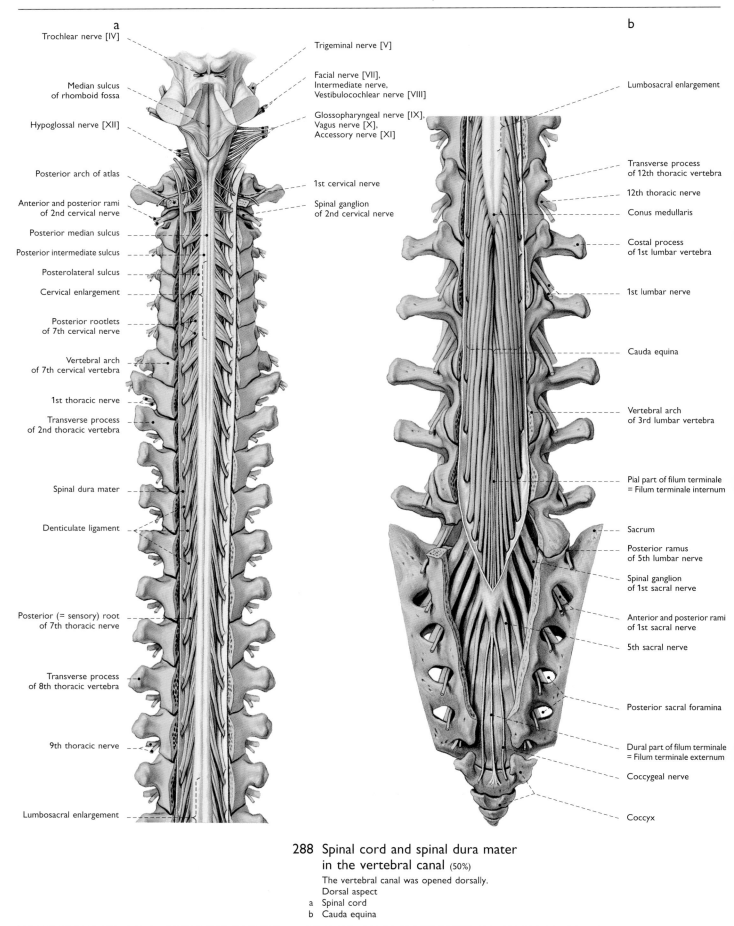

a

Trochlear nerve [IV]

Median sulcus
of rhomboid fossa

Hypoglossal nerve [XII]

Posterior arch of atlas

Anterior and posterior rami
of 2nd cervical nerve

Posterior median sulcus

Posterior intermediate sulcus

Posterolateral sulcus

Cervical enlargement

Posterior rootlets
of 7th cervical nerve

Vertebral arch
of 7th cervical vertebra

1st thoracic nerve

Transverse process
of 2nd thoracic vertebra

Spinal dura mater

Denticulate ligament

Posterior (= sensory) root
of 7th thoracic nerve

Transverse process
of 8th thoracic vertebra

9th thoracic nerve

Lumbosacral enlargement

Trigeminal nerve [V]

Facial nerve [VII],
Intermediate nerve,
Vestibulocochlear nerve [VIII]

Glossopharyngeal nerve [IX],
Vagus nerve [X],
Accessory nerve [XI]

1st cervical nerve

Spinal ganglion
of 2nd cervical nerve

b

Lumbosacral enlargement

Transverse process
of 12th thoracic vertebra

12th thoracic nerve

Conus medullaris

Costal process
of 1st lumbar vertebra

1st lumbar nerve

Cauda equina

Vertebral arch
of 3rd lumbar vertebra

Pial part of filum terminale
= Filum terminale internum

Sacrum

Posterior ramus
of 5th lumbar nerve

Spinal ganglion
of 1st sacral nerve

Anterior and posterior rami
of 1st sacral nerve

5th sacral nerve

Posterior sacral foramina

Dural part of filum terminale
= Filum terminale externum

Coccygeal nerve

Coccyx

**288 Spinal cord and spinal dura mater
in the vertebral canal (50%)**

The vertebral canal was opened dorsally.
Dorsal aspect
a Spinal cord
b Cauda equina

a

Periosteum

Spinal dura mater

Spinal arachnoid mater

Spinal pia mater

Spinal nerve
Posterior (= sensory)
root –
Anterior (= motor)
root –
Spinal ganglion –
Posterior ramus –
Anterior ramus –
White ramus communicans –
Gray ramus communicans –

Epidural space

(Subdural space)

Subarachnoid space
= Leptomeningeal space

Denticulate ligament

Spinal cord
– Gray matter
– White matter
– Anterior median fissure

b

Ligamentum flavum
(cut surface)

Vertebral arch

Posterior spinal vein

Posterior spinal artery

Posterior internal vertebral
venous plexus

Posterior (= sensory) root
of spinal nerve

⟨Interradicular septum⟩

Anterior (= motor) root
of spinal nerve

Anterior spinal veins

Anterior spinal artery

Anterior rootlets
of spinal nerve

Anterolateral sulcus

Anterior median fissure

Epidural space

(Subdural space)

⟨Leptomeningeal septum⟩

Subarachnoid space
= Leptomeningeal space

Spinal pia mater
(cut margin)

Denticulate ligament

Spinal arachnoid mater
(cut margin)

Spinal dura mater
(cut margin)

Spinal branches of posterior
intercostal vein and artery

Spinal ganglion

Anterior internal
vertebral venous plexus

Trunk of spinal nerve

Spinal nerve
– Posterior ramus
– Meningeal ramus
– Anterior ramus
– White ramus communicans
– Gray ramus communicans

Ganglion of sympathetic trunk

Posterior longitudinal ligament

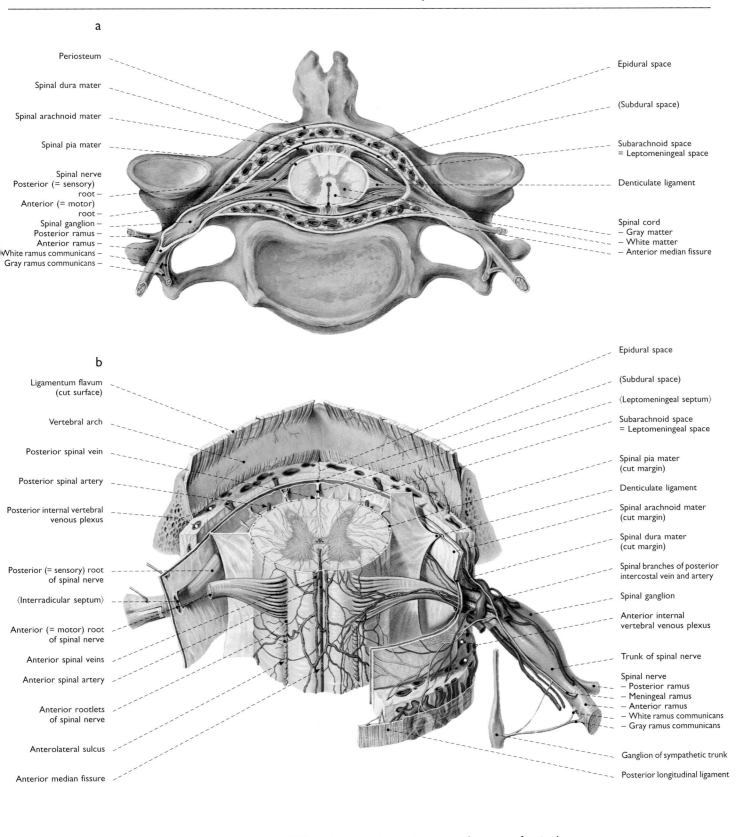

289 Spinal cord, meninges, and roots of spinal nerve
Transverse sections of the spinal cord and the roots of spinal nerve
a Section through the vertebral canal of cervical spine (230%),
superior aspect
b Schematized three-dimensional representation after removal
of vertebral bodies and transverse processes of vertebrae (400%),
ventrocranial aspect

Central Nervous System

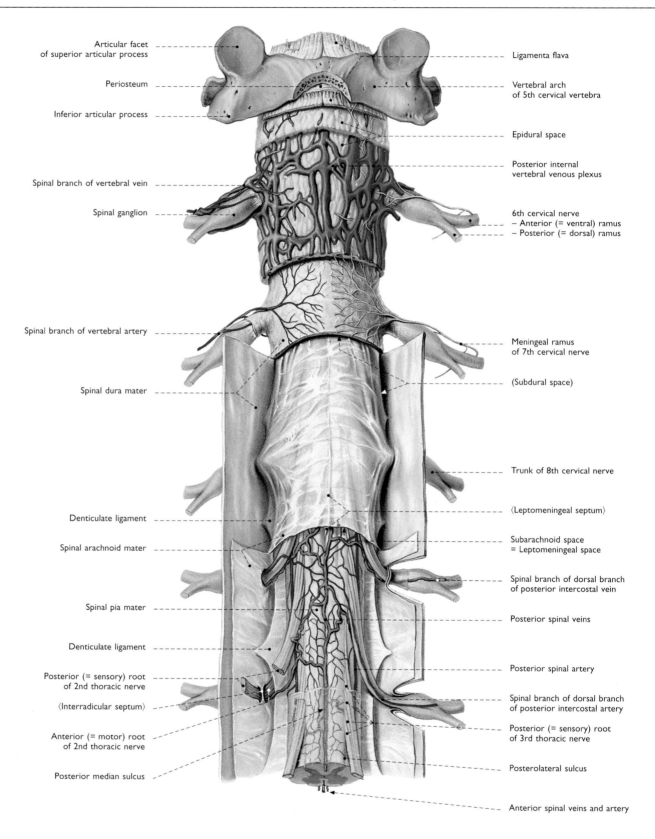

Articular facet
of superior articular process

Periosteum

Inferior articular process

Spinal branch of vertebral vein

Spinal ganglion

Spinal branch of vertebral artery

Spinal dura mater

Denticulate ligament

Spinal arachnoid mater

Spinal pia mater

Denticulate ligament

Posterior (= sensory) root
of 2nd thoracic nerve

⟨Interradicular septum⟩

Anterior (= motor) root
of 2nd thoracic nerve

Posterior median sulcus

Ligamenta flava

Vertebral arch
of 5th cervical vertebra

Epidural space

Posterior internal
vertebral venous plexus

6th cervical nerve
– Anterior (= ventral) ramus
– Posterior (= dorsal) ramus

Meningeal ramus
of 7th cervical nerve

(Subdural space)

Trunk of 8th cervical nerve

⟨Leptomeningeal septum⟩

Subarachnoid space
= Leptomeningeal space

Spinal branch of dorsal branch
of posterior intercostal vein

Posterior spinal veins

Posterior spinal artery

Spinal branch of dorsal branch
of posterior intercostal artery

Posterior (= sensory) root
of 3rd thoracic nerve

Posterolateral sulcus

Anterior spinal veins and artery

290 Spinal cord, meninges, and spinal nerves
The investing structures of the spinal cord and
the blood vessels are demonstrated in layers (150%).
Dorsal aspect

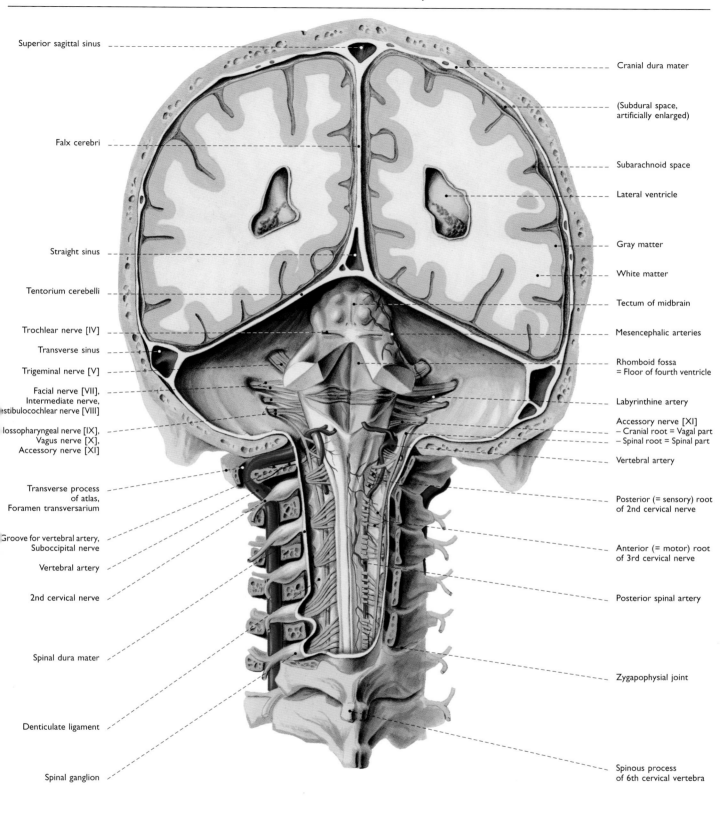

Superior sagittal sinus

Falx cerebri

Straight sinus

Tentorium cerebelli

Trochlear nerve [IV]

Transverse sinus

Trigeminal nerve [V]

Facial nerve [VII],
Intermediate nerve,
Vestibulocochlear nerve [VIII]

Glossopharyngeal nerve [IX],
Vagus nerve [X],
Accessory nerve [XI]

Transverse process
of atlas,
Foramen transversarium

Groove for vertebral artery,
Suboccipital nerve

Vertebral artery

2nd cervical nerve

Spinal dura mater

Denticulate ligament

Spinal ganglion

Cranial dura mater

(Subdural space,
artificially enlarged)

Subarachnoid space

Lateral ventricle

Gray matter

White matter

Tectum of midbrain

Mesencephalic arteries

Rhomboid fossa
= Floor of fourth ventricle

Labyrinthine artery

Accessory nerve [XI]
– Cranial root = Vagal part
– Spinal root = Spinal part

Vertebral artery

Posterior (= sensory) root
of 2nd cervical nerve

Anterior (= motor) root
of 3rd cervical nerve

Posterior spinal artery

Zygapophysial joint

Spinous process
of 6th cervical vertebra

291 Brainstem and dura mater (75%)
Coronal section through the head and neck.
The vertebral arches were removed. Dorsal aspect

a

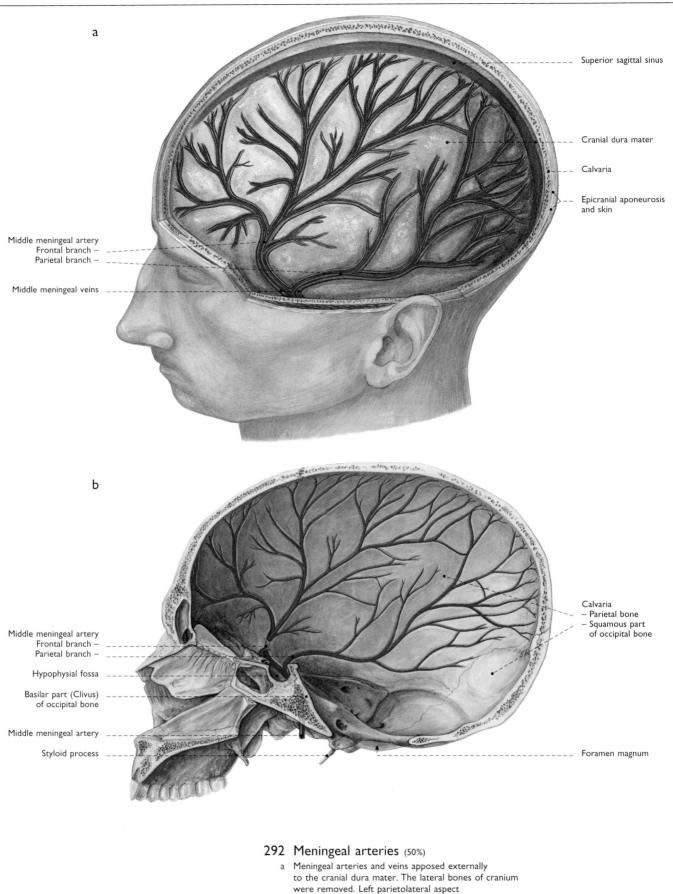

Superior sagittal sinus

Cranial dura mater

Calvaria

Epicranial aponeurosis
and skin

Middle meningeal artery
Frontal branch –
Parietal branch –

Middle meningeal veins

b

Middle meningeal artery
Frontal branch –
Parietal branch –

Hypophysial fossa

Basilar part (Clivus)
of occipital bone

Middle meningeal artery

Styloid process

Calvaria
– Parietal bone
– Squamous part
of occipital bone

Foramen magnum

292 Meningeal arteries (50%)

a Meningeal arteries and veins apposed externally
to the cranial dura mater. The lateral bones of cranium
were removed. Left parietolateral aspect
b Meningeal arteries lying in osseous grooves of the lateral bones
of cranium, medial aspect of the right half of cranium

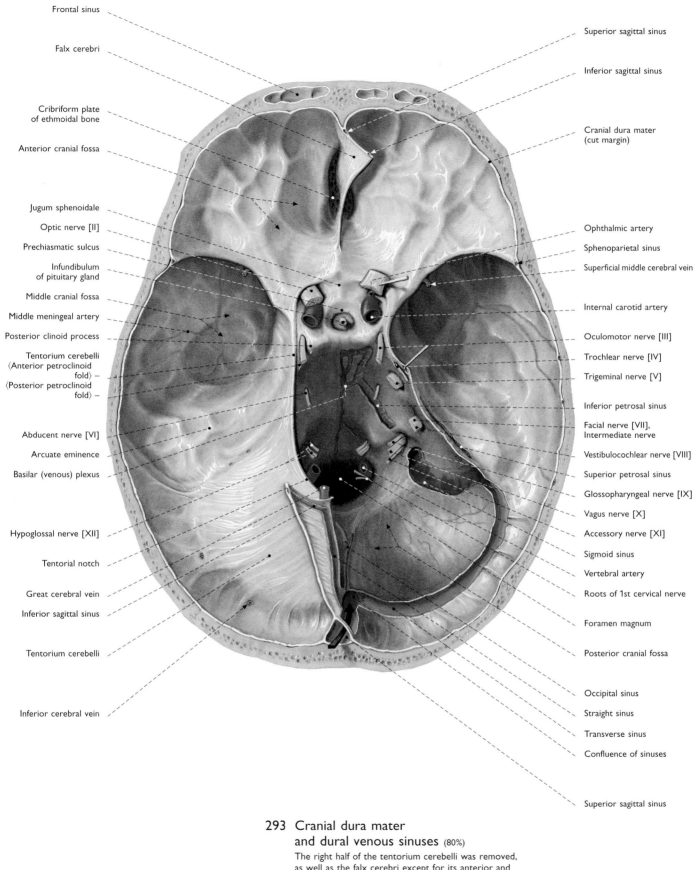

Frontal sinus

Falx cerebri

Cribriform plate
of ethmoidal bone

Anterior cranial fossa

Jugum sphenoidale

Optic nerve [II]

Prechiasmatic sulcus

Infundibulum
of pituitary gland

Middle cranial fossa

Middle meningeal artery

Posterior clinoid process

Tentorium cerebelli
⟨Anterior petroclinoid
fold⟩ —
⟨Posterior petroclinoid
fold⟩ —

Abducent nerve [VI]

Arcuate eminence

Basilar (venous) plexus

Hypoglossal nerve [XII]

Tentorial notch

Great cerebral vein

Inferior sagittal sinus

Tentorium cerebelli

Inferior cerebral vein

Superior sagittal sinus

Inferior sagittal sinus

Cranial dura mater
(cut margin)

Ophthalmic artery

Sphenoparietal sinus

Superficial middle cerebral vein

Internal carotid artery

Oculomotor nerve [III]

Trochlear nerve [IV]

Trigeminal nerve [V]

Inferior petrosal sinus

Facial nerve [VII],
Intermediate nerve

Vestibulocochlear nerve [VIII]

Superior petrosal sinus

Glossopharyngeal nerve [IX]

Vagus nerve [X]

Accessory nerve [XI]

Sigmoid sinus

Vertebral artery

Roots of 1st cervical nerve

Foramen magnum

Posterior cranial fossa

Occipital sinus

Straight sinus

Transverse sinus

Confluence of sinuses

Superior sagittal sinus

**293 Cranial dura mater
and dural venous sinuses** (80%)

The right half of the tentorium cerebelli was removed,
as well as the falx cerebri except for its anterior and
posterior attachments. Some of the dural venous sinuses
were opened. Superior aspect

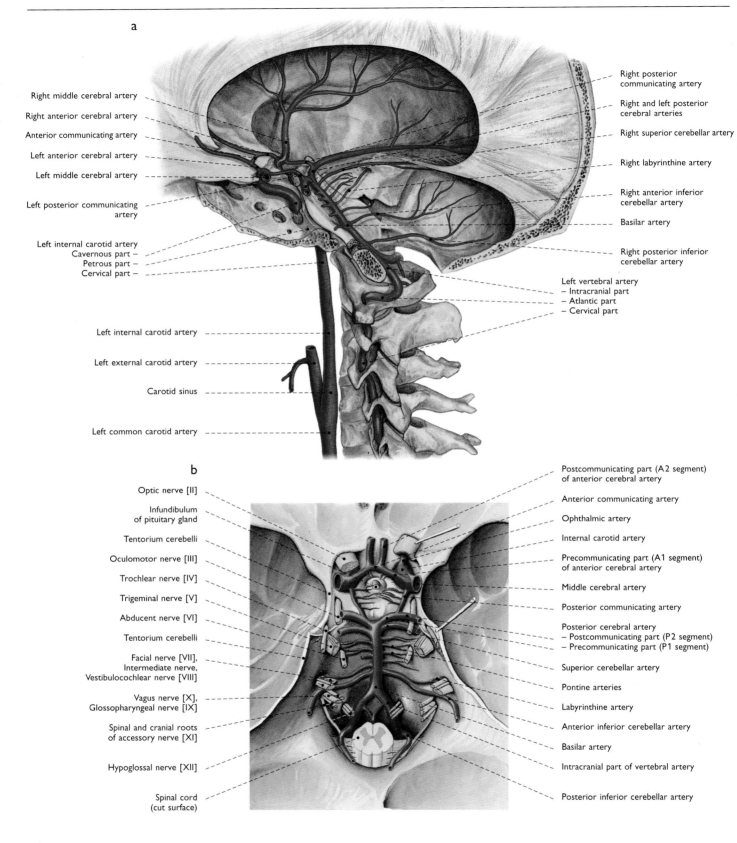

a

Right middle cerebral artery

Right anterior cerebral artery

Anterior communicating artery

Left anterior cerebral artery

Left middle cerebral artery

Left posterior communicating artery

Left internal carotid artery
Cavernous part –
Petrous part –
Cervical part –

Left internal carotid artery

Left external carotid artery

Carotid sinus

Left common carotid artery

Right posterior communicating artery

Right and left posterior cerebral arteries

Right superior cerebellar artery

Right labyrinthine artery

Right anterior inferior cerebellar artery

Basilar artery

Right posterior inferior cerebellar artery

Left vertebral artery
– Intracranial part
– Atlantic part
– Cervical part

b

Optic nerve [II]

Infundibulum of pituitary gland

Tentorium cerebelli

Oculomotor nerve [III]

Trochlear nerve [IV]

Trigeminal nerve [V]

Abducent nerve [VI]

Tentorium cerebelli

Facial nerve [VII], Intermediate nerve, Vestibulocochlear nerve [VIII]

Vagus nerve [X], Glossopharyngeal nerve [IX]

Spinal and cranial roots of accessory nerve [XI]

Hypoglossal nerve [XII]

Spinal cord (cut surface)

Postcommunicating part (A2 segment) of anterior cerebral artery

Anterior communicating artery

Ophthalmic artery

Internal carotid artery

Precommunicating part (A1 segment) of anterior cerebral artery

Middle cerebral artery

Posterior communicating artery

Posterior cerebral artery
– Postcommunicating part (P2 segment)
– Precommunicating part (P1 segment)

Superior cerebellar artery

Pontine arteries

Labyrinthine artery

Anterior inferior cerebellar artery

Basilar artery

Intracranial part of vertebral artery

Posterior inferior cerebellar artery

294 Cranial dura mater and cerebral arterial circle

a Internal carotid and vertebral arteries, cerebral arterial circle (60%), sagittal section to the left of the median plane, parietomedial aspect of the right half

b Cerebral arterial circle (90%), superior aspect

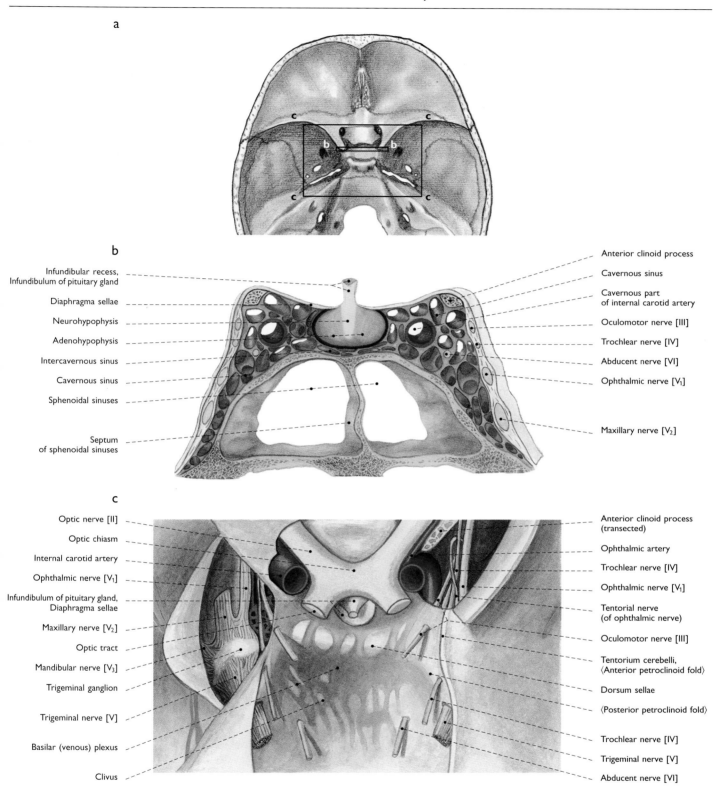

a

b

Infundibular recess,
Infundibulum of pituitary gland

Diaphragma sellae

Neurohypophysis

Adenohypophysis

Intercavernous sinus

Cavernous sinus

Sphenoidal sinuses

Septum
of sphenoidal sinuses

Anterior clinoid process

Cavernous sinus

Cavernous part
of internal carotid artery

Oculomotor nerve [III]

Trochlear nerve [IV]

Abducent nerve [VI]

Ophthalmic nerve [V₁]

Maxillary nerve [V₂]

c

Optic nerve [II]

Optic chiasm

Internal carotid artery

Ophthalmic nerve [V₁]

Infundibulum of pituitary gland,
Diaphragma sellae

Maxillary nerve [V₂]

Optic tract

Mandibular nerve [V₃]

Trigeminal ganglion

Trigeminal nerve [V]

Basilar (venous) plexus

Clivus

Anterior clinoid process
(transected)

Ophthalmic artery

Trochlear nerve [IV]

Ophthalmic nerve [V₁]

Tentorial nerve
(of ophthalmic nerve)

Oculomotor nerve [III]

Tentorium cerebelli,
⟨Anterior petroclinoid fold⟩

Dorsum sellae

⟨Posterior petroclinoid fold⟩

Trochlear nerve [IV]

Trigeminal nerve [V]

Abducent nerve [VI]

295 Cavernous sinus

a Internal surface of the cranial base. The section planes
of figs. b and c are indicated.

b Coronal section through the cavernous sinuses, the pituitary gland,
and the sphenoidal sinuses (300%)

c View of central areas of the cranial base. On the left side,
the cranial dura mater is turned upwards, the cavernous sinus opened,
and the trigeminal ganglion exposed (200%). Superior aspect

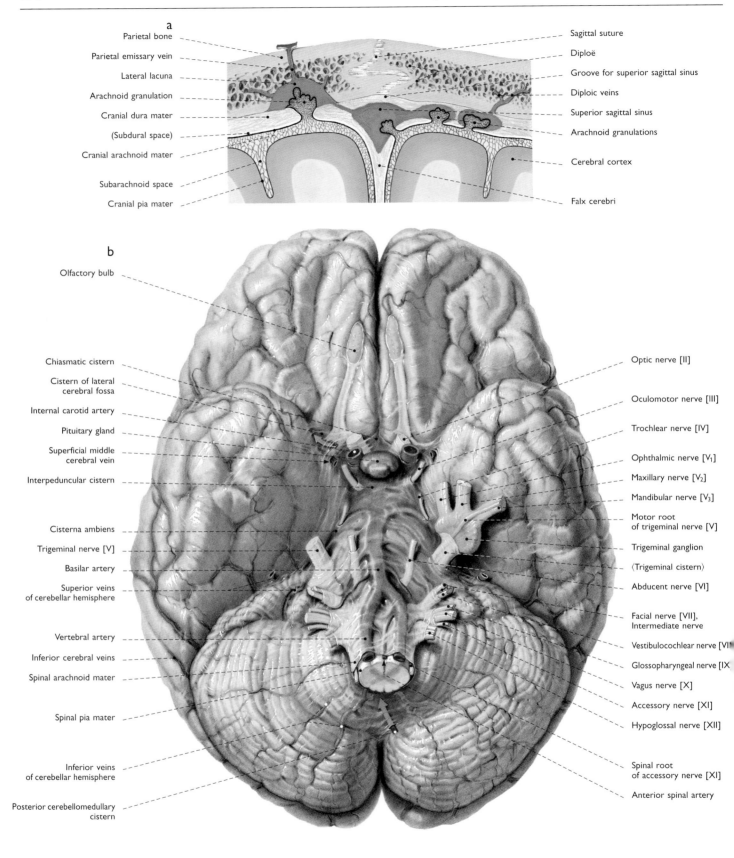

a

Parietal bone
Parietal emissary vein
Lateral lacuna
Arachnoid granulation
Cranial dura mater
(Subdural space)
Cranial arachnoid mater
Subarachnoid space
Cranial pia mater

Sagittal suture
Diploë
Groove for superior sagittal sinus
Diploic veins
Superior sagittal sinus
Arachnoid granulations
Cerebral cortex
Falx cerebri

b

Olfactory bulb
Chiasmatic cistern
Cistern of lateral cerebral fossa
Internal carotid artery
Pituitary gland
Superficial middle cerebral vein
Interpeduncular cistern
Cisterna ambiens
Trigeminal nerve [V]
Basilar artery
Superior veins of cerebellar hemisphere
Vertebral artery
Inferior cerebral veins
Spinal arachnoid mater
Spinal pia mater
Inferior veins of cerebellar hemisphere
Posterior cerebellomedullary cistern

Optic nerve [II]
Oculomotor nerve [III]
Trochlear nerve [IV]
Ophthalmic nerve [V₁]
Maxillary nerve [V₂]
Mandibular nerve [V₃]
Motor root of trigeminal nerve [V]
Trigeminal ganglion
⟨Trigeminal cistern⟩
Abducent nerve [VI]
Facial nerve [VII], Intermediate nerve
Vestibulocochlear nerve [VIII]
Glossopharyngeal nerve [IX]
Vagus nerve [X]
Accessory nerve [XI]
Hypoglossal nerve [XII]
Spinal root of accessory nerve [XI]
Anterior spinal artery

296 Brain and meninges (100%)
a Meninges on the cerebral vault, schematic coronal section
b Brain with leptomeninx (100%), inferior aspect

Frontal pole

Superior cerebral veins

Superficial middle cerebral vein

Superior anastomotic vein

Arachnoid granulations

Longitudinal cerebral fissure

Occipital pole

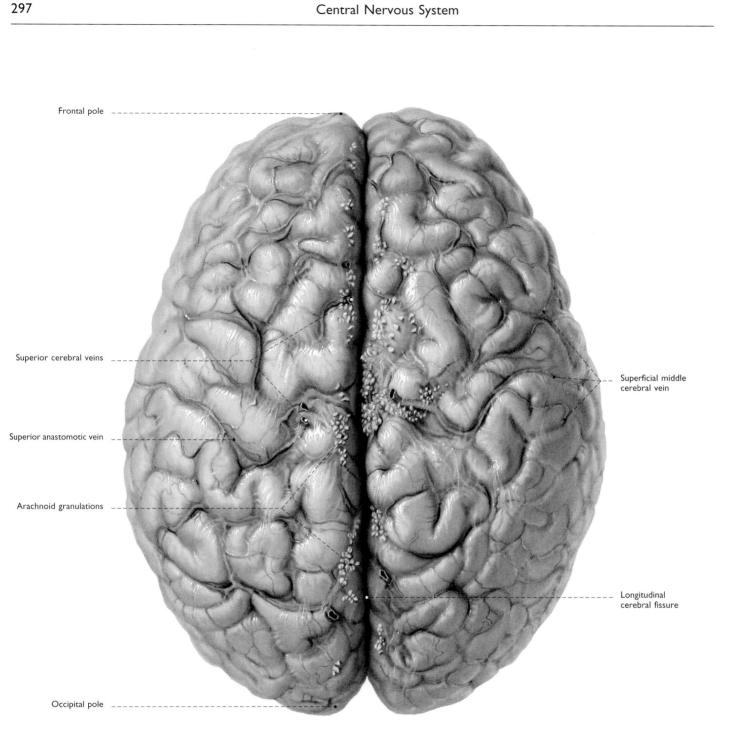

297 Brain with leptomeninx (100%)
Superior aspect

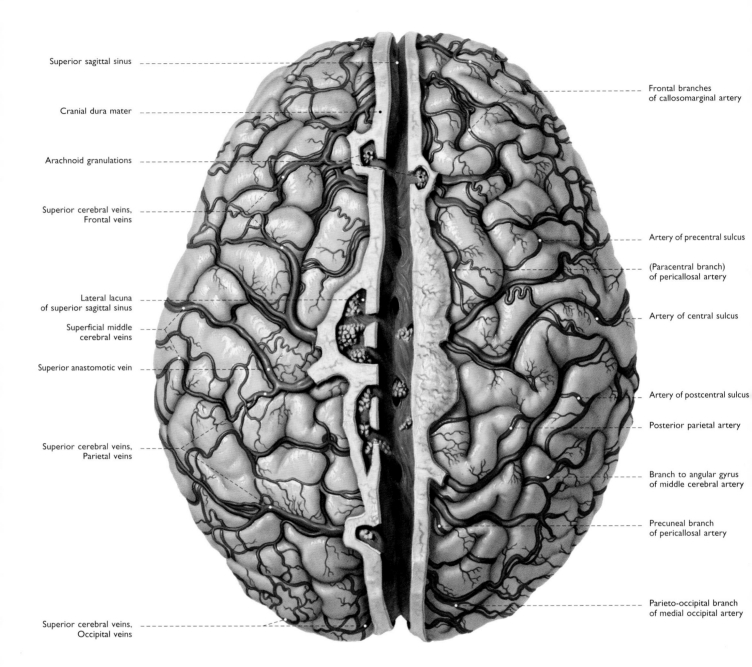

Superior sagittal sinus

Cranial dura mater

Arachnoid granulations

Superior cerebral veins,
Frontal veins

Lateral lacuna
of superior sagittal sinus

Superficial middle
cerebral veins

Superior anastomotic vein

Superior cerebral veins,
Parietal veins

Superior cerebral veins,
Occipital veins

Frontal branches
of callosomarginal artery

Artery of precentral sulcus

(Paracentral branch)
of pericallosal artery

Artery of central sulcus

Artery of postcentral sulcus

Posterior parietal artery

Branch to angular gyrus
of middle cerebral artery

Precuneal branch
of pericallosal artery

Parieto-occipital branch
of medial occipital artery

298 Superficial arteries and veins
of the cerebral hemispheres (100%)
Superior aspect

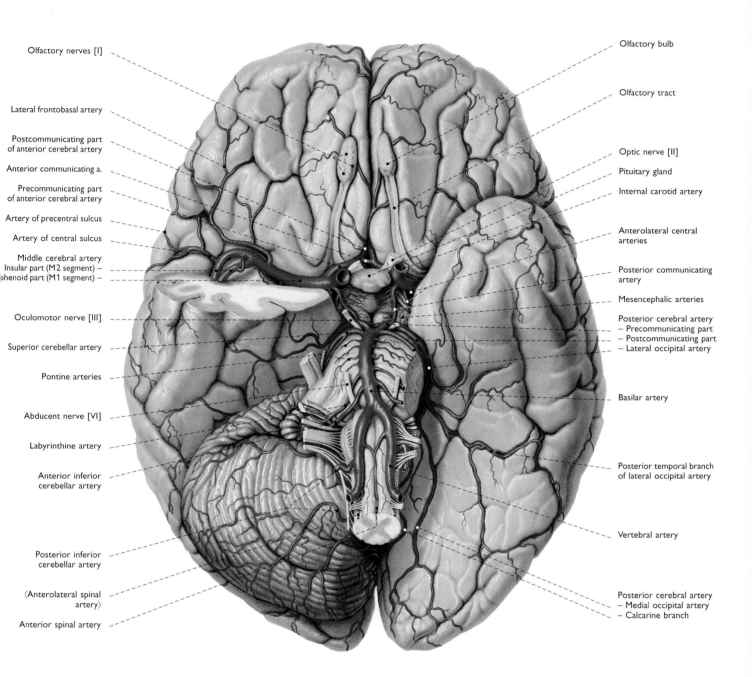

Olfactory nerves [I]

Lateral frontobasal artery

Postcommunicating part
of anterior cerebral artery

Anterior communicating a.

Precommunicating part
of anterior cerebral artery

Artery of precentral sulcus

Artery of central sulcus

Middle cerebral artery
Insular part (M2 segment) –
ᴘhenoid part (M1 segment) –

Oculomotor nerve [III]

Superior cerebellar artery

Pontine arteries

Abducent nerve [VI]

Labyrinthine artery

Anterior inferior
cerebellar artery

Posterior inferior
cerebellar artery

〈Anterolateral spinal
artery〉

Anterior spinal artery

Olfactory bulb

Olfactory tract

Optic nerve [II]

Pituitary gland

Internal carotid artery

Anterolateral central
arteries

Posterior communicating
artery

Mesencephalic arteries

Posterior cerebral artery
– Precommunicating part
– Postcommunicating part
– Lateral occipital artery

Basilar artery

Posterior temporal branch
of lateral occipital artery

Vertebral artery

Posterior cerebral artery
– Medial occipital artery
– Calcarine branch

299 Arteries of the brain (100%)
Inferior aspect

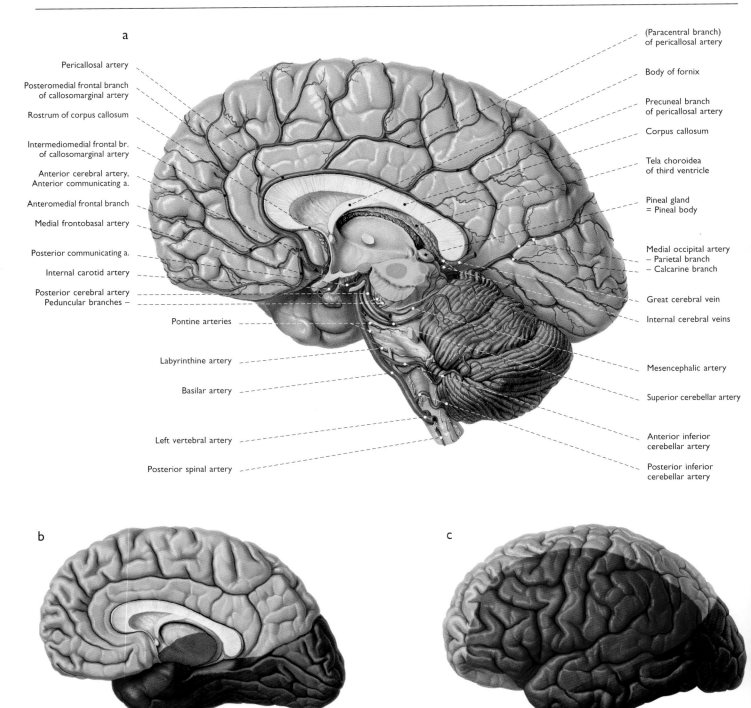

a

Pericallosal artery

Posteromedial frontal branch of callosomarginal artery

Rostrum of corpus callosum

Intermediomedial frontal br. of callosomarginal artery

Anterior cerebral artery, Anterior communicating a.

Anteromedial frontal branch

Medial frontobasal artery

Posterior communicating a.

Internal carotid artery

Posterior cerebral artery Peduncular branches —

Pontine arteries

Labyrinthine artery

Basilar artery

Left vertebral artery

Posterior spinal artery

(Paracentral branch) of pericallosal artery

Body of fornix

Precuneal branch of pericallosal artery

Corpus callosum

Tela choroidea of third ventricle

Pineal gland = Pineal body

Medial occipital artery – Parietal branch – Calcarine branch

Great cerebral vein

Internal cerebral veins

Mesencephalic artery

Superior cerebellar artery

Anterior inferior cerebellar artery

Posterior inferior cerebellar artery

b

c

Anterior cerebral artery

Middle cerebral artery

Posterior cerebral artery

300 Arteries of the brain

a Medial aspect of the right cerebral hemisphere and lateral aspect of the left cerebellar hemisphere (80%)

b, c Circulation areas of the anterior (**yellow**), middle (**red**), and posterior (**brown**) cerebral arteries at the cerebral cortex (50%)

b Medial aspect

c Lateral aspect

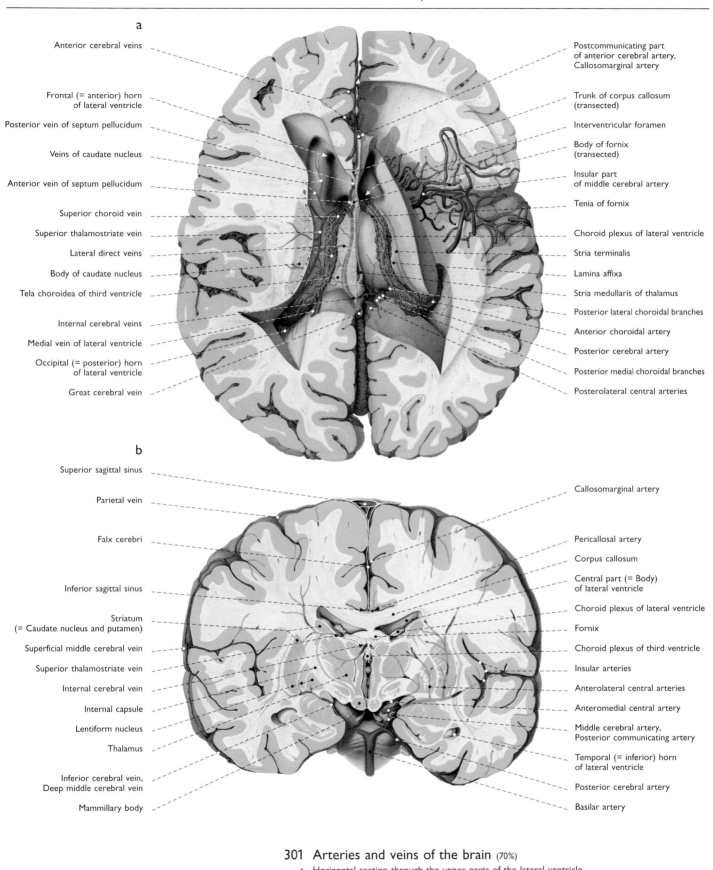

a

Anterior cerebral veins

Frontal (= anterior) horn
of lateral ventricle

Posterior vein of septum pellucidum

Veins of caudate nucleus

Anterior vein of septum pellucidum

Superior choroid vein

Superior thalamostriate vein

Lateral direct veins

Body of caudate nucleus

Tela choroidea of third ventricle

Internal cerebral veins

Medial vein of lateral ventricle

Occipital (= posterior) horn
of lateral ventricle

Great cerebral vein

Postcommunicating part
of anterior cerebral artery,
Callosomarginal artery

Trunk of corpus callosum
(transected)

Interventricular foramen

Body of fornix
(transected)

Insular part
of middle cerebral artery

Tenia of fornix

Choroid plexus of lateral ventricle

Stria terminalis

Lamina affixa

Stria medullaris of thalamus

Posterior lateral choroidal branches

Anterior choroidal artery

Posterior cerebral artery

Posterior medial choroidal branches

Posterolateral central arteries

b

Superior sagittal sinus

Parietal vein

Falx cerebri

Inferior sagittal sinus

Striatum
(= Caudate nucleus and putamen)

Superficial middle cerebral vein

Superior thalamostriate vein

Internal cerebral vein

Internal capsule

Lentiform nucleus

Thalamus

Inferior cerebral vein,
Deep middle cerebral vein

Mammillary body

Callosomarginal artery

Pericallosal artery

Corpus callosum

Central part (= Body)
of lateral ventricle

Choroid plexus of lateral ventricle

Fornix

Choroid plexus of third ventricle

Insular arteries

Anterolateral central arteries

Anteromedial central artery

Middle cerebral artery,
Posterior communicating artery

Temporal (= inferior) horn
of lateral ventricle

Posterior cerebral artery

Basilar artery

301 Arteries and veins of the brain (70%)

a　Horizontal section through the upper parts of the lateral ventricle.
　　The corpus callosum, the fornix, and parts of the cerebral hemispheres
　　were removed. Superior aspect
b　Coronal section at the level of the mammillary bodies,
　　frontal aspect

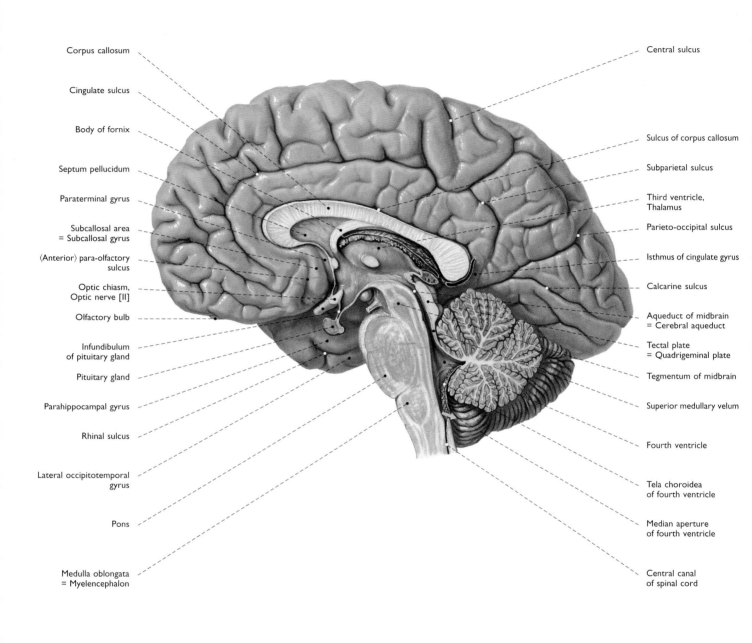

Corpus callosum

Cingulate sulcus

Body of fornix

Septum pellucidum

Paraterminal gyrus

Subcallosal area
= Subcallosal gyrus

⟨Anterior⟩ para-olfactory
sulcus

Optic chiasm,
Optic nerve [II]

Olfactory bulb

Infundibulum
of pituitary gland

Pituitary gland

Parahippocampal gyrus

Rhinal sulcus

Lateral occipitotemporal
gyrus

Pons

Medulla oblongata
= Myelencephalon

Central sulcus

Sulcus of corpus callosum

Subparietal sulcus

Third ventricle,
Thalamus

Parieto-occipital sulcus

Isthmus of cingulate gyrus

Calcarine sulcus

Aqueduct of midbrain
= Cerebral aqueduct

Tectal plate
= Quadrigeminal plate

Tegmentum of midbrain

Superior medullary velum

Fourth ventricle

Tela choroidea
of fourth ventricle

Median aperture
of fourth ventricle

Central canal
of spinal cord

302 Brain (80%)
Median section,
medial aspect of the right hemisphere

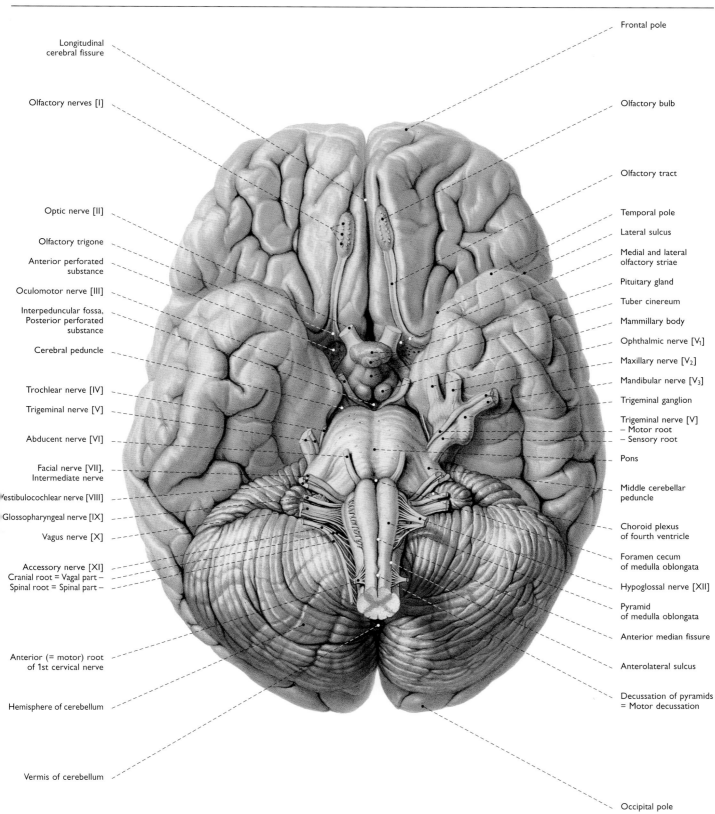

Longitudinal
cerebral fissure

Olfactory nerves [I]

Optic nerve [II]

Olfactory trigone

Anterior perforated
substance

Oculomotor nerve [III]

Interpeduncular fossa,
Posterior perforated
substance

Cerebral peduncle

Trochlear nerve [IV]

Trigeminal nerve [V]

Abducent nerve [VI]

Facial nerve [VII],
Intermediate nerve

Vestibulocochlear nerve [VIII]

Glossopharyngeal nerve [IX]

Vagus nerve [X]

Accessory nerve [XI]
Cranial root = Vagal part –
Spinal root = Spinal part –

Anterior (= motor) root
of 1st cervical nerve

Hemisphere of cerebellum

Vermis of cerebellum

Frontal pole

Olfactory bulb

Olfactory tract

Temporal pole

Lateral sulcus

Medial and lateral
olfactory striae

Pituitary gland

Tuber cinereum

Mammillary body

Ophthalmic nerve [V₁]

Maxillary nerve [V₂]

Mandibular nerve [V₃]

Trigeminal ganglion

Trigeminal nerve [V]
– Motor root
– Sensory root

Pons

Middle cerebellar
peduncle

Choroid plexus
of fourth ventricle

Foramen cecum
of medulla oblongata

Hypoglossal nerve [XII]

Pyramid
of medulla oblongata

Anterior median fissure

Anterolateral sulcus

Decussation of pyramids
= Motor decussation

Occipital pole

303 Brain and cranial nerves (100%)
Inferior aspect

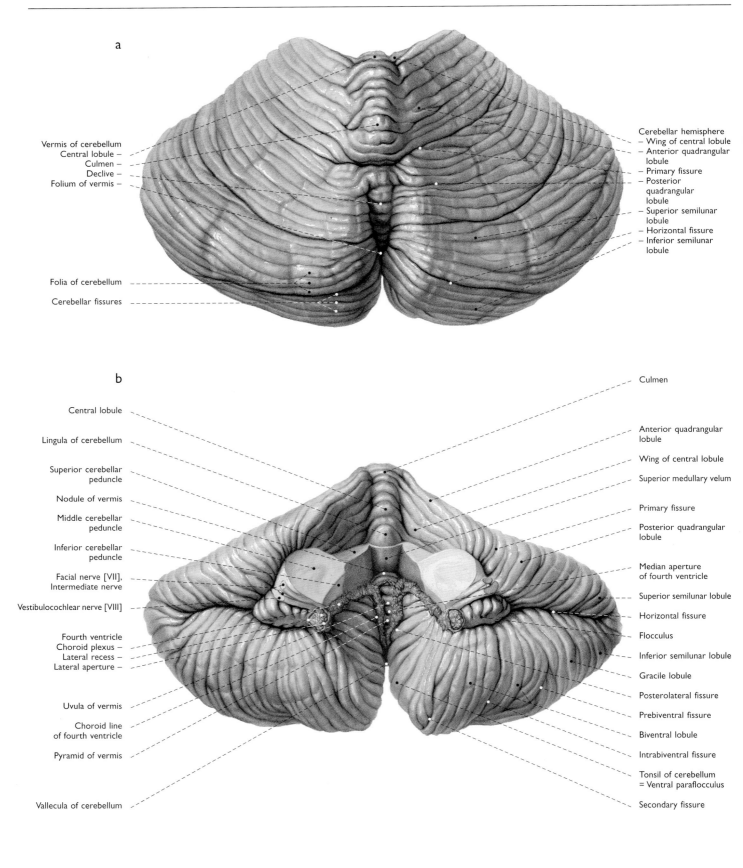

a

Vermis of cerebellum
Central lobule –
Culmen –
Declive –
Folium of vermis –

Folia of cerebellum

Cerebellar fissures

Cerebellar hemisphere
– Wing of central lobule
– Anterior quadrangular lobule
– Primary fissure
– Posterior quadrangular lobule
– Superior semilunar lobule
– Horizontal fissure
– Inferior semilunar lobule

b

Central lobule

Lingula of cerebellum

Superior cerebellar peduncle

Nodule of vermis

Middle cerebellar peduncle

Inferior cerebellar peduncle

Facial nerve [VII], Intermediate nerve

Vestibulocochlear nerve [VIII]

Fourth ventricle
Choroid plexus –
Lateral recess –
Lateral aperture –

Uvula of vermis

Choroid line of fourth ventricle

Pyramid of vermis

Vallecula of cerebellum

Culmen

Anterior quadrangular lobule

Wing of central lobule

Superior medullary velum

Primary fissure

Posterior quadrangular lobule

Median aperture of fourth ventricle

Superior semilunar lobule

Horizontal fissure

Flocculus

Inferior semilunar lobule

Gracile lobule

Posterolateral fissure

Prebiventral fissure

Biventral lobule

Intrabiventral fissure

Tonsil of cerebellum = Ventral paraflocculus

Secondary fissure

304 Cerebellum (120%)
    a   Occipitoparietal aspect
    b   The cerebellar peduncles were transected. Ventral aspect

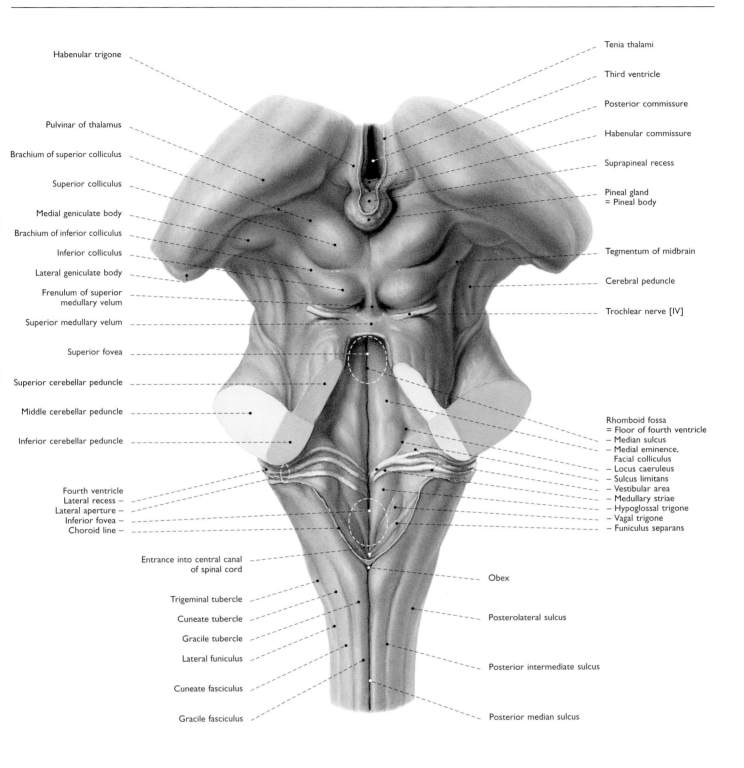

Habenular trigone

Pulvinar of thalamus

Brachium of superior colliculus

Superior colliculus

Medial geniculate body

Brachium of inferior colliculus

Inferior colliculus

Lateral geniculate body

Frenulum of superior
medullary velum

Superior medullary velum

Superior fovea

Superior cerebellar peduncle

Middle cerebellar peduncle

Inferior cerebellar peduncle

Fourth ventricle
Lateral recess –
Lateral aperture –
Inferior fovea –
Choroid line –

Entrance into central canal
of spinal cord

Trigeminal tubercle

Cuneate tubercle

Gracile tubercle

Lateral funiculus

Cuneate fasciculus

Gracile fasciculus

Tenia thalami

Third ventricle

Posterior commissure

Habenular commissure

Suprapineal recess

Pineal gland
= Pineal body

Tegmentum of midbrain

Cerebral peduncle

Trochlear nerve [IV]

Rhomboid fossa
= Floor of fourth ventricle
– Median sulcus
– Medial eminence,
  Facial colliculus
– Locus caeruleus
– Sulcus limitans
– Vestibular area
– Medullary striae
– Hypoglossal trigone
– Vagal trigone
– Funiculus separans

Obex

Posterolateral sulcus

Posterior intermediate sulcus

Posterior median sulcus

305 Brainstem and fourth ventricle (180%)
Dorsal aspect

a

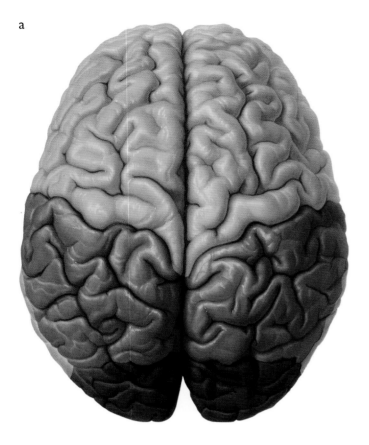

Frontal lobe
Parietal lobe
Occipital lobe
Temporal lobe
Limbic lobe

b

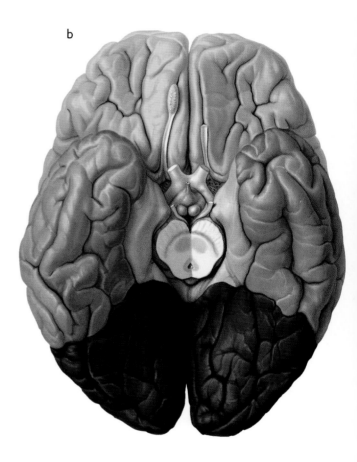

306 Cerebral lobes (70%)
  The diverse cerebral lobes are indicated by different colors.
a Superior aspect
b Inferior aspect

a

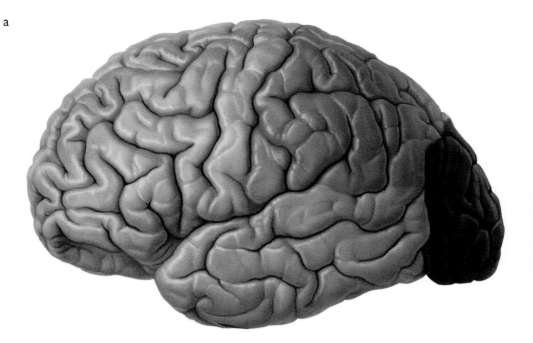

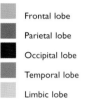

Frontal lobe

Parietal lobe

Occipital lobe

Temporal lobe

Limbic lobe

b

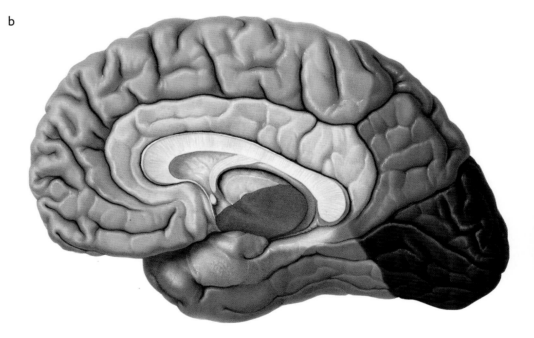

307 Cerebral lobes (80%)

The diverse cerebral lobes are indicated by different colors.

a Left lateral aspect of the left hemisphere

b Medial aspect of the right hemisphere

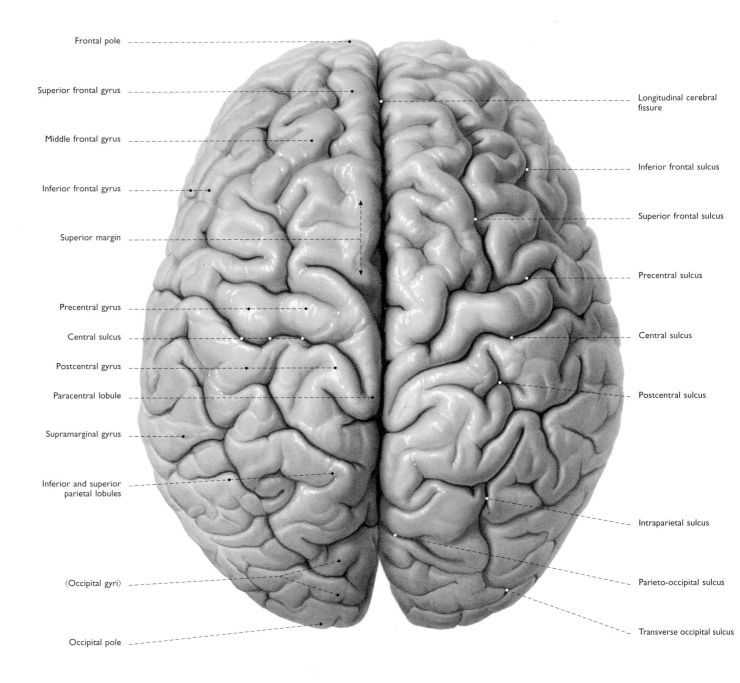

Frontal pole

Superior frontal gyrus

Middle frontal gyrus

Inferior frontal gyrus

Superior margin

Precentral gyrus

Central sulcus

Postcentral gyrus

Paracentral lobule

Supramarginal gyrus

Inferior and superior
parietal lobules

⟨Occipital gyri⟩

Occipital pole

Longitudinal cerebral
fissure

Inferior frontal sulcus

Superior frontal sulcus

Precentral sulcus

Central sulcus

Postcentral sulcus

Intraparietal sulcus

Parieto-occipital sulcus

Transverse occipital sulcus

308 Cerebral hemispheres (100%)
Superior aspect

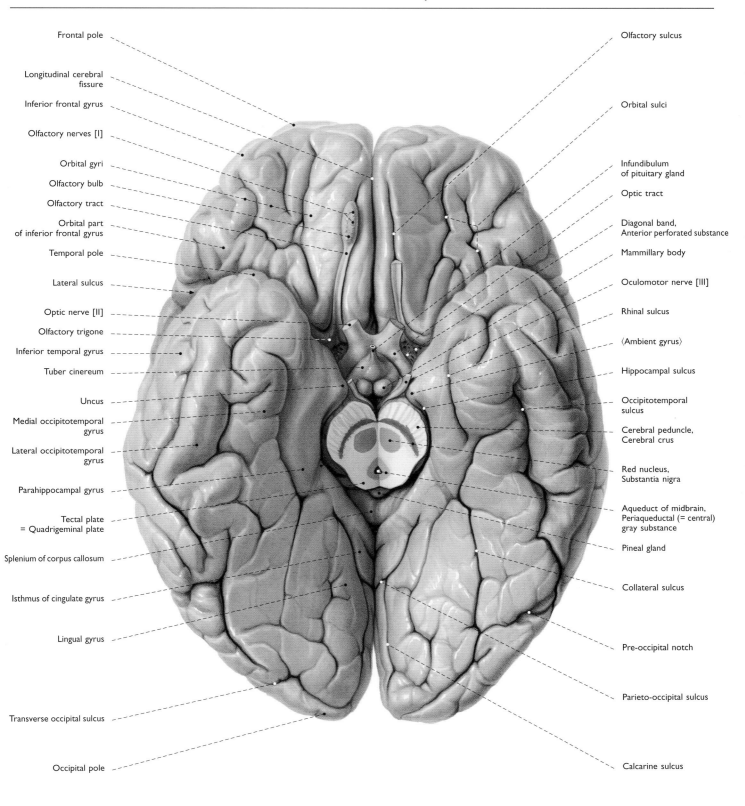

Frontal pole

Longitudinal cerebral fissure

Inferior frontal gyrus

Olfactory nerves [I]

Orbital gyri

Olfactory bulb

Olfactory tract

Orbital part of inferior frontal gyrus

Temporal pole

Lateral sulcus

Optic nerve [II]

Olfactory trigone

Inferior temporal gyrus

Tuber cinereum

Uncus

Medial occipitotemporal gyrus

Lateral occipitotemporal gyrus

Parahippocampal gyrus

Tectal plate = Quadrigeminal plate

Splenium of corpus callosum

Isthmus of cingulate gyrus

Lingual gyrus

Transverse occipital sulcus

Occipital pole

Olfactory sulcus

Orbital sulci

Infundibulum of pituitary gland

Optic tract

Diagonal band, Anterior perforated substance

Mammillary body

Oculomotor nerve [III]

Rhinal sulcus

⟨Ambient gyrus⟩

Hippocampal sulcus

Occipitotemporal sulcus

Cerebral peduncle, Cerebral crus

Red nucleus, Substantia nigra

Aqueduct of midbrain, Periaqueductal (= central) gray substance

Pineal gland

Collateral sulcus

Pre-occipital notch

Parieto-occipital sulcus

Calcarine sulcus

309 Cerebral hemispheres (100%)
The midbrain was transected. Inferior aspect

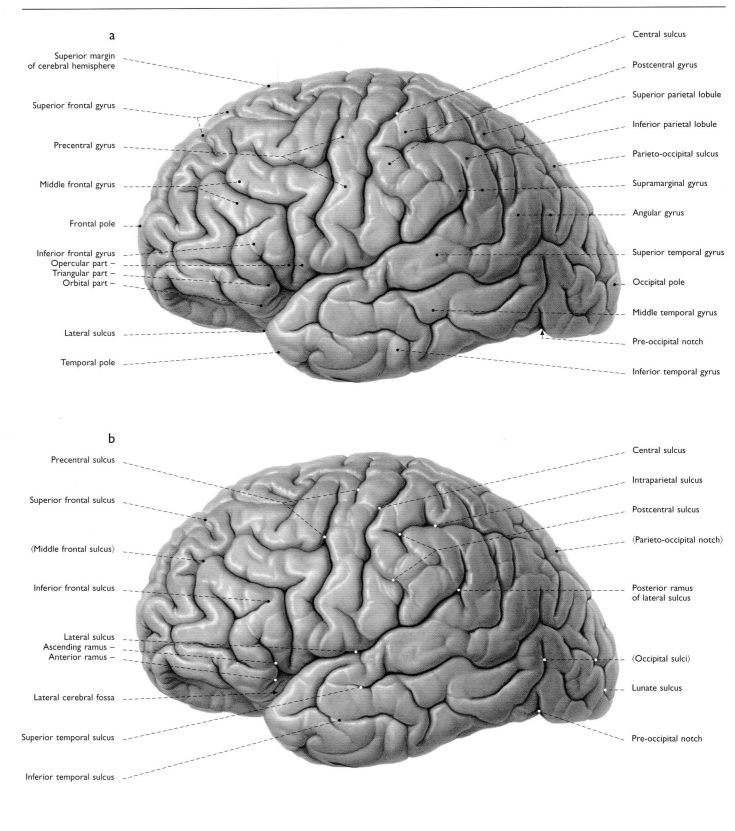

a

Superior margin
of cerebral hemisphere

Superior frontal gyrus

Precentral gyrus

Middle frontal gyrus

Frontal pole

Inferior frontal gyrus
Opercular part –
Triangular part –
Orbital part –

Lateral sulcus

Temporal pole

Central sulcus

Postcentral gyrus

Superior parietal lobule

Inferior parietal lobule

Parieto-occipital sulcus

Supramarginal gyrus

Angular gyrus

Superior temporal gyrus

Occipital pole

Middle temporal gyrus

Pre-occipital notch

Inferior temporal gyrus

b

Precentral sulcus

Superior frontal sulcus

⟨Middle frontal sulcus⟩

Inferior frontal sulcus

Lateral sulcus
Ascending ramus –
Anterior ramus –

Lateral cerebral fossa

Superior temporal sulcus

Inferior temporal sulcus

Central sulcus

Intraparietal sulcus

Postcentral sulcus

⟨Parieto-occipital notch⟩

Posterior ramus
of lateral sulcus

⟨Occipital sulci⟩

Lunate sulcus

Pre-occipital notch

310 Left cerebral hemisphere (80%)
Left lateral aspect
Lettering of the
a   gyri
b   sulci

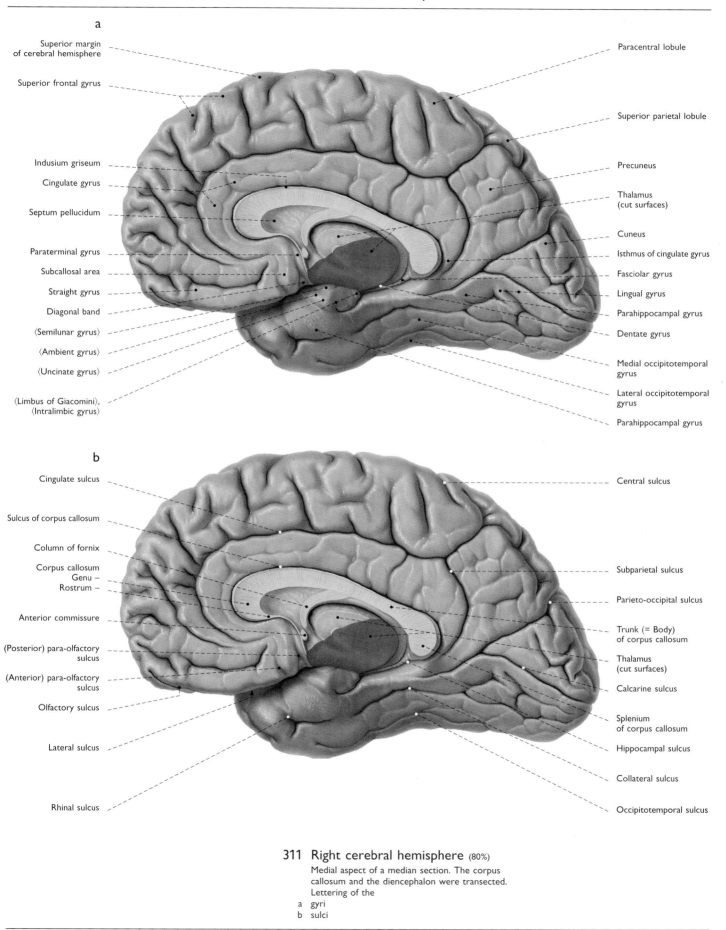

a

Superior margin of cerebral hemisphere

Superior frontal gyrus

Indusium griseum

Cingulate gyrus

Septum pellucidum

Paraterminal gyrus

Subcallosal area

Straight gyrus

Diagonal band

⟨Semilunar gyrus⟩

⟨Ambient gyrus⟩

⟨Uncinate gyrus⟩

⟨Limbus of Giacomini⟩, ⟨Intralimbic gyrus⟩

Paracentral lobule

Superior parietal lobule

Precuneus

Thalamus (cut surfaces)

Cuneus

Isthmus of cingulate gyrus

Fasciolar gyrus

Lingual gyrus

Parahippocampal gyrus

Dentate gyrus

Medial occipitotemporal gyrus

Lateral occipitotemporal gyrus

Parahippocampal gyrus

b

Cingulate sulcus

Sulcus of corpus callosum

Column of fornix

Corpus callosum
Genu –
Rostrum –

Anterior commissure

(Posterior) para-olfactory sulcus

(Anterior) para-olfactory sulcus

Olfactory sulcus

Lateral sulcus

Rhinal sulcus

Central sulcus

Subparietal sulcus

Parieto-occipital sulcus

Trunk (= Body) of corpus callosum

Thalamus (cut surfaces)

Calcarine sulcus

Splenium of corpus callosum

Hippocampal sulcus

Collateral sulcus

Occipitotemporal sulcus

311 Right cerebral hemisphere (80%)
Medial aspect of a median section. The corpus callosum and the diencephalon were transected. Lettering of the
a gyri
b sulci

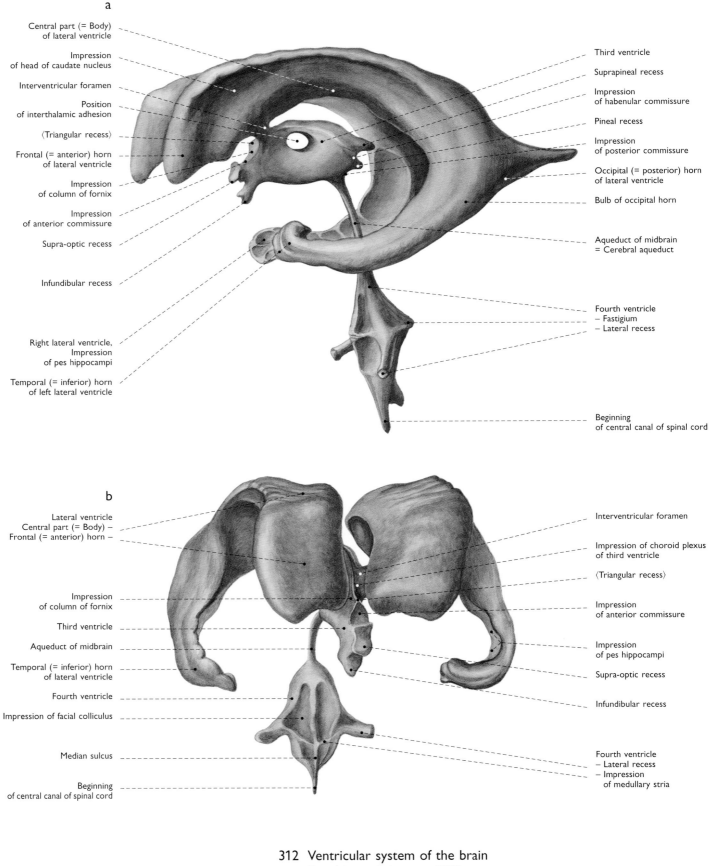

a

Central part (= Body) of lateral ventricle

Impression of head of caudate nucleus

Interventricular foramen

Position of interthalamic adhesion

⟨Triangular recess⟩

Frontal (= anterior) horn of lateral ventricle

Impression of column of fornix

Impression of anterior commissure

Supra-optic recess

Infundibular recess

Right lateral ventricle, Impression of pes hippocampi

Temporal (= inferior) horn of left lateral ventricle

Third ventricle

Suprapineal recess

Impression of habenular commissure

Pineal recess

Impression of posterior commissure

Occipital (= posterior) horn of lateral ventricle

Bulb of occipital horn

Aqueduct of midbrain = Cerebral aqueduct

Fourth ventricle
– Fastigium
– Lateral recess

Beginning of central canal of spinal cord

b

Lateral ventricle
Central part (= Body) –
Frontal (= anterior) horn –

Impression of column of fornix

Third ventricle

Aqueduct of midbrain

Temporal (= inferior) horn of lateral ventricle

Fourth ventricle

Impression of facial colliculus

Median sulcus

Beginning of central canal of spinal cord

Interventricular foramen

Impression of choroid plexus of third ventricle

⟨Triangular recess⟩

Impression of anterior commissure

Impression of pes hippocampi

Supra-optic recess

Infundibular recess

Fourth ventricle
– Lateral recess
– Impression of medullary stria

312 Ventricular system of the brain

Cast of the ventricular system (120%)
a   Left lateral aspect
b   Frontal aspect

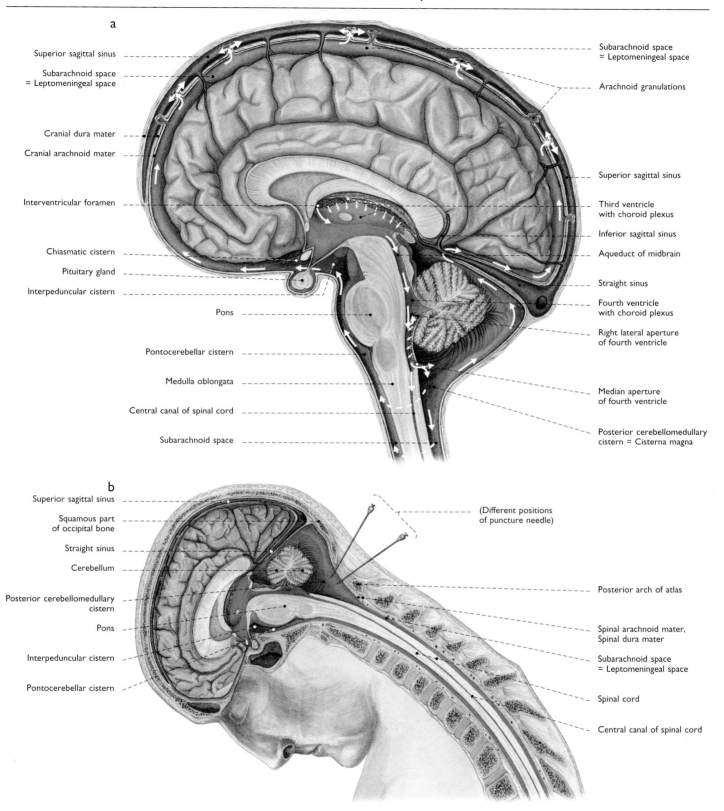

a

Superior sagittal sinus

Subarachnoid space = Leptomeningeal space

Cranial dura mater

Cranial arachnoid mater

Interventricular foramen

Chiasmatic cistern

Pituitary gland

Interpeduncular cistern

Pons

Pontocerebellar cistern

Medulla oblongata

Central canal of spinal cord

Subarachnoid space

Subarachnoid space = Leptomeningeal space

Arachnoid granulations

Superior sagittal sinus

Third ventricle with choroid plexus

Inferior sagittal sinus

Aqueduct of midbrain

Straight sinus

Fourth ventricle with choroid plexus

Right lateral aperture of fourth ventricle

Median aperture of fourth ventricle

Posterior cerebellomedullary cistern = Cisterna magna

b

Superior sagittal sinus

Squamous part of occipital bone

Straight sinus

Cerebellum

Posterior cerebellomedullary cistern

Pons

Interpeduncular cistern

Pontocerebellar cistern

(Different positions of puncture needle)

Posterior arch of atlas

Spinal arachnoid mater, Spinal dura mater

Subarachnoid space = Leptomeningeal space

Spinal cord

Central canal of spinal cord

313 Subarachnoid space, subarachnoid cisterns, and cerebrospinal fluid

a Circulation of the cerebrospinal fluid (60%)
b Puncture of the posterior cerebellomedullary cistern (suboccipital puncture) (35%)

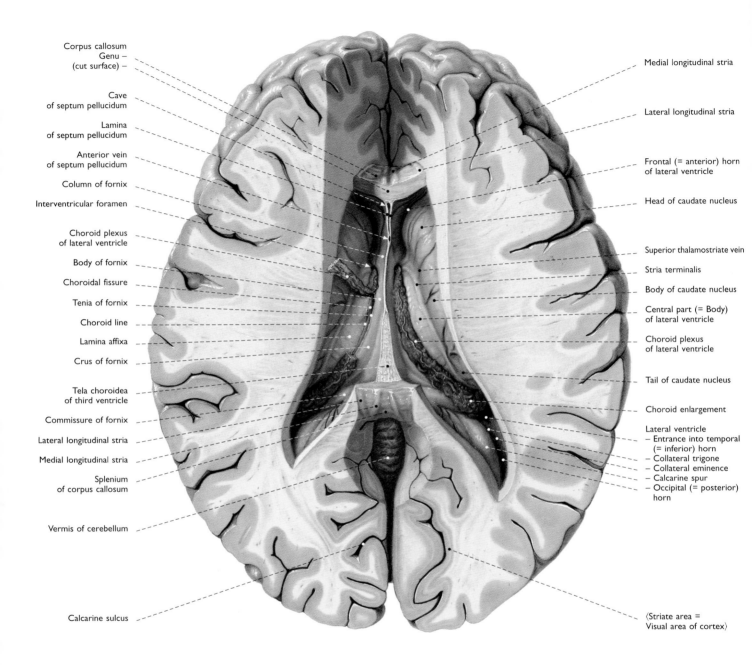

Corpus callosum
Genu —
(cut surface) —

Cave
of septum pellucidum

Lamina
of septum pellucidum

Anterior vein
of septum pellucidum

Column of fornix

Interventricular foramen

Choroid plexus
of lateral ventricle

Body of fornix

Choroidal fissure

Tenia of fornix

Choroid line

Lamina affixa

Crus of fornix

Tela choroidea
of third ventricle

Commissure of fornix

Lateral longitudinal stria

Medial longitudinal stria

Splenium
of corpus callosum

Vermis of cerebellum

Calcarine sulcus

Medial longitudinal stria

Lateral longitudinal stria

Frontal (= anterior) horn
of lateral ventricle

Head of caudate nucleus

Superior thalamostriate vein

Stria terminalis

Body of caudate nucleus

Central part (= Body)
of lateral ventricle

Choroid plexus
of lateral ventricle

Tail of caudate nucleus

Choroid enlargement

Lateral ventricle
— Entrance into temporal
   (= inferior) horn
— Collateral trigone
— Collateral eminence
— Calcarine spur
— Occipital (= posterior)
   horn

⟨Striate area =
Visual area of cortex⟩

314 Lateral ventricles of the brain (100%)
The lateral ventricles were opened from above,
the trunk of the corpus callosum was removed.
Superior aspect of a horizontal section

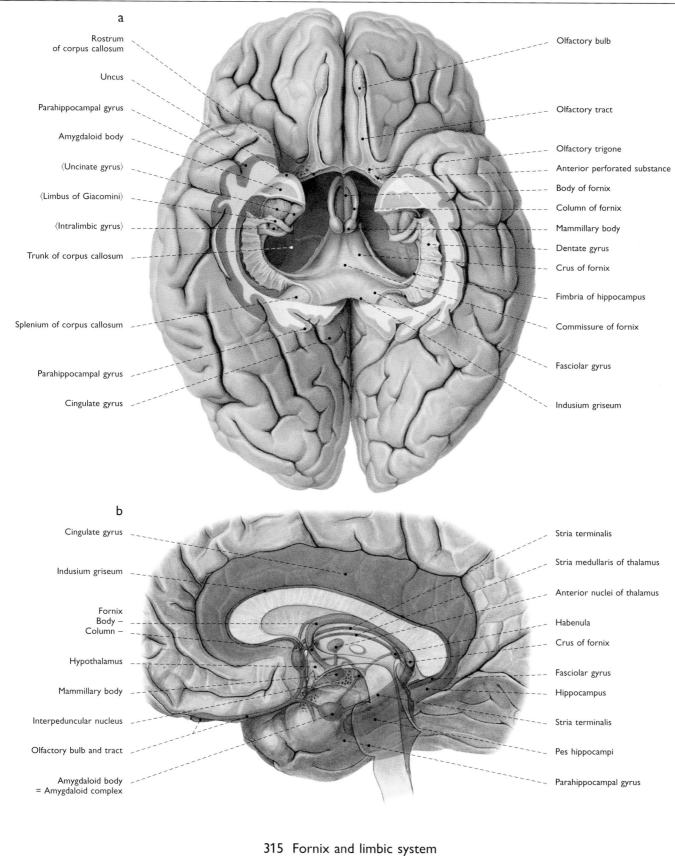

a

Rostrum of corpus callosum

Uncus

Parahippocampal gyrus

Amygdaloid body

⟨Uncinate gyrus⟩

⟨Limbus of Giacomini⟩

⟨Intralimbic gyrus⟩

Trunk of corpus callosum

Splenium of corpus callosum

Parahippocampal gyrus

Cingulate gyrus

Olfactory bulb

Olfactory tract

Olfactory trigone

Anterior perforated substance

Body of fornix

Column of fornix

Mammillary body

Dentate gyrus

Crus of fornix

Fimbria of hippocampus

Commissure of fornix

Fasciolar gyrus

Indusium griseum

b

Cingulate gyrus

Indusium griseum

Fornix
Body –
Column –

Hypothalamus

Mammillary body

Interpeduncular nucleus

Olfactory bulb and tract

Amygdaloid body
= Amygdaloid complex

Stria terminalis

Stria medullaris of thalamus

Anterior nuclei of thalamus

Habenula

Crus of fornix

Fasciolar gyrus

Hippocampus

Stria terminalis

Pes hippocampi

Parahippocampal gyrus

315 Fornix and limbic system

a The undersurface of the brain was partly dissected
in order to show the fornix and some other elements
of the limbic system (80%). Inferior aspect

b The structures of the limbic system are emphasized
by brown color (100%). Medial aspect of the right hemisphere

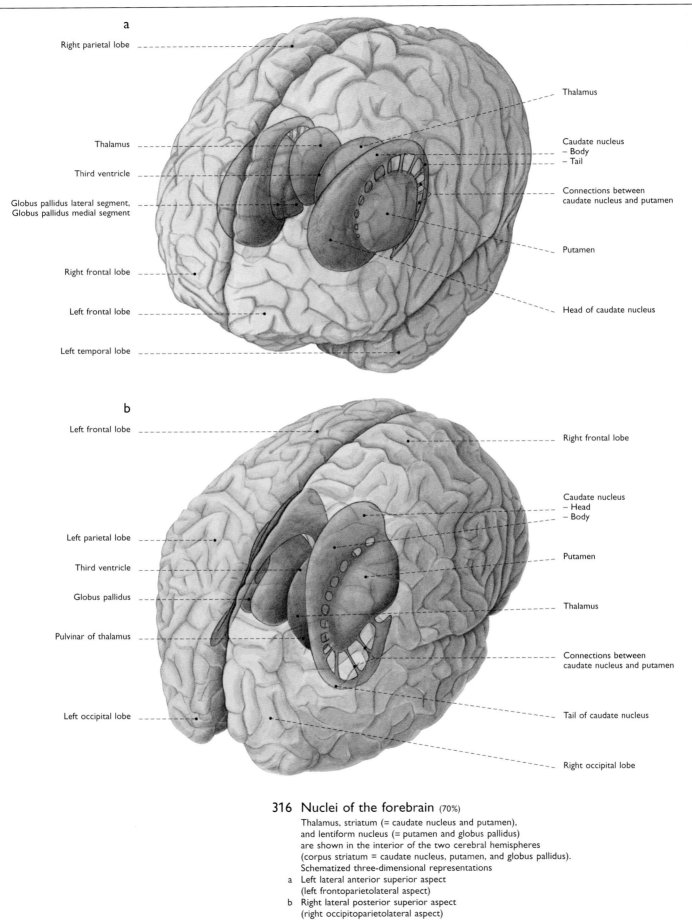

a

Right parietal lobe

Thalamus

Thalamus

Third ventricle

Globus pallidus lateral segment,
Globus pallidus medial segment

Right frontal lobe

Left frontal lobe

Left temporal lobe

Caudate nucleus
– Body
– Tail

Connections between
caudate nucleus and putamen

Putamen

Head of caudate nucleus

b

Left frontal lobe

Left parietal lobe

Third ventricle

Globus pallidus

Pulvinar of thalamus

Left occipital lobe

Right frontal lobe

Caudate nucleus
– Head
– Body

Putamen

Thalamus

Connections between
caudate nucleus and putamen

Tail of caudate nucleus

Right occipital lobe

316 Nuclei of the forebrain (70%)

Thalamus, striatum (= caudate nucleus and putamen),
and lentiform nucleus (= putamen and globus pallidus)
are shown in the interior of the two cerebral hemispheres
(corpus striatum = caudate nucleus, putamen, and globus pallidus).
Schematized three-dimensional representations
a   Left lateral anterior superior aspect
    (left frontoparietolateral aspect)
b   Right lateral posterior superior aspect
    (right occipitoparietolateral aspect)

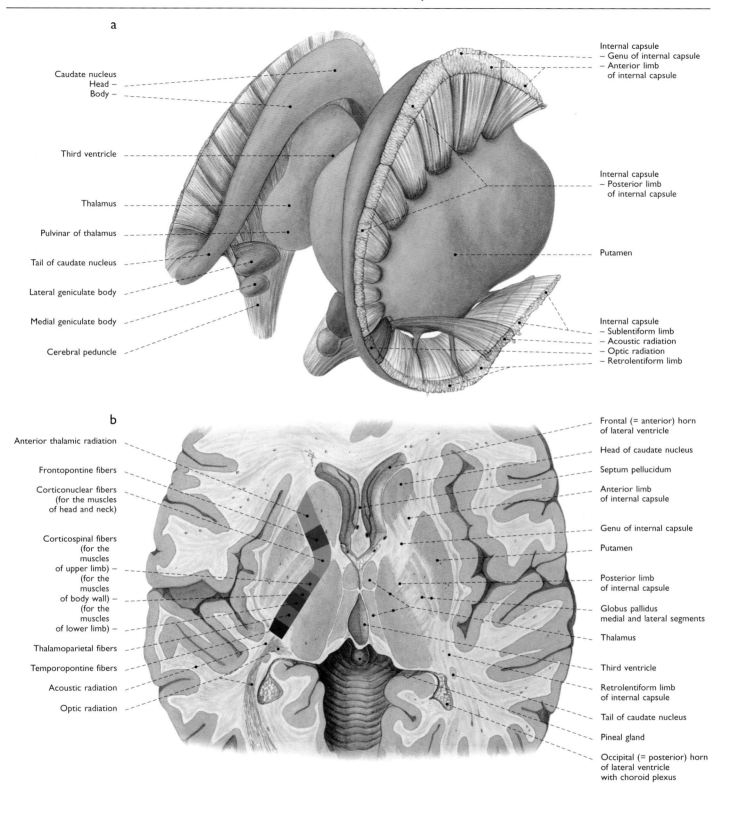

a

Caudate nucleus
Head –
Body –

Third ventricle

Thalamus

Pulvinar of thalamus

Tail of caudate nucleus

Lateral geniculate body

Medial geniculate body

Cerebral peduncle

Internal capsule
– Genu of internal capsule
– Anterior limb
  of internal capsule

Internal capsule
– Posterior limb
  of internal capsule

Putamen

Internal capsule
– Sublentiform limb
– Acoustic radiation
– Optic radiation
– Retrolentiform limb

b

Anterior thalamic radiation

Frontopontine fibers

Corticonuclear fibers
(for the muscles
of head and neck)

Corticospinal fibers
(for the
muscles
of upper limb) –
(for the
muscles
of body wall) –
(for the
muscles
of lower limb) –

Thalamoparietal fibers

Temporopontine fibers

Acoustic radiation

Optic radiation

Frontal (= anterior) horn
of lateral ventricle

Head of caudate nucleus

Septum pellucidum

Anterior limb
of internal capsule

Genu of internal capsule

Putamen

Posterior limb
of internal capsule

Globus pallidus
medial and lateral segments

Thalamus

Third ventricle

Retrolentiform limb
of internal capsule

Tail of caudate nucleus

Pineal gland

Occipital (= posterior) horn
of lateral ventricle
with choroid plexus

317　Nuclei of the forebrain and internal capsule

a　Thalamus, striatum (= caudate nucleus and putamen),
　and lentiform nucleus (= putamen and globus pallidus)
　with the internal capsule (250%), right occipitolateral aspect
b　Horizontal section through the internal capsule and adjacent nuclei
　(corpus striatum = caudate nucleus, putamen, and globus pallidus).
　On the left side, the main components of the internal capsule
　are marked by different colors (90%). Superior aspect

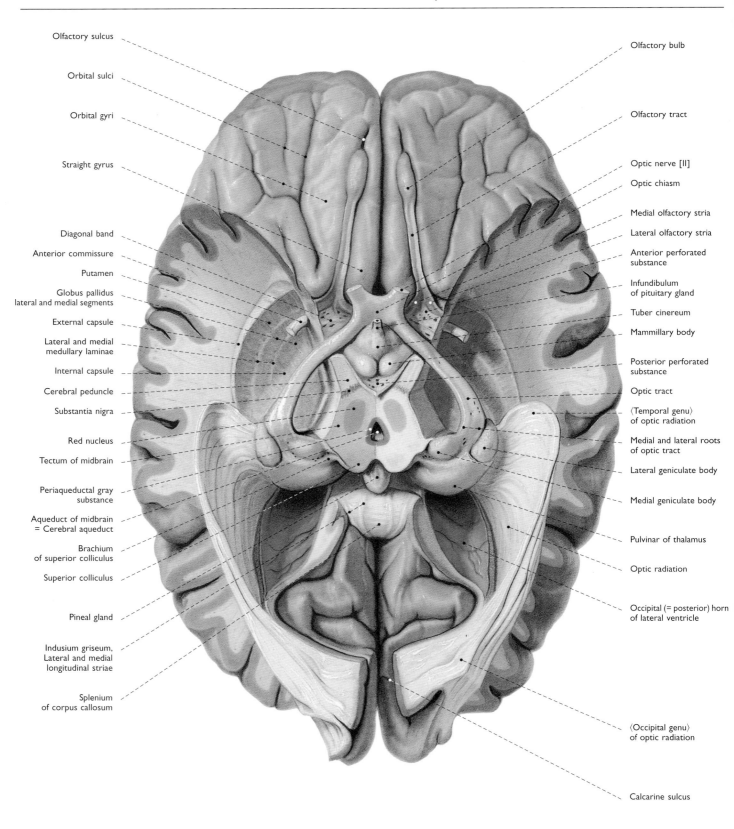

Olfactory sulcus

Orbital sulci

Orbital gyri

Straight gyrus

Diagonal band

Anterior commissure

Putamen

Globus pallidus
lateral and medial segments

External capsule

Lateral and medial
medullary laminae

Internal capsule

Cerebral peduncle

Substantia nigra

Red nucleus

Tectum of midbrain

Periaqueductal gray
substance

Aqueduct of midbrain
= Cerebral aqueduct

Brachium
of superior colliculus

Superior colliculus

Pineal gland

Indusium griseum,
Lateral and medial
longitudinal striae

Splenium
of corpus callosum

Olfactory bulb

Olfactory tract

Optic nerve [II]

Optic chiasm

Medial olfactory stria

Lateral olfactory stria

Anterior perforated
substance

Infundibulum
of pituitary gland

Tuber cinereum

Mammillary body

Posterior perforated
substance

Optic tract

⟨Temporal genu⟩
of optic radiation

Medial and lateral roots
of optic tract

Lateral geniculate body

Medial geniculate body

Pulvinar of thalamus

Optic radiation

Occipital (= posterior) horn
of lateral ventricle

⟨Occipital genu⟩
of optic radiation

Calcarine sulcus

318 Visual pathway (100%)
Dissection showing the optic radiation. The fibers pass
from the optic chiasm to the visual area of cortex.
Inferior aspect

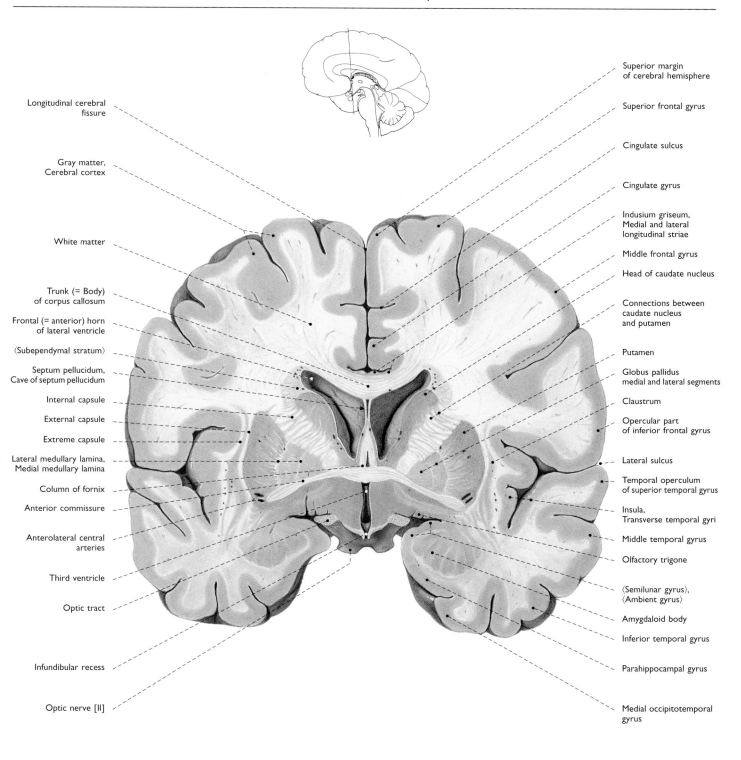

Longitudinal cerebral fissure

Gray matter, Cerebral cortex

White matter

Trunk (= Body) of corpus callosum

Frontal (= anterior) horn of lateral ventricle

⟨Subependymal stratum⟩

Septum pellucidum, Cave of septum pellucidum

Internal capsule

External capsule

Extreme capsule

Lateral medullary lamina, Medial medullary lamina

Column of fornix

Anterior commissure

Anterolateral central arteries

Third ventricle

Optic tract

Infundibular recess

Optic nerve [II]

Superior margin of cerebral hemisphere

Superior frontal gyrus

Cingulate sulcus

Cingulate gyrus

Indusium griseum, Medial and lateral longitudinal striae

Middle frontal gyrus

Head of caudate nucleus

Connections between caudate nucleus and putamen

Putamen

Globus pallidus medial and lateral segments

Claustrum

Opercular part of inferior frontal gyrus

Lateral sulcus

Temporal operculum of superior temporal gyrus

Insula, Transverse temporal gyri

Middle temporal gyrus

Olfactory trigone

⟨Semilunar gyrus⟩, ⟨Ambient gyrus⟩

Amygdaloid body

Inferior temporal gyrus

Parahippocampal gyrus

Medial occipitotemporal gyrus

319 Brain (100%)
Coronal section in the plane of the anterior commissure, occipital aspect

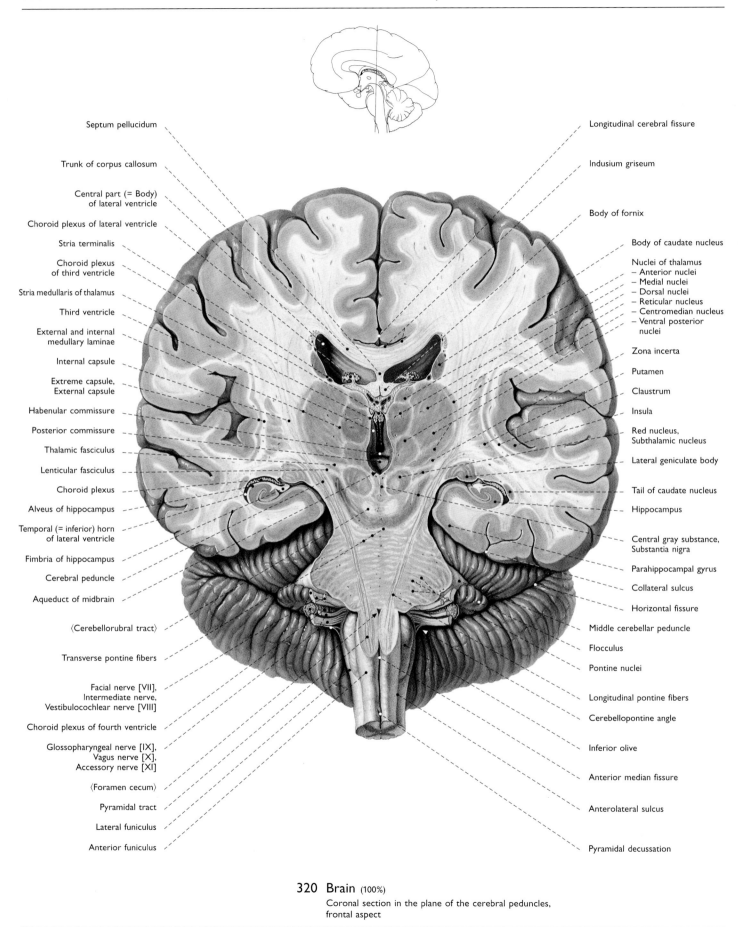

Septum pellucidum

Trunk of corpus callosum

Central part (= Body) of lateral ventricle

Choroid plexus of lateral ventricle

Stria terminalis

Choroid plexus of third ventricle

Stria medullaris of thalamus

Third ventricle

External and internal medullary laminae

Internal capsule

Extreme capsule, External capsule

Habenular commissure

Posterior commissure

Thalamic fasciculus

Lenticular fasciculus

Choroid plexus

Alveus of hippocampus

Temporal (= inferior) horn of lateral ventricle

Fimbria of hippocampus

Cerebral peduncle

Aqueduct of midbrain

⟨Cerebellorubral tract⟩

Transverse pontine fibers

Facial nerve [VII], Intermediate nerve, Vestibulocochlear nerve [VIII]

Choroid plexus of fourth ventricle

Glossopharyngeal nerve [IX], Vagus nerve [X], Accessory nerve [XI]

⟨Foramen cecum⟩

Pyramidal tract

Lateral funiculus

Anterior funiculus

Longitudinal cerebral fissure

Indusium griseum

Body of fornix

Body of caudate nucleus

Nuclei of thalamus
– Anterior nuclei
– Medial nuclei
– Dorsal nuclei
– Reticular nucleus
– Centromedian nucleus
– Ventral posterior nuclei

Zona incerta

Putamen

Claustrum

Insula

Red nucleus, Subthalamic nucleus

Lateral geniculate body

Tail of caudate nucleus

Hippocampus

Central gray substance, Substantia nigra

Parahippocampal gyrus

Collateral sulcus

Horizontal fissure

Middle cerebellar peduncle

Flocculus

Pontine nuclei

Longitudinal pontine fibers

Cerebellopontine angle

Inferior olive

Anterior median fissure

Anterolateral sulcus

Pyramidal decussation

320 Brain (100%)
Coronal section in the plane of the cerebral peduncles, frontal aspect

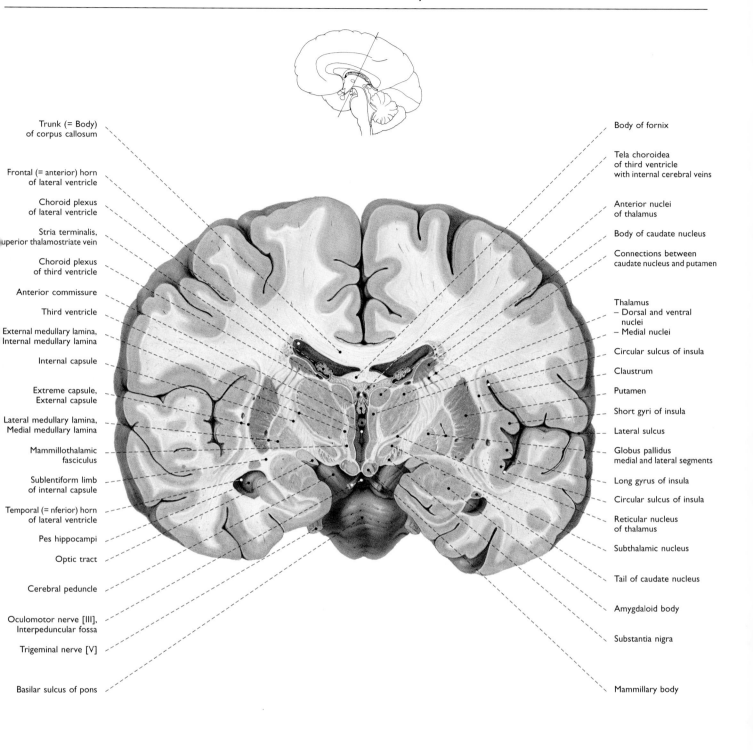

Trunk (= Body)
of corpus callosum

Frontal (= anterior) horn
of lateral ventricle

Choroid plexus
of lateral ventricle

Stria terminalis,
uperior thalamostriate vein

Choroid plexus
of third ventricle

Anterior commissure

Third ventricle

External medullary lamina,
Internal medullary lamina

Internal capsule

Extreme capsule,
External capsule

Lateral medullary lamina,
Medial medullary lamina

Mammillothalamic
fasciculus

Sublentiform limb
of internal capsule

Temporal (= nferior) horn
of lateral ventricle

Pes hippocampi

Optic tract

Cerebral peduncle

Oculomotor nerve [III],
Interpeduncular fossa

Trigeminal nerve [V]

Basilar sulcus of pons

Body of fornix

Tela choroidea
of third ventricle
with internal cerebral veins

Anterior nuclei
of thalamus

Body of caudate nucleus

Connections between
caudate nucleus and putamen

Thalamus
– Dorsal and ventral
nuclei
– Medial nuclei

Circular sulcus of insula

Claustrum

Putamen

Short gyri of insula

Lateral sulcus

Globus pallidus
medial and lateral segments

Long gyrus of insula

Circular sulcus of insula

Reticular nucleus
of thalamus

Subthalamic nucleus

Tail of caudate nucleus

Amygdaloid body

Substantia nigra

Mammillary body

321 Brain (100%)
Coronal section in the plane of the mammillary bodies,
frontal aspect

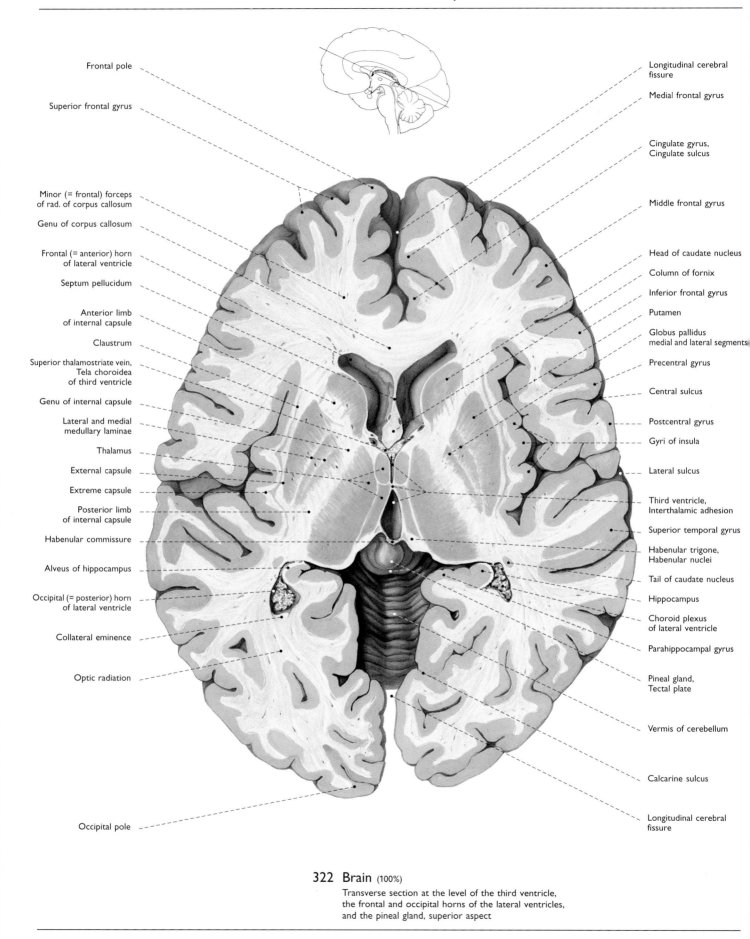

Frontal pole

Superior frontal gyrus

Minor (= frontal) forceps
of rad. of corpus callosum

Genu of corpus callosum

Frontal (= anterior) horn
of lateral ventricle

Septum pellucidum

Anterior limb
of internal capsule

Claustrum

Superior thalamostriate vein,
Tela choroidea
of third ventricle

Genu of internal capsule

Lateral and medial
medullary laminae

Thalamus

External capsule

Extreme capsule

Posterior limb
of internal capsule

Habenular commissure

Alveus of hippocampus

Occipital (= posterior) horn
of lateral ventricle

Collateral eminence

Optic radiation

Occipital pole

Longitudinal cerebral
fissure

Medial frontal gyrus

Cingulate gyrus,
Cingulate sulcus

Middle frontal gyrus

Head of caudate nucleus

Column of fornix

Inferior frontal gyrus

Putamen

Globus pallidus
medial and lateral segments

Precentral gyrus

Central sulcus

Postcentral gyrus

Gyri of insula

Lateral sulcus

Third ventricle,
Interthalamic adhesion

Superior temporal gyrus

Habenular trigone,
Habenular nuclei

Tail of caudate nucleus

Hippocampus

Choroid plexus
of lateral ventricle

Parahippocampal gyrus

Pineal gland,
Tectal plate

Vermis of cerebellum

Calcarine sulcus

Longitudinal cerebral
fissure

322 Brain (100%)

Transverse section at the level of the third ventricle,
the frontal and occipital horns of the lateral ventricles,
and the pineal gland, superior aspect

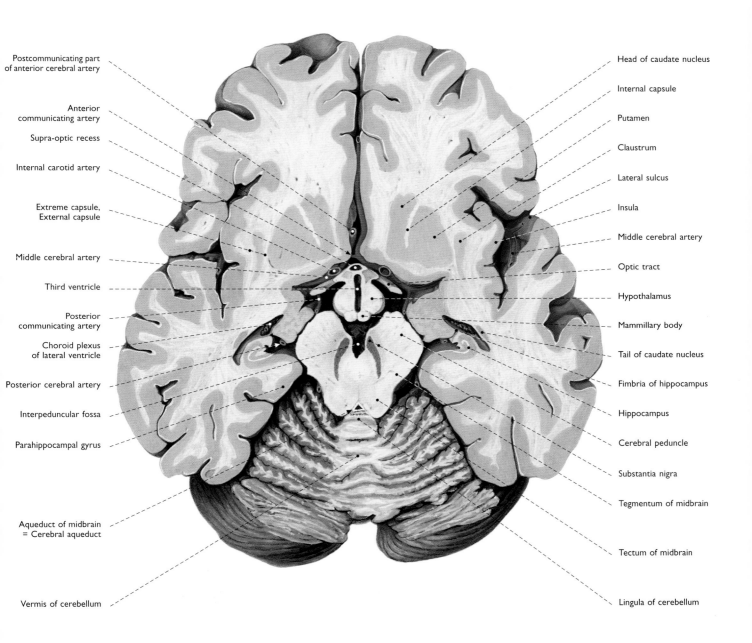

Postcommunicating part
of anterior cerebral artery

Anterior
communicating artery

Supra-optic recess

Internal carotid artery

Extreme capsule,
External capsule

Middle cerebral artery

Third ventricle

Posterior
communicating artery

Choroid plexus
of lateral ventricle

Posterior cerebral artery

Interpeduncular fossa

Parahippocampal gyrus

Aqueduct of midbrain
= Cerebral aqueduct

Vermis of cerebellum

Head of caudate nucleus

Internal capsule

Putamen

Claustrum

Lateral sulcus

Insula

Middle cerebral artery

Optic tract

Hypothalamus

Mammillary body

Tail of caudate nucleus

Fimbria of hippocampus

Hippocampus

Cerebral peduncle

Substantia nigra

Tegmentum of midbrain

Tectum of midbrain

Lingula of cerebellum

323 Brain (100%)
Transverse section at the level of the midbrain
and the mammillary bodies, superior aspect

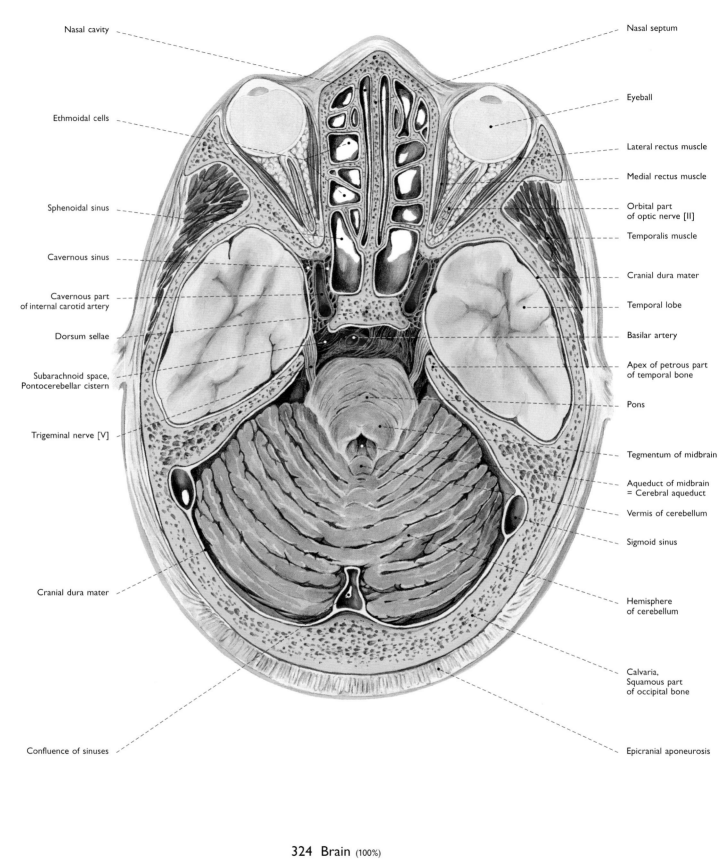

Nasal cavity

Ethmoidal cells

Sphenoidal sinus

Cavernous sinus

Cavernous part
of internal carotid artery

Dorsum sellae

Subarachnoid space,
Pontocerebellar cistern

Trigeminal nerve [V]

Cranial dura mater

Confluence of sinuses

Nasal septum

Eyeball

Lateral rectus muscle

Medial rectus muscle

Orbital part
of optic nerve [II]

Temporalis muscle

Cranial dura mater

Temporal lobe

Basilar artery

Apex of petrous part
of temporal bone

Pons

Tegmentum of midbrain

Aqueduct of midbrain
= Cerebral aqueduct

Vermis of cerebellum

Sigmoid sinus

Hemisphere
of cerebellum

Calvaria,
Squamous part
of occipital bone

Epicranial aponeurosis

324 Brain (100%)

Transverse section through the head and brain at the level
of the optic nerves, the dorsum sellae, the pons, and
the aqueduct of midbrain, superior aspect

Visual Organ and Orbital Cavity

a

Eyebrow

Superior (= upper) eyelid
⟨Supratarsal part⟩ –
⟨Tarsal part⟩ –

Medial palpebral commissure,
Medial angle of eye

Palpebronasal fold
= Medial canthic fold

Lateral palpebral commissure,
Lateral angle of eye

⟨Lateral palpebral raphe⟩

Palpebral fissure

Eyelashes

Inferior (= lower) eyelid

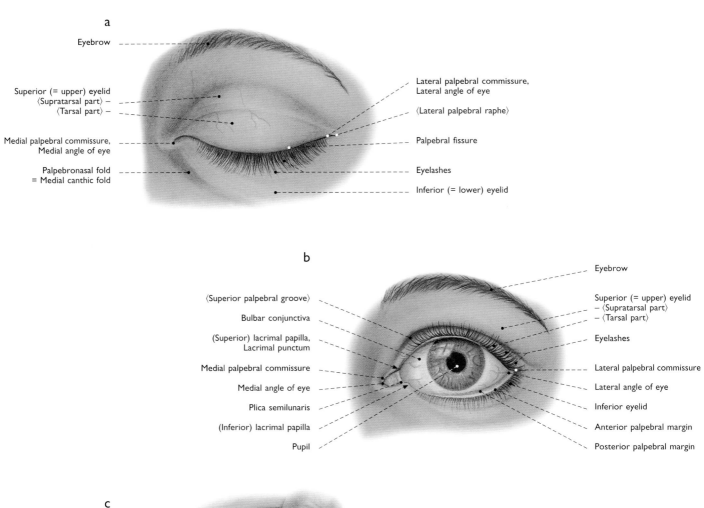

b

⟨Superior palpebral groove⟩

Bulbar conjunctiva

(Superior) lacrimal papilla,
Lacrimal punctum

Medial palpebral commissure

Medial angle of eye

Plica semilunaris

(Inferior) lacrimal papilla

Pupil

Eyebrow

Superior (= upper) eyelid
– ⟨Supratarsal part⟩
– ⟨Tarsal part⟩

Eyelashes

Lateral palpebral commissure

Lateral angle of eye

Inferior eyelid

Anterior palpebral margin

Posterior palpebral margin

c

Pupil

Iris

(Superior) lacrimal papilla

Plica semilunaris

Medial palpebral commissure

Lacrimal caruncle

(Inferior) lacrimal papilla

Inferior conjunctival fornix

Anterior palpebral margin

Posterior palpebral margin

Bulbar conjunctiva

Lateral palpebral commissure

Anterior palpebral margin

Posterior palpebral margin

Palpebral conjunctiva

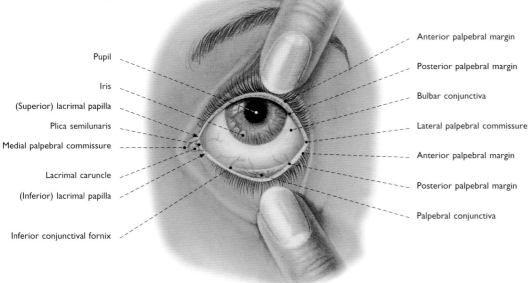

326 Eye (110%)
Frontal aspect
a Eyelids of the left eye closed
b Eyelids of the left eye open
c Upper eyelid of the left eye drawn upwards
   and lower eyelid slightly everted

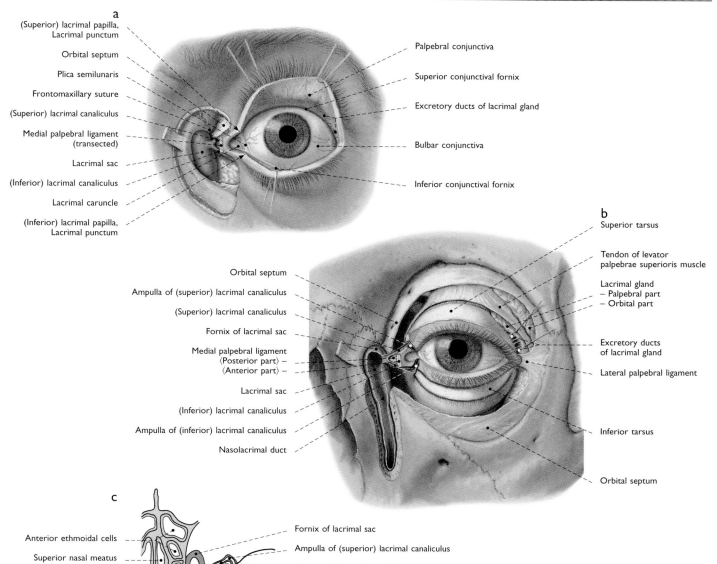

a
(Superior) lacrimal papilla, Lacrimal punctum
Orbital septum
Plica semilunaris
Frontomaxillary suture
(Superior) lacrimal canaliculus
Medial palpebral ligament (transected)
Lacrimal sac
(Inferior) lacrimal canaliculus
Lacrimal caruncle
(Inferior) lacrimal papilla, Lacrimal punctum

Palpebral conjunctiva
Superior conjunctival fornix
Excretory ducts of lacrimal gland
Bulbar conjunctiva
Inferior conjunctival fornix

b
Superior tarsus
Tendon of levator palpebrae superioris muscle
Lacrimal gland
– Palpebral part
– Orbital part
Excretory ducts of lacrimal gland
Lateral palpebral ligament
Inferior tarsus
Orbital septum

Orbital septum
Ampulla of (superior) lacrimal canaliculus
(Superior) lacrimal canaliculus
Fornix of lacrimal sac
Medial palpebral ligament
⟨Posterior part⟩ –
⟨Anterior part⟩ –
Lacrimal sac
(Inferior) lacrimal canaliculus
Ampulla of (inferior) lacrimal canaliculus
Nasolacrimal duct

c
Anterior ethmoidal cells
Superior nasal meatus
Bony nasal septum
Middle nasal concha
Middle nasal meatus
Inferior nasal concha
Inferior nasal meatus
Hard palate

Fornix of lacrimal sac
Ampulla of (superior) lacrimal canaliculus
(Superior) lacrimal canaliculus
Lacrimal puncta
Lacrimal sac
Nasolacrimal canal
Nasolacrimal duct
Maxillary sinus
Lacrimal fold
Opening of nasolacrimal duct

327 Lacrimal apparatus

Frontal aspect

a The lacrimal apparatus of the left eye was exposed. The upper eyelid was pulled upwards by sutures, while the lower eyelid was drawn slightly downwards (90%).

b Left orbit and its contents after removal of facial muscles. The tendon of the levator palpebrae superioris muscle was divided by a bow-shaped incision. The lacrimal apparatus was exposed and opened (90%).

c Schematized coronal section through the lacrimal ducts, the nasolacrimal duct, and the nasal cavity of the left side (100%)

a

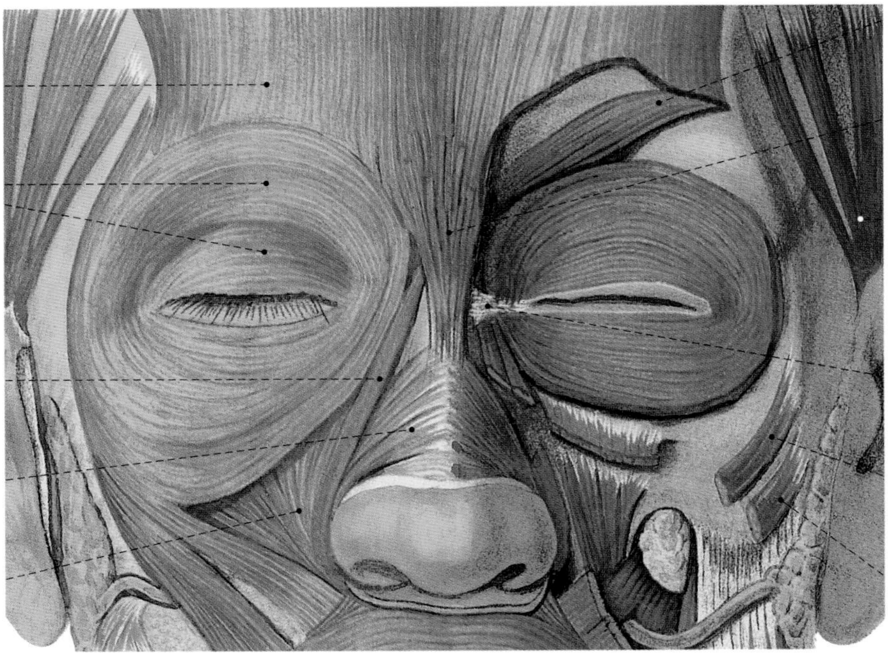

Frontal belly
of occipitofrontalis muscle

Orbicularis oculi muscle
Orbital part –
Palpebral part –

Levator labii superioris
alaeque nasi muscle

Nasalis muscle

Levator labii superioris muscle

Depressor supercilii muscle

Procerus muscle

Auricularis superior muscle

Medial palpebral ligament

Zygomaticus minor muscle
(cut)

Zygomaticus major muscle
(cut)

b

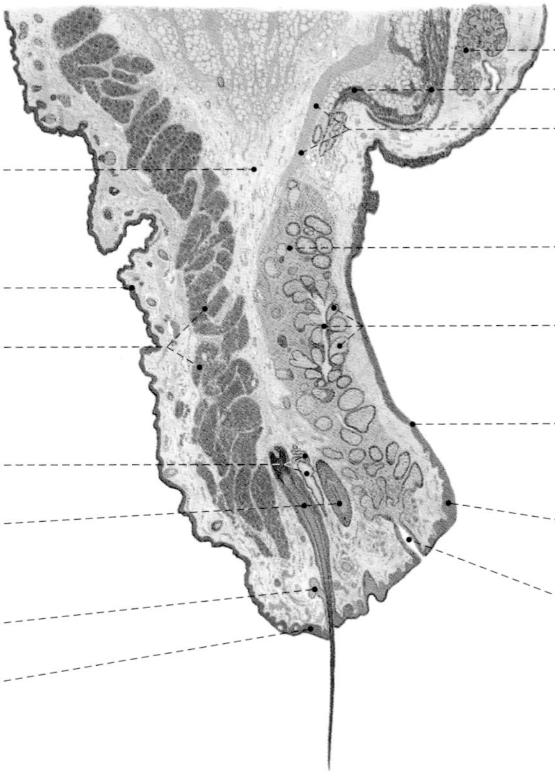

Connective tissue

Anterior surface of eyelid

Palpebral part
of orbicularis oculi muscle

Ciliary glands

Eyelashes

Sebaceous glands

Anterior palpebral margin

(Accessory lacrimal gland)

Superior tarsal muscle

Tendon of levator
palpebrae superioris muscle

Superior tarsus

Tarsal glands
with excretory duct

Posterior surface of eyelid

Posterior palpebral margin

Excretory duct of tarsal gland

328 Orbital region
a Facial muscles around the eyes (75%),
frontal aspect
b Sagittal section through the upper eyelid (700%),
medial aspect

a

Anterior cranial fossa

Frontal sinus

Orbit

Anterior ethmoidal cells

Bony nasal cavity

Infra-orbital canal

Maxillary hiatus

Maxillary sinus

Alveolar process
of maxilla

Molar tooth

Crista galli

Orbit
– Orbital surface of frontal bone
– Superior orbital fissure
– Orbital plate
of ethmoidal bone
– Orbital surface of maxilla
– Inferior orbital fissure

Middle nasal concha

Bony nasal septum

Inferior nasal concha

Palatine process
of maxilla

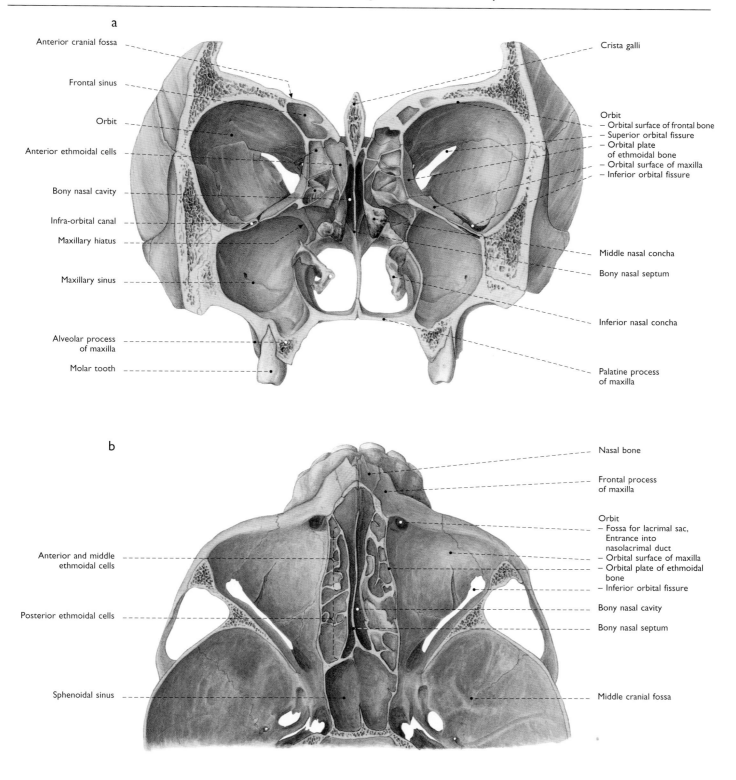

b

Anterior and middle
ethmoidal cells

Posterior ethmoidal cells

Sphenoidal sinus

Nasal bone

Frontal process
of maxilla

Orbit
– Fossa for lacrimal sac,
Entrance into
nasolacrimal duct
– Orbital surface of maxilla
– Orbital plate of ethmoidal
bone
– Inferior orbital fissure

Bony nasal cavity

Bony nasal septum

Middle cranial fossa

329 Orbital cavity (85%)
Sections through the osseous walls of the orbits,
the nasal cavity, and paranasal sinuses
a Coronal section through the viscerocranium,
frontal aspect
b Transverse section through the middle parts of the orbits,
the nasal cavity, the ethmoidal labyrinth, and sphenoidal sinuses,
superior aspect

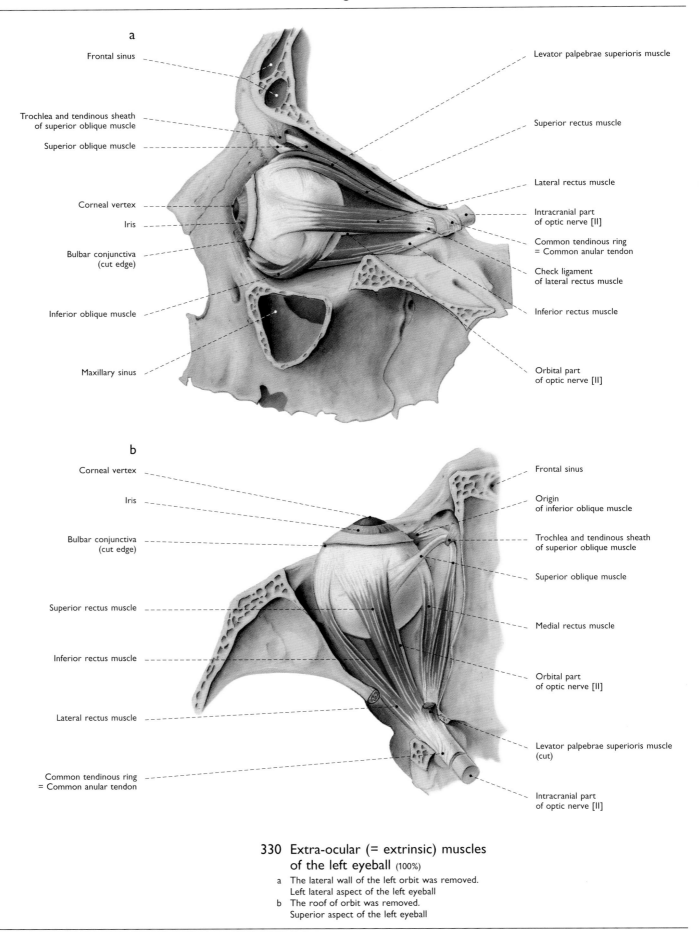

a

Frontal sinus

Trochlea and tendinous sheath
of superior oblique muscle

Superior oblique muscle

Corneal vertex

Iris

Bulbar conjunctiva
(cut edge)

Inferior oblique muscle

Maxillary sinus

Levator palpebrae superioris muscle

Superior rectus muscle

Lateral rectus muscle

Intracranial part
of optic nerve [II]

Common tendinous ring
= Common anular tendon

Check ligament
of lateral rectus muscle

Inferior rectus muscle

Orbital part
of optic nerve [II]

b

Corneal vertex

Iris

Bulbar conjunctiva
(cut edge)

Superior rectus muscle

Inferior rectus muscle

Lateral rectus muscle

Common tendinous ring
= Common anular tendon

Frontal sinus

Origin
of inferior oblique muscle

Trochlea and tendinous sheath
of superior oblique muscle

Superior oblique muscle

Medial rectus muscle

Orbital part
of optic nerve [II]

Levator palpebrae superioris muscle
(cut)

Intracranial part
of optic nerve [II]

330 Extra-ocular (= extrinsic) muscles
of the left eyeball (100%)
a The lateral wall of the left orbit was removed.
   Left lateral aspect of the left eyeball
b The roof of orbit was removed.
   Superior aspect of the left eyeball

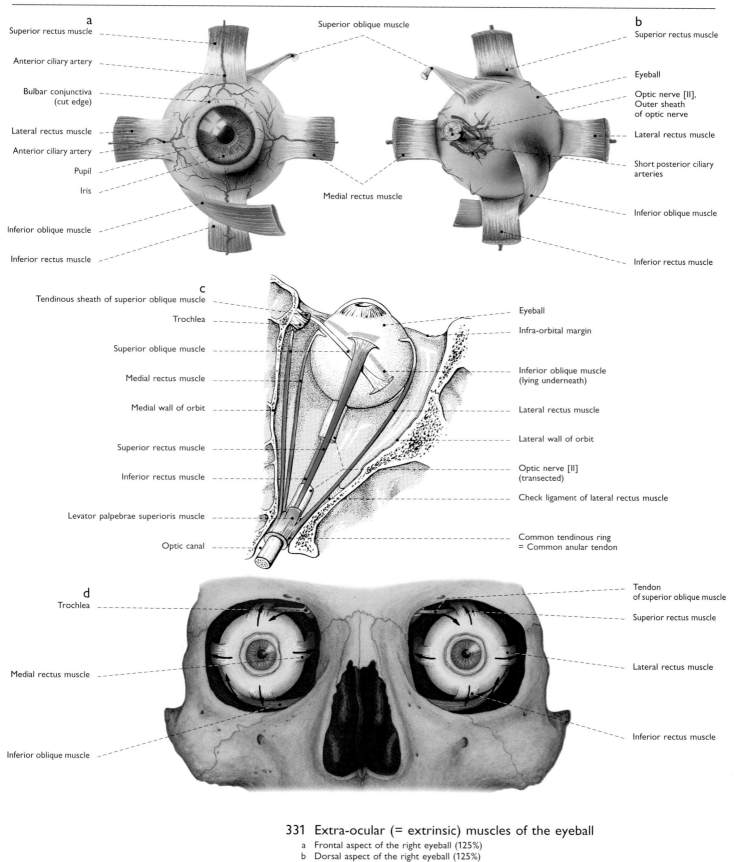

a

Superior rectus muscle

Anterior ciliary artery

Bulbar conjunctiva (cut edge)

Lateral rectus muscle

Anterior ciliary artery

Pupil

Iris

Inferior oblique muscle

Inferior rectus muscle

Superior oblique muscle

Medial rectus muscle

b

Superior rectus muscle

Eyeball

Optic nerve [II], Outer sheath of optic nerve

Lateral rectus muscle

Short posterior ciliary arteries

Inferior oblique muscle

Inferior rectus muscle

c

Tendinous sheath of superior oblique muscle

Trochlea

Superior oblique muscle

Medial rectus muscle

Medial wall of orbit

Superior rectus muscle

Inferior rectus muscle

Levator palpebrae superioris muscle

Optic canal

Eyeball

Infra-orbital margin

Inferior oblique muscle (lying underneath)

Lateral rectus muscle

Lateral wall of orbit

Optic nerve [II] (transected)

Check ligament of lateral rectus muscle

Common tendinous ring = Common anular tendon

d

Trochlea

Medial rectus muscle

Inferior oblique muscle

Tendon of superior oblique muscle

Superior rectus muscle

Lateral rectus muscle

Inferior rectus muscle

331 Extra-ocular (= extrinsic) muscles of the eyeball

a Frontal aspect of the right eyeball (125%)
b Dorsal aspect of the right eyeball (125%)
c Superior aspect of the right orbit (100%), schematic representation
d Frontal aspect of the osseous orbits and both eyeballs with muscle insertions (80%). The arrows indicate ocular movements after muscle contraction.

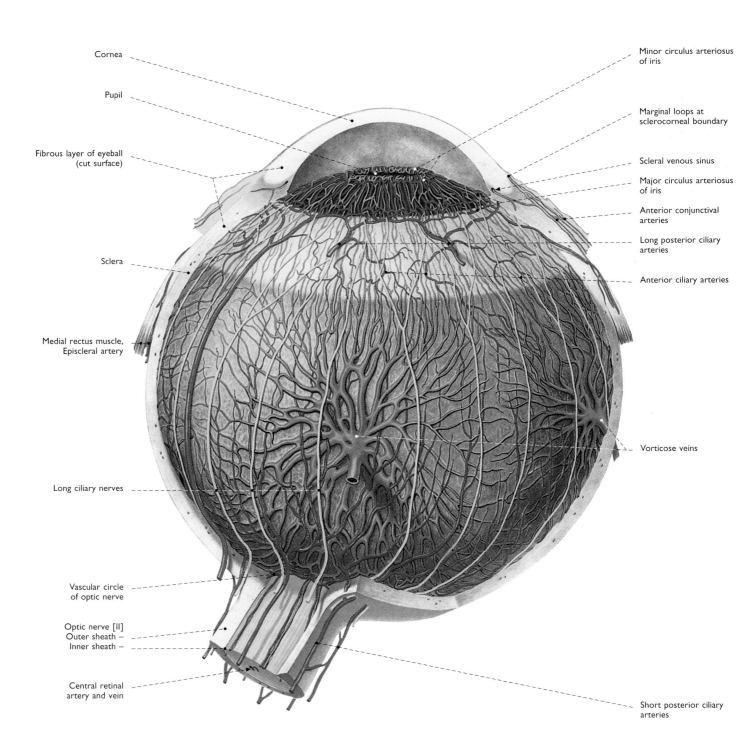

Cornea

Pupil

Fibrous layer of eyeball
(cut surface)

Sclera

Medial rectus muscle,
Episcleral artery

Long ciliary nerves

Vascular circle
of optic nerve

Optic nerve [II]
Outer sheath –
Inner sheath –

Central retinal
artery and vein

Minor circulus arteriosus
of iris

Marginal loops at
sclerocorneal boundary

Scleral venous sinus

Major circulus arteriosus
of iris

Anterior conjunctival
arteries

Long posterior ciliary
arteries

Anterior ciliary arteries

Vorticose veins

Short posterior ciliary
arteries

332 Blood vessels of the eyeball (500%)
The upper part of the fibrous layer of the right eyeball
was removed nearly completely. Superior aspect

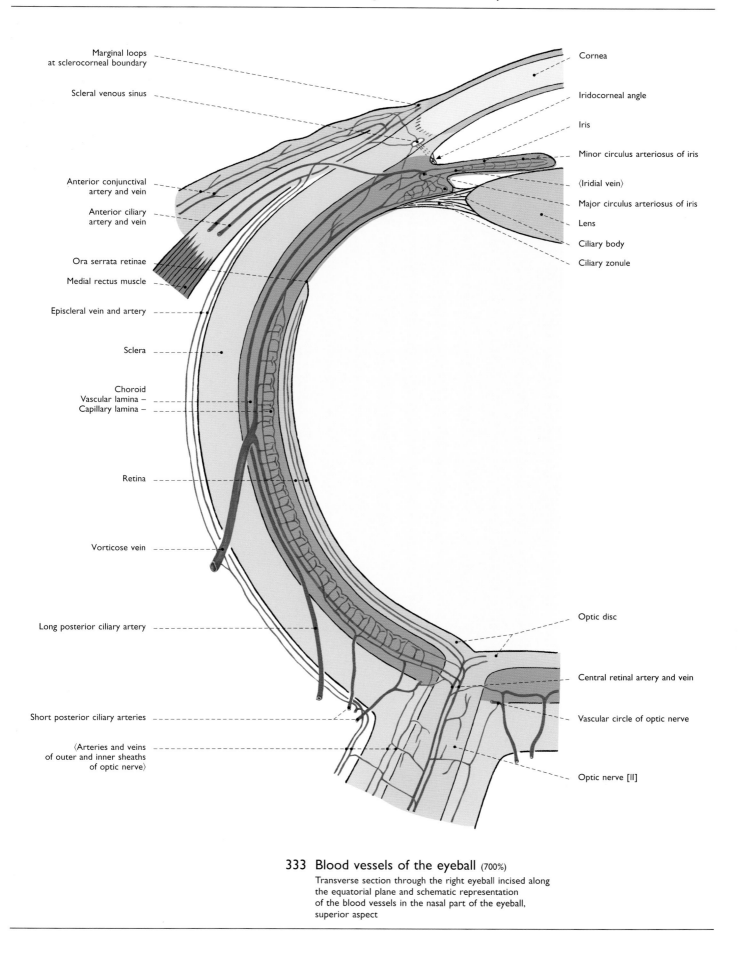

Marginal loops at sclerocorneal boundary

Scleral venous sinus

Anterior conjunctival artery and vein

Anterior ciliary artery and vein

Ora serrata retinae

Medial rectus muscle

Episcleral vein and artery

Sclera

Choroid
Vascular lamina –
Capillary lamina –

Retina

Vorticose vein

Long posterior ciliary artery

Short posterior ciliary arteries

⟨Arteries and veins of outer and inner sheaths of optic nerve⟩

Cornea

Iridocorneal angle

Iris

Minor circulus arteriosus of iris

⟨Iridial vein⟩

Major circulus arteriosus of iris

Lens

Ciliary body

Ciliary zonule

Optic disc

Central retinal artery and vein

Vascular circle of optic nerve

Optic nerve [II]

333 Blood vessels of the eyeball (700%)
Transverse section through the right eyeball incised along the equatorial plane and schematic representation of the blood vessels in the nasal part of the eyeball, superior aspect

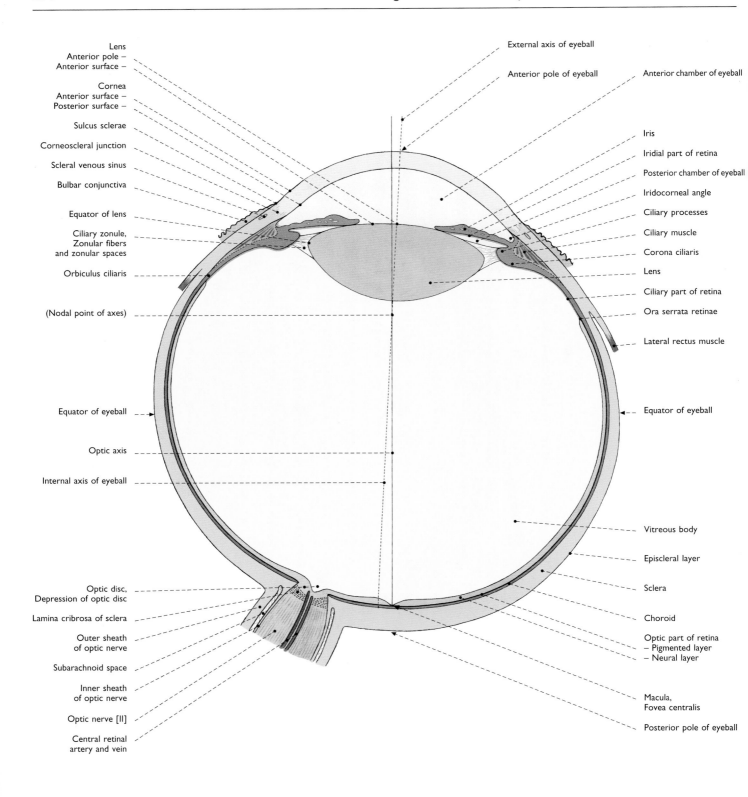

Lens
Anterior pole −
Anterior surface −

Cornea
Anterior surface −
Posterior surface −

Sulcus sclerae

Corneoscleral junction

Scleral venous sinus

Bulbar conjunctiva

Equator of lens

Ciliary zonule,
Zonular fibers
and zonular spaces

Orbiculus ciliaris

(Nodal point of axes)

Equator of eyeball

Optic axis

Internal axis of eyeball

Optic disc,
Depression of optic disc

Lamina cribrosa of sclera

Outer sheath
of optic nerve

Subarachnoid space

Inner sheath
of optic nerve

Optic nerve [II]

Central retinal
artery and vein

External axis of eyeball

Anterior pole of eyeball

Anterior chamber of eyeball

Iris

Iridial part of retina

Posterior chamber of eyeball

Iridocorneal angle

Ciliary processes

Ciliary muscle

Corona ciliaris

Lens

Ciliary part of retina

Ora serrata retinae

Lateral rectus muscle

Equator of eyeball

Vitreous body

Episcleral layer

Sclera

Choroid

Optic part of retina
− Pigmented layer
− Neural layer

Macula,
Fovea centralis

Posterior pole of eyeball

334 Eyeball (500%)
Schematized transverse section through the right eyeball
of an adult with normal vision, superior aspect

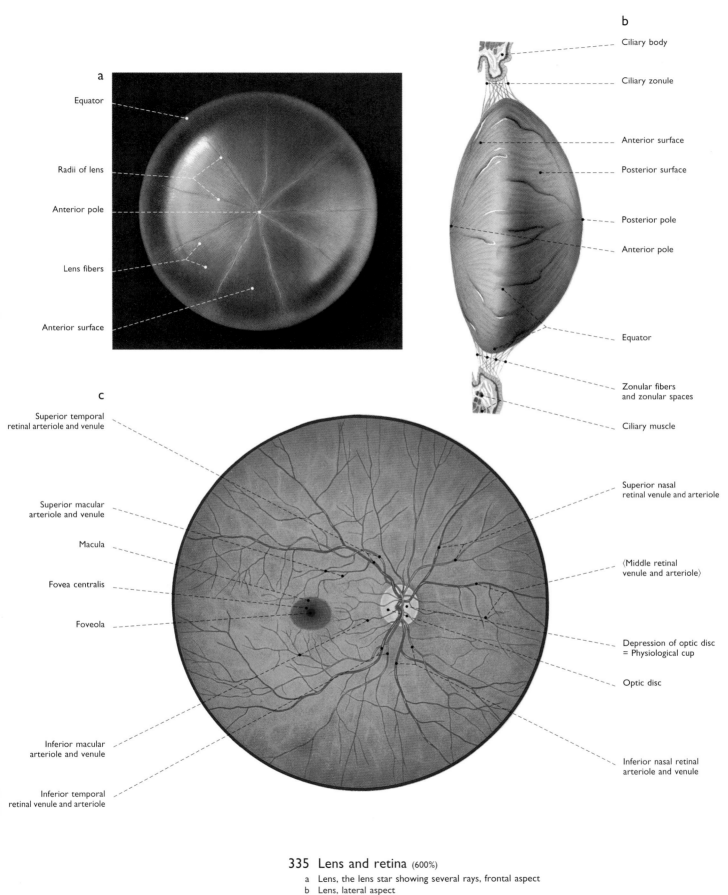

a

Equator

Radii of lens

Anterior pole

Lens fibers

Anterior surface

b

Ciliary body

Ciliary zonule

Anterior surface

Posterior surface

Posterior pole

Anterior pole

Equator

Zonular fibers
and zonular spaces

Ciliary muscle

c

Superior temporal
retinal arteriole and venule

Superior macular
arteriole and venule

Macula

Fovea centralis

Foveola

Inferior macular
arteriole and venule

Inferior temporal
retinal venule and arteriole

Superior nasal
retinal venule and arteriole

⟨Middle retinal
venule and arteriole⟩

Depression of optic disc
= Physiological cup

Optic disc

Inferior nasal retinal
arteriole and venule

335 Lens and retina (600%)

a   Lens, the lens star showing several rays, frontal aspect
b   Lens, lateral aspect
c   Retina of the right eye, ophthalmoscopic view

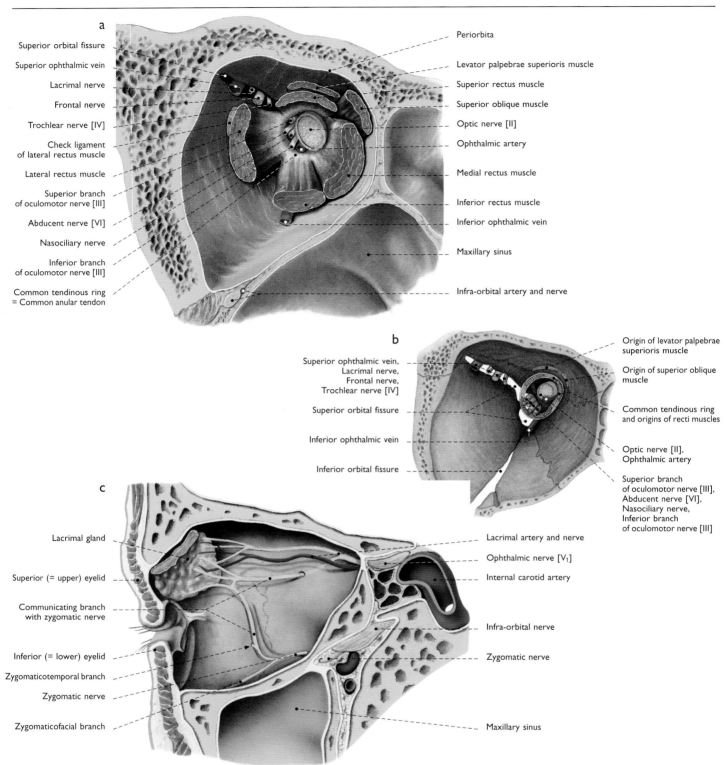

a
Superior orbital fissure — — Periorbita
Superior ophthalmic vein — — Levator palpebrae superioris muscle
Lacrimal nerve — — Superior rectus muscle
Frontal nerve — — Superior oblique muscle
Trochlear nerve [IV] — — Optic nerve [II]
Check ligament — — Ophthalmic artery
of lateral rectus muscle
Lateral rectus muscle — — Medial rectus muscle
Superior branch — — Inferior rectus muscle
of oculomotor nerve [III]
Abducent nerve [VI] — — Inferior ophthalmic vein
Nasociliary nerve
Inferior branch — — Maxillary sinus
of oculomotor nerve [III]
Common tendinous ring — — Infra-orbital artery and nerve
= Common anular tendon

b
Superior ophthalmic vein, — — Origin of levator palpebrae
Lacrimal nerve, superioris muscle
Frontal nerve, — — Origin of superior oblique
Trochlear nerve [IV] muscle
Superior orbital fissure — — Common tendinous ring
and origins of recti muscles
Inferior ophthalmic vein — — Optic nerve [II],
Ophthalmic artery
Inferior orbital fissure — — Superior branch
of oculomotor nerve [III],
Abducent nerve [VI],
Nasociliary nerve,
Inferior branch
of oculomotor nerve [III]

c
Lacrimal gland — — Lacrimal artery and nerve
— Ophthalmic nerve [V₁]
Superior (= upper) eyelid — — Internal carotid artery
Communicating branch
with zygomatic nerve
— Infra-orbital nerve
Inferior (= lower) eyelid — — Zygomatic nerve
Zygomaticotemporal branch
Zygomatic nerve
Zygomaticofacial branch — — Maxillary sinus

336 Right orbital cavity
a Coronal section. The common tendinous ring and the origins
of the extra-ocular muscles are shown (150%). Frontal aspect
b Coronal section. The extra-ocular muscles were transected close
to their origins, the nerves and blood vessels were cut close
to their points of entry into or exit from the orbit, respectively (150%).
Frontal aspect
c Sagittal section. The lateral wall of the right orbit is shown
after removal of the eyeball (100%). Medial aspect

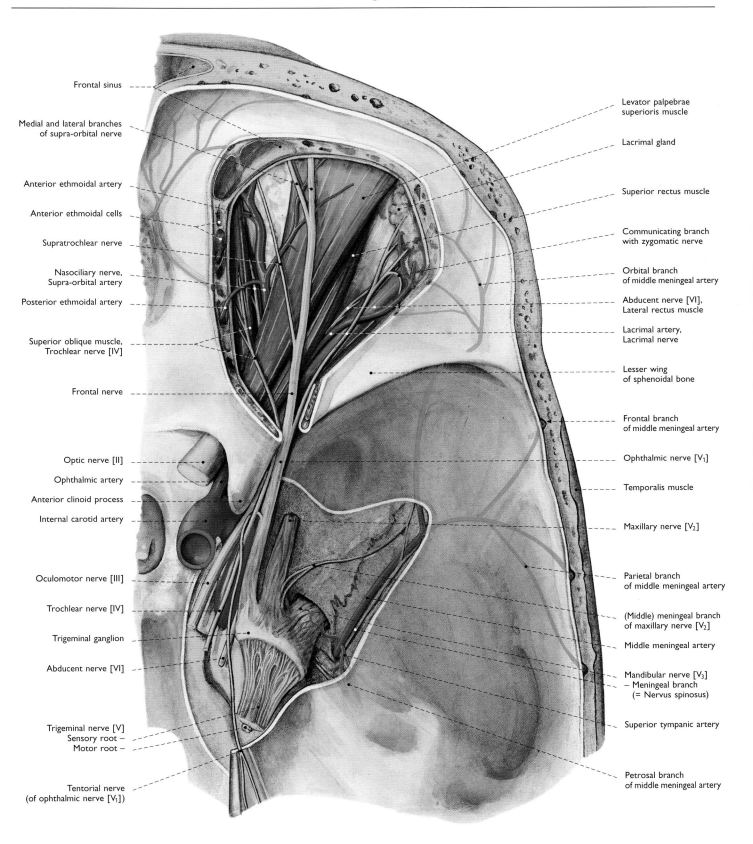

Frontal sinus

Medial and lateral branches of supra-orbital nerve

Anterior ethmoidal artery

Anterior ethmoidal cells

Supratrochlear nerve

Nasociliary nerve, Supra-orbital artery

Posterior ethmoidal artery

Superior oblique muscle, Trochlear nerve [IV]

Frontal nerve

Optic nerve [II]

Ophthalmic artery

Anterior clinoid process

Internal carotid artery

Oculomotor nerve [III]

Trochlear nerve [IV]

Trigeminal ganglion

Abducent nerve [VI]

Trigeminal nerve [V]
Sensory root –
Motor root –

Tentorial nerve
(of ophthalmic nerve [V₁])

Levator palpebrae superioris muscle

Lacrimal gland

Superior rectus muscle

Communicating branch with zygomatic nerve

Orbital branch of middle meningeal artery

Abducent nerve [VI], Lateral rectus muscle

Lacrimal artery, Lacrimal nerve

Lesser wing of sphenoidal bone

Frontal branch of middle meningeal artery

Ophthalmic nerve [V₁]

Temporalis muscle

Maxillary nerve [V₂]

Parietal branch of middle meningeal artery

(Middle) meningeal branch of maxillary nerve [V₂]

Middle meningeal artery

Mandibular nerve [V₃]
– Meningeal branch
(= Nervus spinosus)

Superior tympanic artery

Petrosal branch of middle meningeal artery

337 Right orbital cavity and middle cranial fossa (200%)
The roof of orbit and the cranial dura mater beyond the right orbit
were removed, the right cavernous sinus was opened. Superior aspect

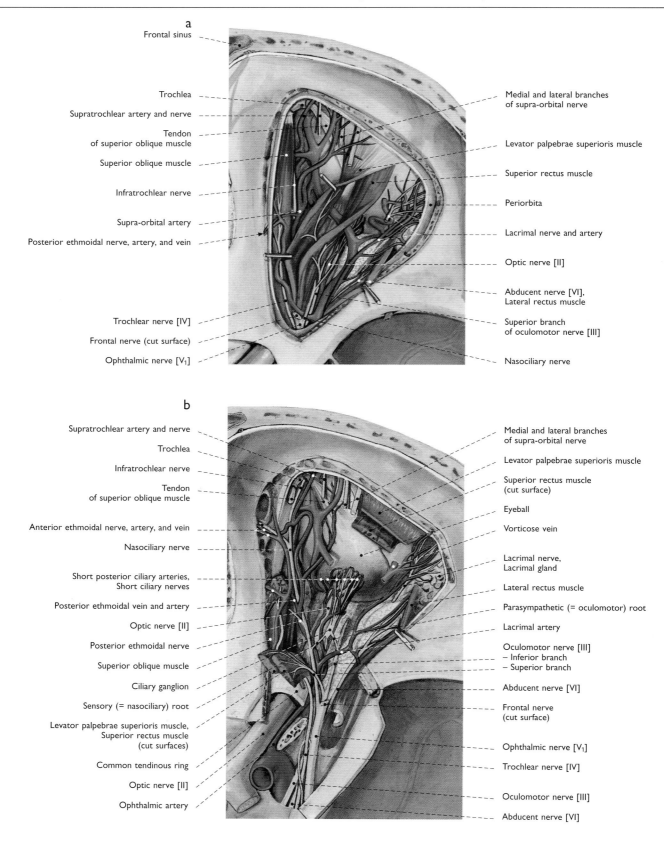

a
Frontal sinus

Trochlea
Supratrochlear artery and nerve
Tendon
of superior oblique muscle
Superior oblique muscle

Infratrochlear nerve

Supra-orbital artery
Posterior ethmoidal nerve, artery, and vein

Trochlear nerve [IV]
Frontal nerve (cut surface)
Ophthalmic nerve [V₁]

Medial and lateral branches
of supra-orbital nerve

Levator palpebrae superioris muscle

Superior rectus muscle

Periorbita

Lacrimal nerve and artery

Optic nerve [II]

Abducent nerve [VI],
Lateral rectus muscle

Superior branch
of oculomotor nerve [III]

Nasociliary nerve

b

Supratrochlear artery and nerve
Trochlea
Infratrochlear nerve
Tendon
of superior oblique muscle

Anterior ethmoidal nerve, artery, and vein
Nasociliary nerve

Short posterior ciliary arteries,
Short ciliary nerves
Posterior ethmoidal vein and artery
Optic nerve [II]
Posterior ethmoidal nerve
Superior oblique muscle
Ciliary ganglion
Sensory (= nasociliary) root
Levator palpebrae superioris muscle,
Superior rectus muscle
(cut surfaces)
Common tendinous ring
Optic nerve [II]
Ophthalmic artery

Medial and lateral branches
of supra-orbital nerve
Levator palpebrae superioris muscle
Superior rectus muscle
(cut surface)
Eyeball

Vorticose vein

Lacrimal nerve,
Lacrimal gland
Lateral rectus muscle
Parasympathetic (= oculomotor) root

Lacrimal artery

Oculomotor nerve [III]
– Inferior branch
– Superior branch
Abducent nerve [VI]
Frontal nerve
(cut surface)

Ophthalmic nerve [V₁]

Trochlear nerve [IV]

Oculomotor nerve [III]

Abducent nerve [VI]

338 Right orbital cavity (140%)
The roof of orbit was removed. Superior aspect
a Superficial dissection
b Deeper dissection just above the optic nerve. The upper
extra-ocular muscles were transected and removed.

a

Frontal sinus (opened)

Trochlea

Tendon of superior oblique muscle

Superior oblique muscle

Medial rectus muscle

Infratrochlear nerve

Anterior ethmoidal artery, nerve

Optic nerve [II] (cut surface)

Inferior rectus muscle

Inferior branch of oculomotor nerve [III]

Trochlear nerve [IV]

Posterior ethmoidal nerve and artery

Oculomotor nerve [III]

Nasociliary nerve

Common tendinous ring

Ophthalmic artery

Optic nerve [II] (transected)

Internal carotid artery

Levator palpebrae superioris m. (cut)

Superior rectus muscle (cut)

Lacrimal gland

Eyeball

Lacrimal nerve and artery

Lateral rectus muscle

Short posterior ciliary arteries

Long ciliary nerve

Short ciliary nerves

Abducent nerve [VI]

Ciliary ganglion

Parasympathetic (= oculomotor) root

Sensory (= nasociliary) root

Abducent nerve [VI]

Superior ophthalmic vein

Frontal nerve (cut)

Middle cranial fossa

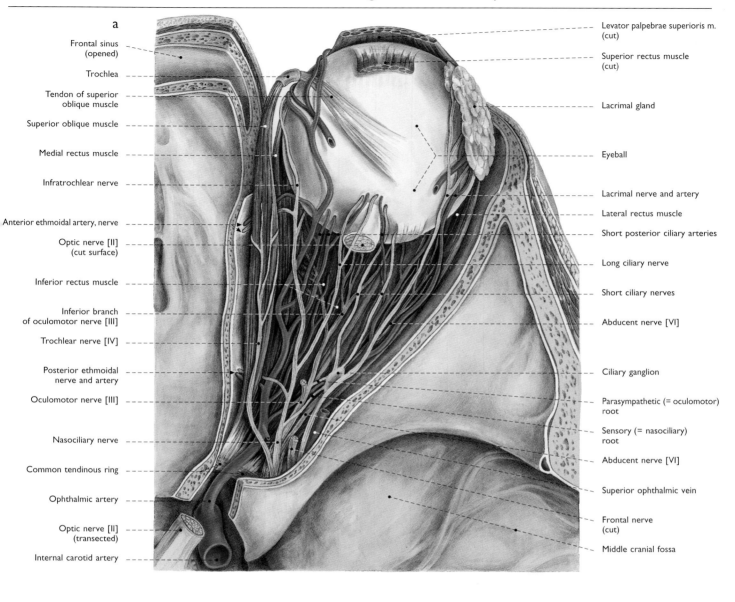

b

Ciliary ganglion
Parasympathetic (= oculomotor) root –
Sympathetic root –
Sensory (= nasociliary) root –

Trigeminal nerve [V]

Ophthalmic nerve [V₁]

Internal carotid artery with internal carotid plexus

Oculomotor nerve [III]

Ciliary ganglion

Short ciliary nerves

Long ciliary nerve

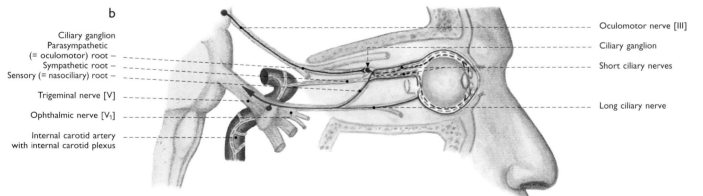

339 Right orbital cavity

a Deep dissection. The roof of orbit was completely removed, the upper extra-ocular muscles and the optic nerve were transected and removed (150%). Superior aspect

b Ciliary ganglion and autonomic innervation of the eyeball (80%), schematic representation, right lateral aspect

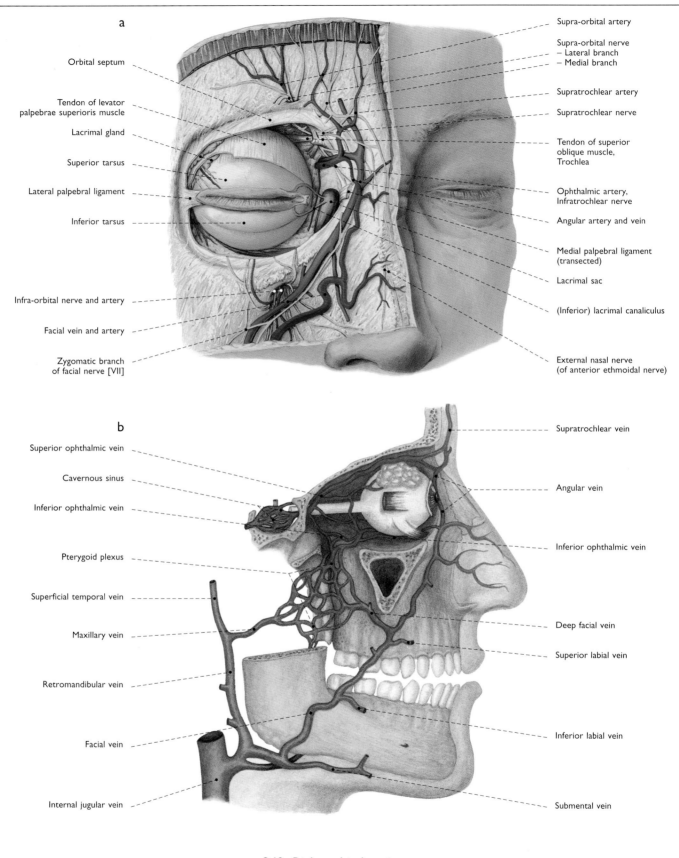

a

Orbital septum

Tendon of levator palpebrae superioris muscle

Lacrimal gland

Superior tarsus

Lateral palpebral ligament

Inferior tarsus

Infra-orbital nerve and artery

Facial vein and artery

Zygomatic branch of facial nerve [VII]

Supra-orbital artery

Supra-orbital nerve
– Lateral branch
– Medial branch

Supratrochlear artery

Supratrochlear nerve

Tendon of superior oblique muscle, Trochlea

Ophthalmic artery, Infratrochlear nerve

Angular artery and vein

Medial palpebral ligament (transected)

Lacrimal sac

(Inferior) lacrimal canaliculus

External nasal nerve (of anterior ethmoidal nerve)

b

Superior ophthalmic vein

Cavernous sinus

Inferior ophthalmic vein

Pterygoid plexus

Superficial temporal vein

Maxillary vein

Retromandibular vein

Facial vein

Internal jugular vein

Supratrochlear vein

Angular vein

Inferior ophthalmic vein

Deep facial vein

Superior labial vein

Inferior labial vein

Submental vein

340 Right orbital region

a   Gross anatomy of the right orbital region (90%)
b   Drainage of the orbital veins. The lateral wall of the right orbit and parts of the ramus of mandible were removed, the right maxillary sinus is opened (80%). Right lateral aspect

Vestibulocochlear Organ

a

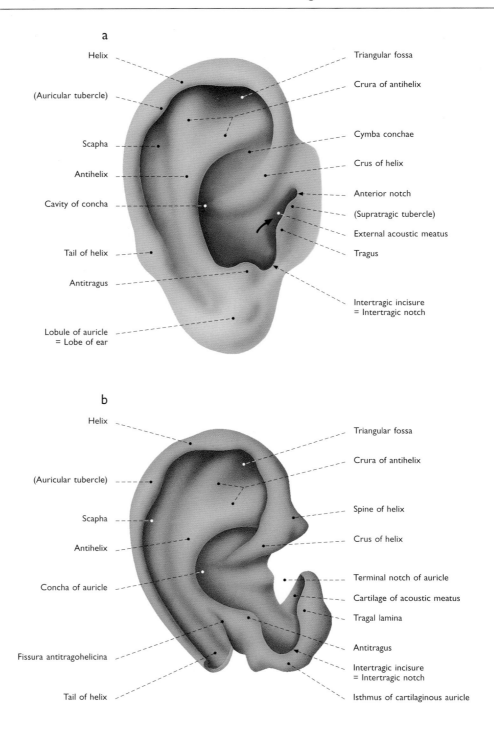

Helix ----------- Triangular fossa

(Auricular tubercle) ----------- Crura of antihelix

Scapha ----------- Cymba conchae

Antihelix ----------- Crus of helix

Cavity of concha ----------- Anterior notch

----------- (Supratragic tubercle)

----------- External acoustic meatus

Tail of helix ----------- Tragus

Antitragus -----------

----------- Intertragic incisure
= Intertragic notch

Lobule of auricle
= Lobe of ear -----------

b

Helix ----------- Triangular fossa

----------- Crura of antihelix

(Auricular tubercle) ----------- Spine of helix

Scapha ----------- Crus of helix

Antihelix -----------

Concha of auricle ----------- Terminal notch of auricle

----------- Cartilage of acoustic meatus

----------- Tragal lamina

----------- Antitragus

Fissura antitragohelicina ----------- Intertragic incisure
= Intertragic notch

Tail of helix ----------- Isthmus of cartilaginous auricle

342 External ear (110%)
Right external ear, lateral aspect
a Auricle = pinna
b Auricular cartilage

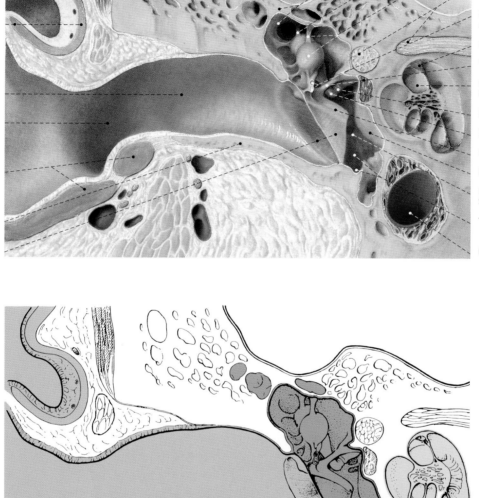

a

Squamous part
of temporal bone

Temporalis muscle

Auricular cartilage

(Bony) external
acoustic meatus

Cartilaginous external
acoustic meatus

Cartilage of acoustic meatus

Tympanic part
of temporal bone

Tympanic membrane

Tegmental wall,
Epitympanic recess

Superior ligament of incus

Superior ligament of malleus,
Head of malleus

Facial nerve [VII]

Chorda tympani,
Anterior fold of malleus

Cochlear duct

Tensor tympani muscle

Scala vestibuli,
Scala tympani

Head of stapes

Long limb of incus

Handle of malleus

Promontory

Round window,
Secondary tympanic membrane

Tympanic cavity,
Hypotympanic recess

Internal carotid artery

b

External ear   Middle ear   Internal ear

343 Ear (300%)

a, b   Coronal section through the external, middle,
and internal ear of the right side, frontal aspect

b   The parts of the external, middle, and internal ear
are indicated by different colors.

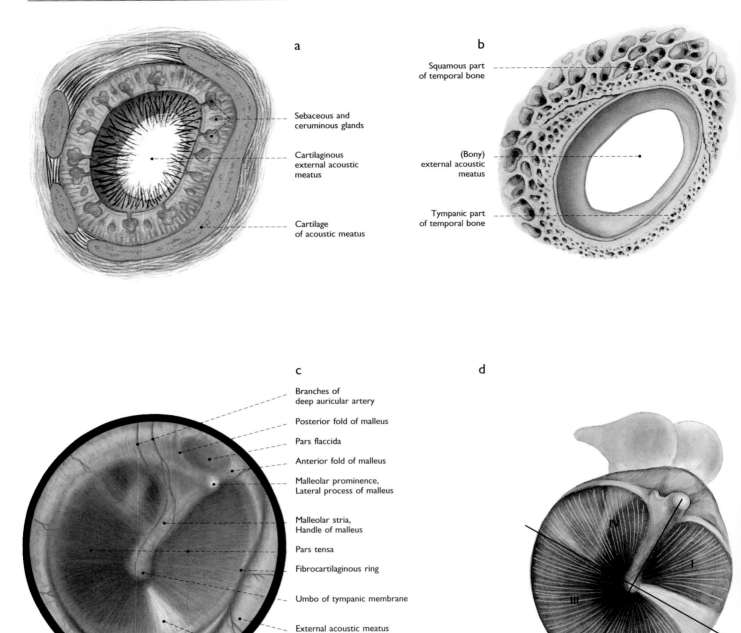

a

Sebaceous and
ceruminous glands

Cartilaginous
external acoustic
meatus

Cartilage
of acoustic meatus

b

Squamous part
of temporal bone

(Bony)
external acoustic
meatus

Tympanic part
of temporal bone

c

Branches of
deep auricular artery

Posterior fold of malleus

Pars flaccida

Anterior fold of malleus

Malleolar prominence,
Lateral process of malleus

Malleolar stria,
Handle of malleus

Pars tensa

Fibrocartilaginous ring

Umbo of tympanic membrane

External acoustic meatus

(Cone of reflected light)

d

**344 External acoustic meatus and
tympanic membrane of the right ear**

Lateral aspect
a, b   Cross sections through the
    a   cartilaginous and
    b   noncartilaginous parts
        of the right external acoustic meatus (500%)
    c   View of the right tympanic membrane (= eardrum)
        as seen through an otoscope in a living subject (700%)
    d   Division of the right tympanic membrane into quadrants,
        schematic representation

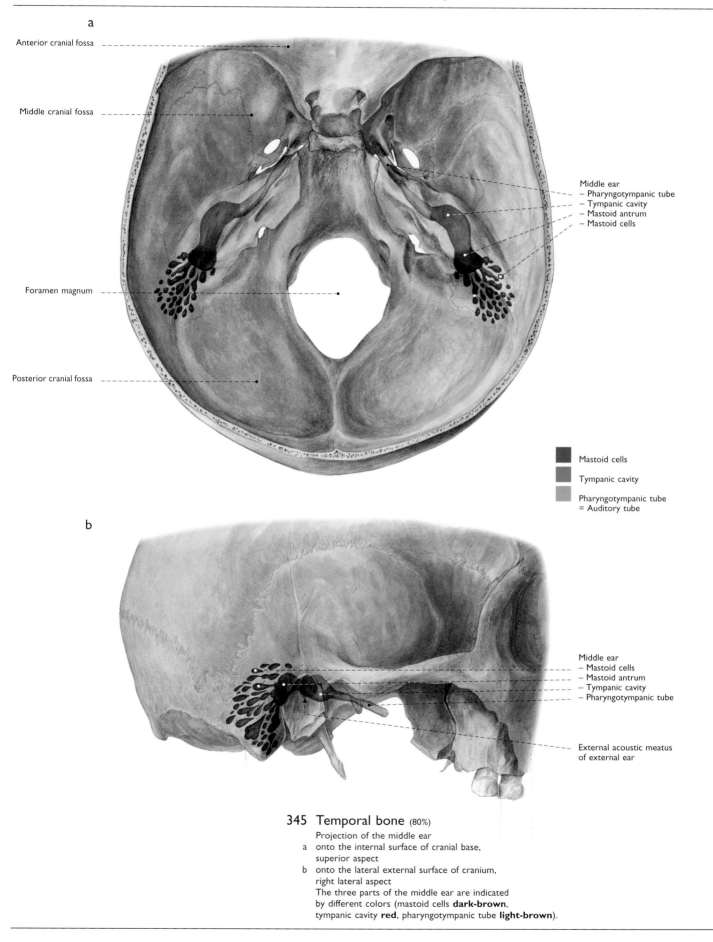

a

Anterior cranial fossa

Middle cranial fossa

Middle ear
– Pharyngotympanic tube
– Tympanic cavity
– Mastoid antrum
– Mastoid cells

Foramen magnum

Posterior cranial fossa

Mastoid cells

Tympanic cavity

Pharyngotympanic tube
= Auditory tube

b

Middle ear
– Mastoid cells
– Mastoid antrum
– Tympanic cavity
– Pharyngotympanic tube

External acoustic meatus
of external ear

345 Temporal bone (80%)
Projection of the middle ear
a onto the internal surface of cranial base,
   superior aspect
b onto the lateral external surface of cranium,
   right lateral aspect
   The three parts of the middle ear are indicated
   by different colors (mastoid cells **dark-brown**,
   tympanic cavity **red**, pharyngotympanic tube **light-brown**).

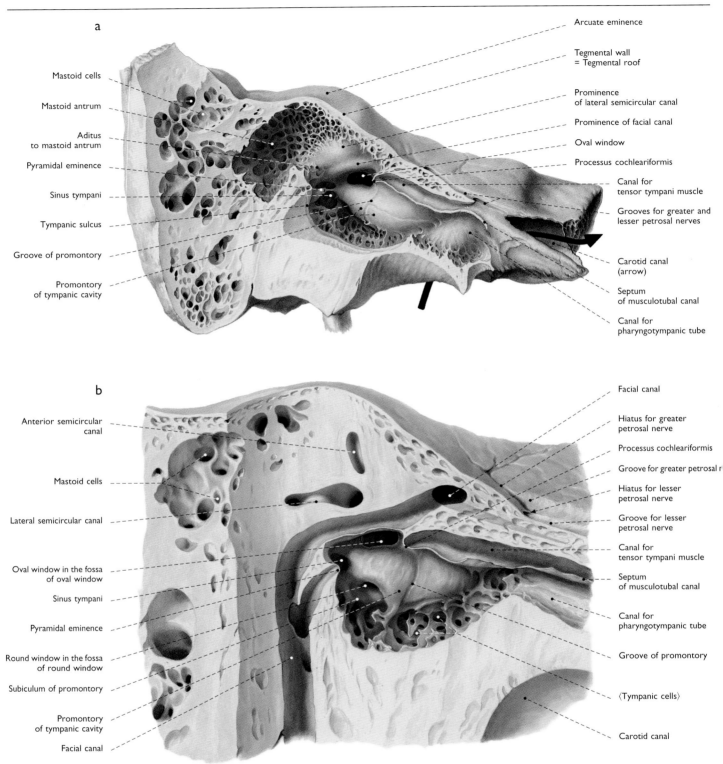

a

Mastoid cells

Mastoid antrum

Aditus
to mastoid antrum

Pyramidal eminence

Sinus tympani

Tympanic sulcus

Groove of promontory

Promontory
of tympanic cavity

Arcuate eminence

Tegmental wall
= Tegmental roof

Prominence
of lateral semicircular canal

Prominence of facial canal

Oval window

Processus cochleariformis

Canal for
tensor tympani muscle

Grooves for greater and
lesser petrosal nerves

Carotid canal
(arrow)

Septum
of musculotubal canal

Canal for
pharyngotympanic tube

b

Anterior semicircular
canal

Mastoid cells

Lateral semicircular canal

Oval window in the fossa
of oval window

Sinus tympani

Pyramidal eminence

Round window in the fossa
of round window

Subiculum of promontory

Promontory
of tympanic cavity

Facial canal

Facial canal

Hiatus for greater
petrosal nerve

Processus cochleariformis

Groove for greater petrosal r

Hiatus for lesser
petrosal nerve

Groove for lesser
petrosal nerve

Canal for
tensor tympani muscle

Septum
of musculotubal canal

Canal for
pharyngotympanic tube

Groove of promontory

⟨Tympanic cells⟩

Carotid canal

346   Right tympanic cavity

Frontolateral aspect of the labyrinthine (= medial) wall
of tympanic cavity. The oval window was opened by removing
the stapes. The mucous membrane was taken away.

a   Vertical section through the petrous part of temporal bone
parallel to its longitudinal axis (200%)

b   Section through the tympanic cavity parallel to its labyrinthine wall.
The facial canal was opened and the facial nerve removed (500%).

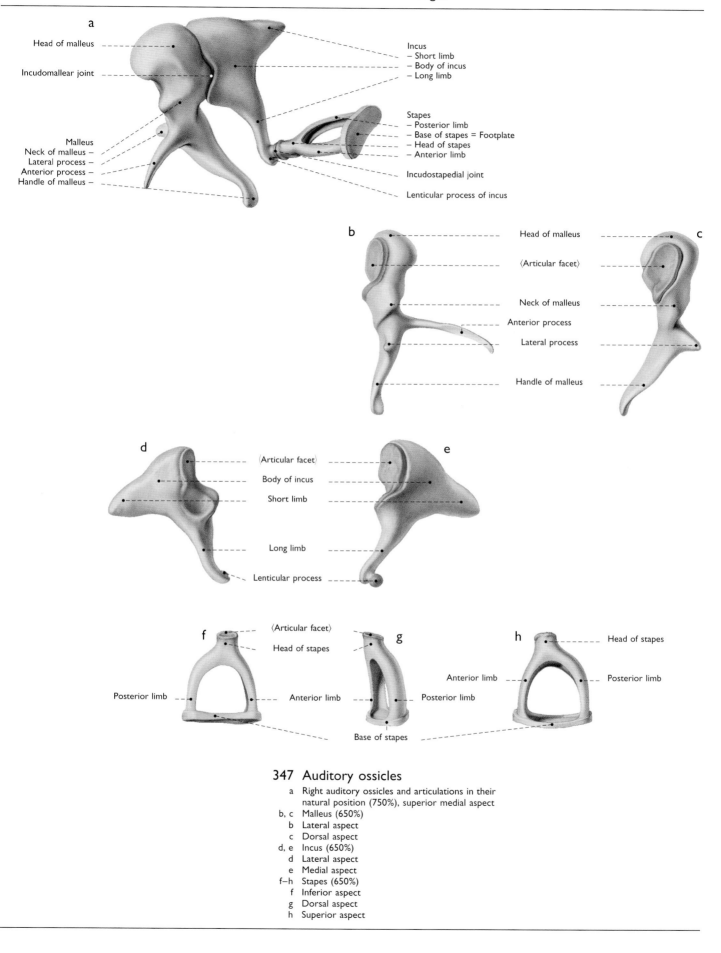

a

Head of malleus

Incudomallear joint

Incus
– Short limb
– Body of incus
– Long limb

Stapes
– Posterior limb
– Base of stapes = Footplate
– Head of stapes
– Anterior limb

Incudostapedial joint

Lenticular process of incus

Malleus
Neck of malleus –
Lateral process –
Anterior process –
Handle of malleus –

b

Head of malleus

⟨Articular facet⟩

Neck of malleus

Anterior process

Lateral process

Handle of malleus

c

d

⟨Articular facet⟩

Body of incus

Short limb

Long limb

Lenticular process

e

f

⟨Articular facet⟩

Head of stapes

Posterior limb

Anterior limb

g

Posterior limb

Base of stapes

Anterior limb

h

Head of stapes

Posterior limb

347 Auditory ossicles

a Right auditory ossicles and articulations in their
   natural position (750%), superior medial aspect
b, c Malleus (650%)
   b Lateral aspect
   c Dorsal aspect
d, e Incus (650%)
   d Lateral aspect
   e Medial aspect
f–h Stapes (650%)
   f Inferior aspect
   g Dorsal aspect
   h Superior aspect

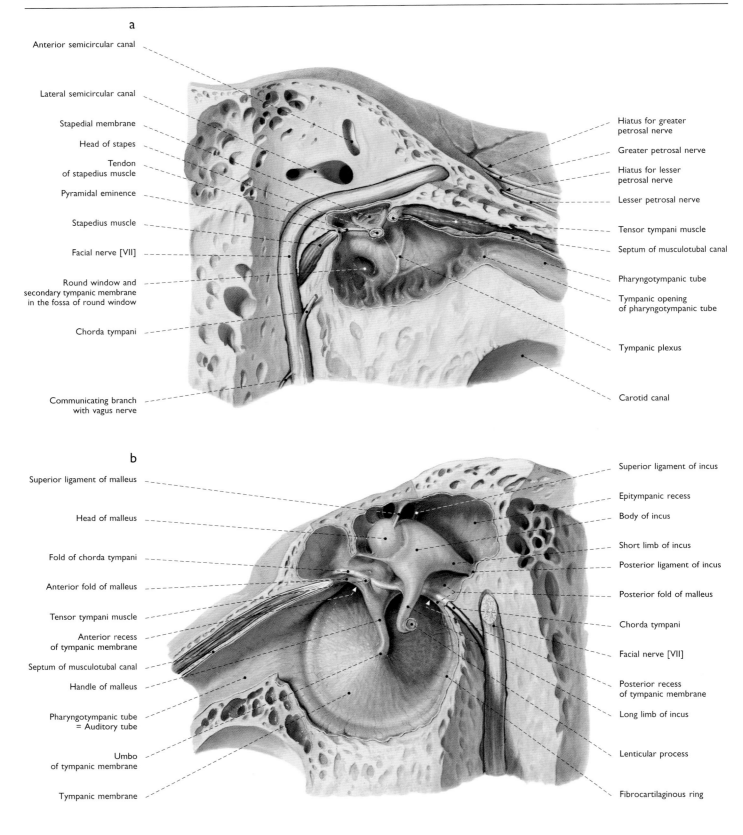

a

Anterior semicircular canal

Lateral semicircular canal

Stapedial membrane

Head of stapes

Tendon
of stapedius muscle

Pyramidal eminence

Stapedius muscle

Facial nerve [VII]

Round window and
secondary tympanic membrane
in the fossa of round window

Chorda tympani

Communicating branch
with vagus nerve

Hiatus for greater
petrosal nerve

Greater petrosal nerve

Hiatus for lesser
petrosal nerve

Lesser petrosal nerve

Tensor tympani muscle

Septum of musculotubal canal

Pharyngotympanic tube

Tympanic opening
of pharyngotympanic tube

Tympanic plexus

Carotid canal

b

Superior ligament of malleus

Head of malleus

Fold of chorda tympani

Anterior fold of malleus

Tensor tympani muscle

Anterior recess
of tympanic membrane

Septum of musculotubal canal

Handle of malleus

Pharyngotympanic tube
= Auditory tube

Umbo
of tympanic membrane

Tympanic membrane

Superior ligament of incus

Epitympanic recess

Body of incus

Short limb of incus

Posterior ligament of incus

Posterior fold of malleus

Chorda tympani

Facial nerve [VII]

Posterior recess
of tympanic membrane

Long limb of incus

Lenticular process

Fibrocartilaginous ring

348 Right tympanic cavity (450%)

a Section through the tympanic cavity parallel to its labyrinthine (= medial) wall,
frontolateral aspect of the labyrinthine wall
b Section through the tympanic cavity parallel to its membranous (= lateral) wall,
medial aspect of the membranous wall

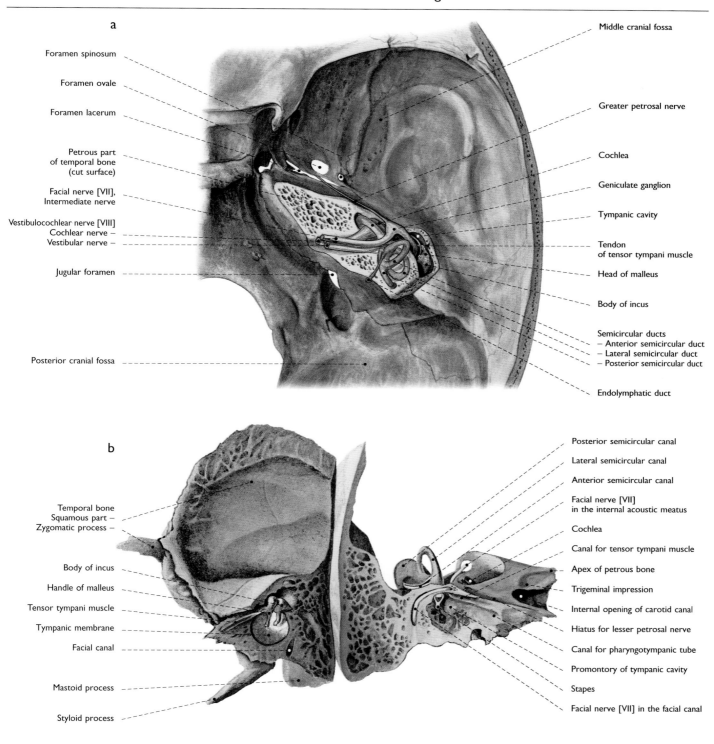

a

Foramen spinosum

Foramen ovale

Foramen lacerum

Petrous part
of temporal bone
(cut surface)

Facial nerve [VII],
Intermediate nerve

Vestibulocochlear nerve [VIII]
Cochlear nerve —
Vestibular nerve —

Jugular foramen

Posterior cranial fossa

Middle cranial fossa

Greater petrosal nerve

Cochlea

Geniculate ganglion

Tympanic cavity

Tendon
of tensor tympani muscle

Head of malleus

Body of incus

Semicircular ducts
— Anterior semicircular duct
— Lateral semicircular duct
— Posterior semicircular duct

Endolymphatic duct

b

Temporal bone
Squamous part —
Zygomatic process —

Body of incus

Handle of malleus

Tensor tympani muscle

Tympanic membrane

Facial canal

Mastoid process

Styloid process

Posterior semicircular canal

Lateral semicircular canal

Anterior semicircular canal

Facial nerve [VII]
in the internal acoustic meatus

Cochlea

Canal for tensor tympani muscle

Apex of petrous bone

Trigeminal impression

Internal opening of carotid canal

Hiatus for lesser petrosal nerve

Canal for pharyngotympanic tube

Promontory of tympanic cavity

Stapes

Facial nerve [VII] in the facial canal

349 Temporal bone and facial canal (100%)

a Transverse section through the petrous part of the right temporal bone
in the plane of the internal acoustic meatus. Exposure of the cochlear duct,
the semicircular ducts, and the facial and vestibulocochlear nerves.
The facial canal and the tympanic cavity were opened. Superior aspect

b Vertical section through the middle ear, the facial canal, and the mastoid cells
of the petrous part of the right temporal bone. The lateral part was turned outwards.
The semicircular canals and the cochlea were exposed. Frontolateral aspect

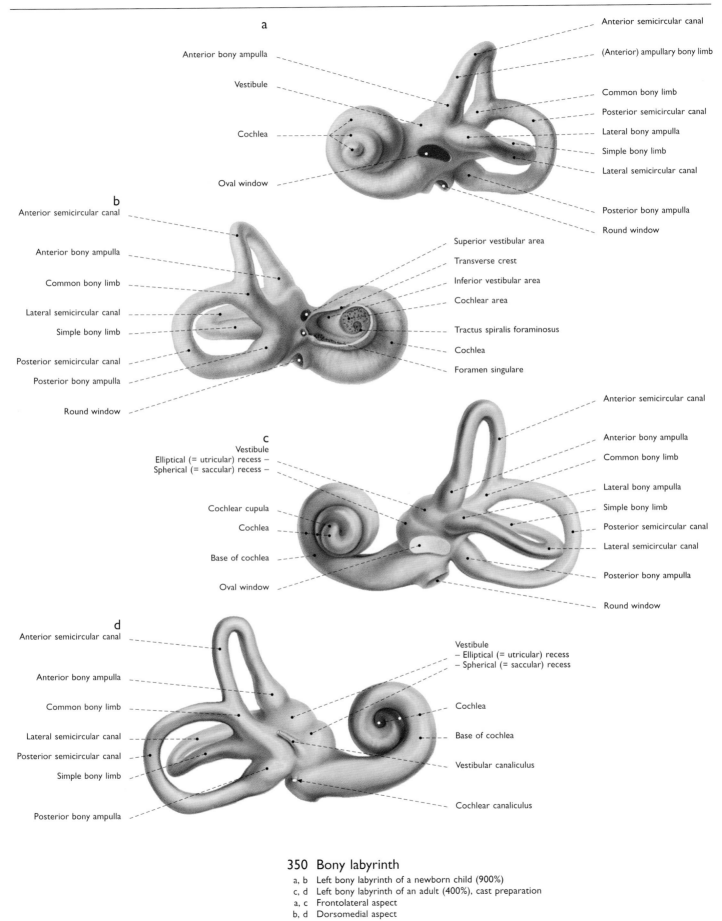

a

Anterior bony ampulla

Vestibule

Cochlea

Oval window

Anterior semicircular canal

(Anterior) ampullary bony limb

Common bony limb

Posterior semicircular canal

Lateral bony ampulla

Simple bony limb

Lateral semicircular canal

Posterior bony ampulla

Round window

b

Anterior semicircular canal

Anterior bony ampulla

Common bony limb

Lateral semicircular canal

Simple bony limb

Posterior semicircular canal

Posterior bony ampulla

Round window

Superior vestibular area

Transverse crest

Inferior vestibular area

Cochlear area

Tractus spiralis foraminosus

Cochlea

Foramen singulare

c

Vestibule
Elliptical (= utricular) recess –
Spherical (= saccular) recess –

Cochlear cupula

Cochlea

Base of cochlea

Oval window

Anterior semicircular canal

Anterior bony ampulla

Common bony limb

Lateral bony ampulla

Simple bony limb

Posterior semicircular canal

Lateral semicircular canal

Posterior bony ampulla

Round window

d

Anterior semicircular canal

Anterior bony ampulla

Common bony limb

Lateral semicircular canal

Posterior semicircular canal

Simple bony limb

Posterior bony ampulla

Vestibule
– Elliptical (= utricular) recess
– Spherical (= saccular) recess

Cochlea

Base of cochlea

Vestibular canaliculus

Cochlear canaliculus

350 Bony labyrinth

a, b   Left bony labyrinth of a newborn child (900%)
c, d   Left bony labyrinth of an adult (400%), cast preparation
a, c   Frontolateral aspect
b, d   Dorsomedial aspect

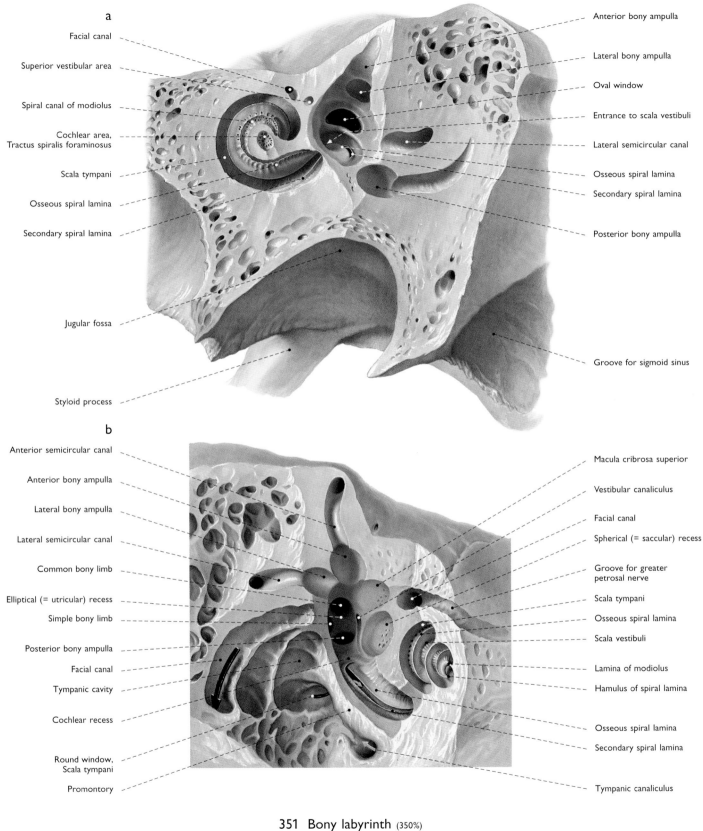

a

Facial canal

Superior vestibular area

Spiral canal of modiolus

Cochlear area,
Tractus spiralis foraminosus

Scala tympani

Osseous spiral lamina

Secondary spiral lamina

Jugular fossa

Styloid process

Anterior bony ampulla

Lateral bony ampulla

Oval window

Entrance to scala vestibuli

Lateral semicircular canal

Osseous spiral lamina

Secondary spiral lamina

Posterior bony ampulla

Groove for sigmoid sinus

b

Anterior semicircular canal

Anterior bony ampulla

Lateral bony ampulla

Lateral semicircular canal

Common bony limb

Elliptical (= utricular) recess

Simple bony limb

Posterior bony ampulla

Facial canal

Tympanic cavity

Cochlear recess

Round window,
Scala tympani

Promontory

Macula cribrosa superior

Vestibular canaliculus

Facial canal

Spherical (= saccular) recess

Groove for greater
petrosal nerve

Scala tympani

Osseous spiral lamina

Scala vestibuli

Lamina of modiolus

Hamulus of spiral lamina

Osseous spiral lamina

Secondary spiral lamina

Tympanic canaliculus

351 Bony labyrinth (350%)
Vestibule, cochlea, and parts of the semicircular canals
of the petrous part of the right temporal bone
a revealed and opened by dissection of the petrous bone from medial,
dorsomedial aspect
b exposed and opened by dissection of the petrous bone from lateral,
frontolateral aspect

a

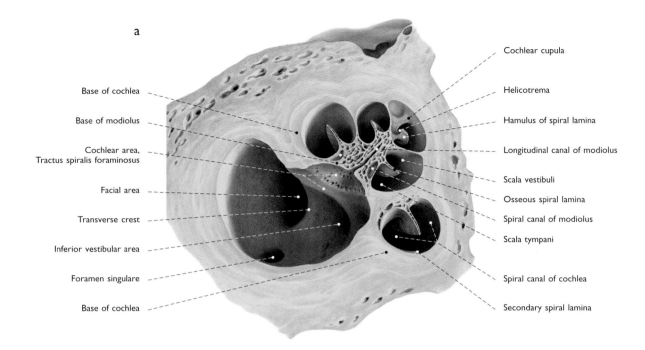

Base of cochlea — Cochlear cupula

Base of modiolus — Helicotrema

Cochlear area, Tractus spiralis foraminosus — Hamulus of spiral lamina

Facial area — Longitudinal canal of modiolus

Transverse crest — Scala vestibuli

Inferior vestibular area — Osseous spiral lamina

Foramen singulare — Spiral canal of modiolus

Base of cochlea — Scala tympani

Spiral canal of cochlea

Secondary spiral lamina

b

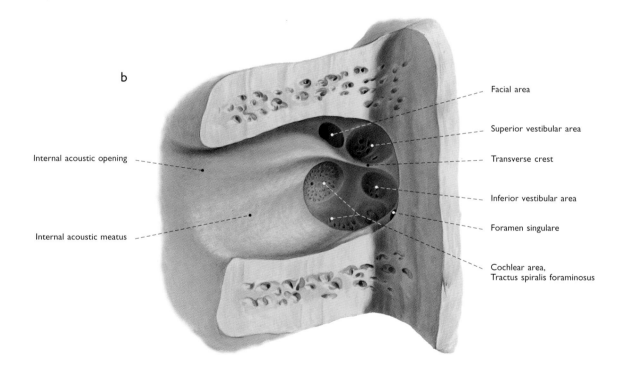

Internal acoustic opening — Facial area

Superior vestibular area

Transverse crest

Inferior vestibular area

Internal acoustic meatus — Foramen singulare

Cochlear area, Tractus spiralis foraminosus

352 Internal ear (500%)

a   The cochlea and the fundus of the internal acoustic meatus were opened by an axial bisection through the right cochlea. Superior aspect

b   Internal acoustic meatus after removal of the dorsomedial wall of the right internal acoustic meatus, medial aspect

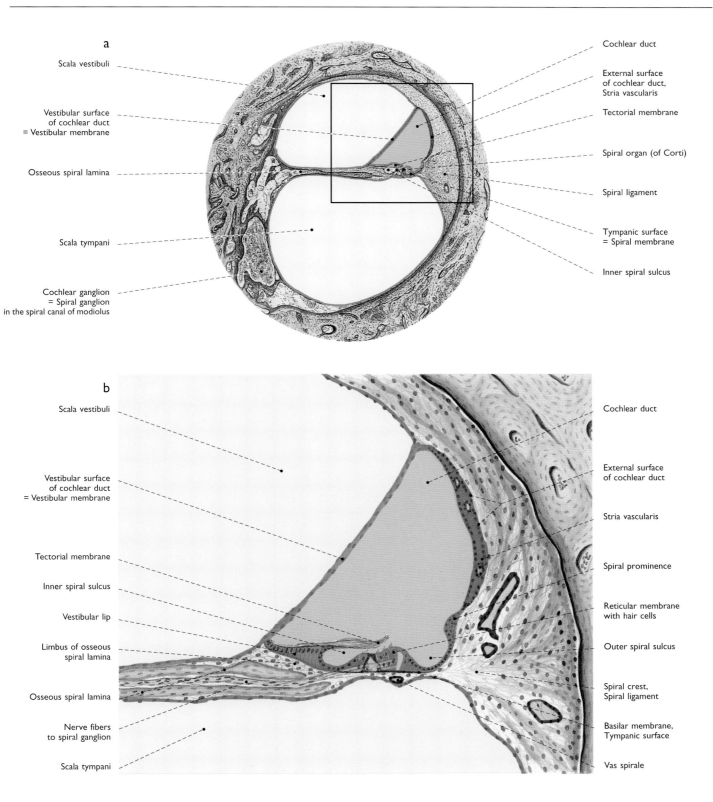

a

Scala vestibuli

Vestibular surface
of cochlear duct
= Vestibular membrane

Osseous spiral lamina

Scala tympani

Cochlear ganglion
= Spiral ganglion
in the spiral canal of modiolus

Cochlear duct

External surface
of cochlear duct,
Stria vascularis

Tectorial membrane

Spiral organ (of Corti)

Spiral ligament

Tympanic surface
= Spiral membrane

Inner spiral sulcus

b

Scala vestibuli

Vestibular surface
of cochlear duct
= Vestibular membrane

Tectorial membrane

Inner spiral sulcus

Vestibular lip

Limbus of osseous
spiral lamina

Osseous spiral lamina

Nerve fibers
to spiral ganglion

Scala tympani

Cochlear duct

External surface
of cochlear duct

Stria vascularis

Spiral prominence

Reticular membrane
with hair cells

Outer spiral sulcus

Spiral crest,
Spiral ligament

Basilar membrane,
Tympanic surface

Vas spirale

353 Cochlea, cochlear duct, and spiral organ of Corti

a Transverse section through one of the coils of cochlea (2000%)
b Enlargement of an area similar to the rectangle in fig. a showing
a cross section through the cochlear duct (8000%)

Subject Index